P9-CQJ-654

Peeking at Peak Oil

Peeking at Peak Oil

Kjell Aleklett
With Michael Lardelli
Illustrated by Olle Qvennerstedt

Kjell Aleklett
Professor of Physics
Uppsala University
Department of Earth Sciences
752 36 Uppsala, Sweden
kjell.aleklett@geo.uu.se

ISBN 978-1-4614-3423-8 ISBN 978-1-4614-3424-5 (eBook)
DOI 10.1007/978-1-4614-3424-5
Springer NewYork Heidelberg Dordrecht London

Library of Congress Control Number: 2012936647

Printed on acid-free paper

Springer is part of Springer Science+Business Media (www.springer.com)

To my wife, daughters, and grandchildren

About the Author, Illustrator, and Translator

Kjell Aleklett

Kjell Aleklett is Professor of Physics at Uppsala University in Sweden where he leads the Uppsala Global Energy Systems Group (UGES). He holds a doctorate in nuclear physics from the University of Gothenburg, Sweden, and worked as a postdoctoral fellow and staff scientist from 1977 to 1985 at the Natural Science Laboratory at Studsvik, Sweden. In 1978–1979 and again in 1983, he was invited to work with Nobel Prize winner Glenn T. Seaborg at the Lawrence Berkeley Laboratory of the University of California, Berkeley, United States. His collaboration with Seaborg spanned 20 years. He was appointed as an associate professor at Uppsala University in 1986 and promoted to full professor in 2000. His interest in the world's energy supply began in 1994 when he acted as energy advisor to the deputy prime minister of Sweden. He subsequently changed his field of research from nuclear physics to the depletion of oil, gas, and coal and its global consequences in 2002. Together with Colin Campbell he organized the *First International Workshop on Oil Depletion* in May 2002 at Uppsala University.

It was in connection with this workshop that ASPO, the Association for the Study of Peak Oil & Gas, was established. Since 2003 he has been president of ASPO International (official website: www.peakoil.net). In 2005 the Prime Minister of Sweden, Göran Persson, became aware of Peak Oil and the research of UGES. He decided to appoint an Oil Commission for which Kjell Aleklett gave the introductory seminar. In that year Kjell Aleklett was also asked to give testimony on Peak Oil before the U.S. House of Representatives Subcommittee on Energy and Air Quality. In 2007 he was asked by the OECD to write a report on "Peak-Oil and the Evolving Strategies of Oil Importing and Exporting Countries: Facing the Hard Truth about an Import Decline for the OECD Countries." In 2009 he testified on Peak Oil for the Australian Senate Standing Committees on Rural and Regional Affairs and Transport. Kjell Aleklett frequently lectures and gives interviews on Peak Oil at community, national, and international events. His hosts have included international transport and oil corporations, governments, and security agencies.

Olle Qvennerstedt

Olle Qvennerstedt was educated as an illustrator and graphic artist at Berghs Reklamskola in Stockholm from 1961 to 1964. Since then he has been active as an illustrator for, among others, Sweden's foremost broadsheet newspaper *Dagens Nyheter* and numerous advertising agencies. Between 1977 and 1981 Olle also worked in animated film. In 1981 he took the step into freelance work with a broad spectrum of commissions and since the early 1990s he has been active as an independent artist. He has now presented over 70 exhibitions of painting, drawing, and graphic art both in Sweden and internationally. In the mid-1990s Olle began producing evocative and light-hearted illustrations for Kjell Aleklett's lectures, and was the obvious choice to illustrate *Peeking at Peak Oil*.

Michael Lardelli

Michael Lardelli received his doctorate in developmental genetics from the Council for National Academic Awards of the United Kingdom in 1991. He then worked in Sweden as a postdoctoral fellow and assistant professor before returning to Australia in 1997. He currently teaches genetics and investigates the molecular mechanisms underlying Alzheimer's disease at the University of Adelaide. He has been engaged in spreading awareness about Peak Oil and other resource issues since 2004 and has translated Kjell Aleklett's blog, *Aleklett's Energy Mix,* into English since 2008. To provide a more nuanced text, Professor Aleklett wrote the first drafts of *Peeking at Peak Oil* in his mother tongue of Swedish for subsequent translation by Michael.

Prologue

We live in a world that can no longer function without oil. Our dependence on oil has become so great that we can justifiably state we are addicted to it. We know that oil was formed under highly unusual and uncommon circumstances during the past 500 million years. Most of the world's extractable oil was discovered between 1945 and 1970. We know where on our Earth it is still possible to find new oilfields, however, the amount those new oilfields will yield will be limited compared to those oilfields already discovered.

During the past 100 years, detailed information on oil reserves and oil production has mostly been kept confidential by oil companies and national and international organizations. Over the past 10 years independent and university-based researchers have presented research that has made some of this information available to the wider community. Among the leaders in this work have been the researchers at the Uppsala Global Energy Systems group (UGES) at Uppsala University in Sweden. One aim of this book is to summarize the research findings of UGES in an easily understood manner.

To understand and prepare ourselves for a future that is not "business as usual" we must comprehend certain basic principles about oil. We must know where oil occurs and how much can be "produced" (extracted from underground). One of the most important things to understand is that, in any year, we can only produce a certain limited proportion of the oil that exists underground. This knowledge leads inexorably to the conclusion that there is a point in history when oil production reaches a maximum possible rate, Peak Oil, before declining. We must also understand what "unconventional" oil is and how it, together with new technology, can influence our future.

The western lifestyle is an oil addiction, and China and other developing nations also want their share of this drug. In 1950, the world had 2.5 billion inhabitants. Now there are 7.0 billion and, following current trends, we will have an additional 2.5 billion by 2050. All of these new world citizens will want to have their share of the world's oil at the same time as those concerned about climate change tell us we must stop using oil completely. Living in a world where oil production has peaked will mean competition (and maybe conflict) over the remaining oil resources where the unsuccessful will go without. You may choose to remain ignorant of your future and so yield control of it to others. However, if you want to understand where the world is heading then you need to be "Peeking at Peak Oil."

Acknowledgments

The path to completion of the book *Peeking at Peak Oil* has been long and sometimes difficult. First I would like to thank all of those who inspired me and then supported me as I began my research into Peak Oil. Without Colin Campbell there would have been no beginning and without support from Professors Bo Höistad and Mats Leijon there would presumably have been no research on Peak Oil at Uppsala University. Moral support from Professors Erkki Brändas, Leif Karlsson, Sven Kullander, and Svante Svensson has also been important.

Of equal importance has been the support I have received from my entire ASPO family around the world. I could name a great many people but to avoid the risk of overlooking someone I will simply say instead a huge THANK YOU to you all. Those of you with whom I have had especially close involvement know who you are and know that you are on my acknowledgment list.

This book is based on the research conducted by the Uppsala Global Energy Systems group (UGES) in the Department of Physics and Astronomy (from January 2012, Department of Earth Sciences), Uppsala University, Sweden. A crucial part of our research activity has been the work of my four doctoral students between 2003 and August 2011. I am very grateful to Mikael Höök, Kristofer Jakobsson, Fredrik Robelius, and Bengt Söderbergh. During the same period we have supervised 12 Diploma students and all of them have played important parts in our research work. It was especially gratifying when I was able to appoint our group's first Diploma student Anders Sivertsson.

A male colleague of mine has described the process of preparing a book as being like an extended childbirth. Of course, as men we have no experience of actually delivering a child but becoming a father is also a long process from when one receives the happy news that a child is expected until one proudly holds that child in one's arms. For me, bringing the book *Peeking at Peak Oil* into the world has been a little like becoming a parent but it has taken somewhat longer than 9 months.

Work on this book began in November 2010 when Michael Lardelli, Olle Qvennerstedt, and I got together in Australia at what we called "Camp Peak Oil Adelaide." The fact that Michael had been a postdoctoral scientist in Sweden for 6 years and had learned Swedish meant that I could write in my mother tongue and so concentrate on the book's contents. As a scientist he was able to elegantly interpret my descriptions of our research into English. I would never have signed the contract with the publisher, Springer, if I had not had Michael with me. When I say thank you there is a great deal more behind the words than gratitude.

Since the mid-1990s Olle has helped me illustrate my research presentations and many have envied the fact that I have such a friend. That Olle was willing to prepare the illustrations for this book was also a precondition to my signature on the contract with Springer. During "Camp Peak Oil Adelaide" Olle extended his offer to include all the diagrams and graphs and this gave the book a completely new character with (in my opinion) a very approachable charm. My thank you to Olle also encompasses so much more than the simple meaning of those words. Olle, I am immensely proud to have your beautiful artwork in my book.

As an embryonic concept grows and takes form there are many midwives who check its progress. *Peeking at Peak Oil* has taken form under the watchful eyes of a number of skillful professionals who have checked its contents to ensure it developed correctly. Primarily I would like to name Colin Campbell, Bob Hirsch, and Jean Laherrère who each helped check particular chapters in the book. Colin has also carefully read the entire text. Others who have given their time to examine and critique this work have been my friends and colleagues Göran Nyman, Leif Karlsson, and Simon Snowden.

When we established Camp Peak Oil Adelaide, Bob Couch put his house at our disposal. Bob also describes himself as a "nitpicker" and without his final checking of the text the list of errors that slipped through into print would have been much longer. Many thanks for your generosity and help!

My wife Ann-Cathrine has literally lived with *Peeking at Peak Oil* during the entire time it took to write it. As a teacher with an appreciation of the written word she has, of course, given me much good advice along the way. She became the book's expectant mother and deserves so many thanks! The rest of my family – my daughters Malin, Lovisa, and Kristin and their families – have also patiently tolerated my obsession with the Peak Oil topic and with this book. Many thanks to you all! It was during one of our many discussions about Peak Oil that Kristin's boyfriend David Kadish conceived the book's title of *Peeking at Peak Oil* for which I am very grateful.

Of course, I also owe a debt of gratitude to Dr. Liesbeth Mol, Editorial Director Physics at Springer who contacted me and came to Uppsala to persuade me to write *Peeking at Peak Oil,* and David J. Packer, Springer New York, who has been the executive editor of the book.

March 2012 Kjell Aleklett

Contents

Abbreviations

1P	Proven reserve
2P	Proven and probable reserves
3P	Proven, probable, and possible reserves
ADCO	Abu Dhabi Company for Onshore Oil Operation
API	American Petroleum Institute
ASPO	Association for the Study of Peak Oil and Gas
BPSR	BP Statistical Review of World Energy
CCS	Carbon Capture and Storage
CDIAC	Carbon Dioxide Information Analysis Center
CIA	U.S. Central Intelligence Agency
CTL	Coal-to-Liquids
CP	Cumulative Production
CERA	IHS Cambridge Energy Research Associates
CEO	Chief Executive Officer
CO_2	Carbon dioxide
DoE	U.S. Department of Energy
DRRR	Depletion of Remaining Recoverable Resources
EIA	U.S. Energy Information Administration
EU	European Union
GDP	Gross Domestic Product
GFF	Global Futures Forum
GOM	Gulf of Mexico
GTL	Gas-to-Liquids
IEA	International Energy Agency
IIASA	International Institute for Applied Systems Analysis
IOC	International Oil Company
IPCC	United Nations Intergovernmental Panel on Climate Change
ITF	International Transport Forum
MUST	Sweden's Military Intelligence and Security Service
NGL	Natural Gas Liquids
NOC	National Oil Company

NYMEX	New York Mercantile Exchange
OECD	Organization for Economic Co-operation and Development
OGJ	Oil and Gas Journal
OOIP	Oil Originally In Place
OPEC	Organization of the Petroleum Exporting Countries
PPP	Purchasing Power Parity
RRR	Remaining Recoverable Resources
SRES	Special Report on Emission Scenarios
SSA	Sub-Saharan Africa
UAE	United Arab Emirates
USGS	U.S. Geological Survey
URR	Ultimately Recoverable Resources
UGES	Uppsala Global Energy Systems
WEO	World Energy Outlook

Units

b	Barrels – 159 liters
boe	Barrel of oil equivalents
G	Billion (giga)
Gb	Billions of barrels, Gigabarrels
Gb/y	Billions of barrels per year, Gigabarrels per year
Gb/a	Gigabarrels per annum
J	Joule
k	Thousand (kilo)
kWh	Thousands Watthours
L	Liter
M	Million (mega)
Mb/d	Million barrels per day
ppm	Parts per million
T	Thousand billion (tera)
Tcf	Trillions of cubic feet
Tcm	Trillions of cubic meter
Wh	Watthour
ZJ	Zettajoule (1Z = 1,000,000,000,000,000,000,000)

Chapter 1

Introduction

Our everyday lives are completely dependent upon energy. Normally, we do not give a thought as to how the food on our table, the comfortable temperature of our household, our daily travel, or nearly everything else around us can be related to different forms of energy. We started modestly, using twigs and branches to fuel fires for use in preparing cooked food. Modern celebrity chefs still use the heat from burning wood to cook food but more often they use heat from burning natural gas or from electricity generated in coal-burning power stations. Meals today are prepared from ingredients obtained from all over the world and transported to us using oil. We also need oil for our personal transportation, for heating our homes, and as a raw material for plastics and other chemical products. In fact, our dependence on oil for production and transportation of food and other essentials from far away means that we now cannot live without it.

For thousands of years of human history we survived using only the renewable energy from the sun. Solar energy is captured in biomass (e.g., grain and wood), drives the winds, and produces the rain that fills rivers. We ate the grain, burned the wood, captured the wind, and used the flowing waters to build up our societies. Very recently, in the past 200 years, we learned how to tap the ancient solar energy stored in coal, oil, and natural gas. It took millions of years to capture and store this energy. During the nineteenth century coal was our most important fuel. However, oil rose to predominant importance during the twentieth century and now, in the early twenty-first century, natural gas is an increasingly significant source of energy. We still use renewable forms of energy but we rely on over 10 times as much energy from fossil fuels.

K. Aleklett *Peeking at Peak Oil*, DOI 10.1007/978-1-4614-3424-5_1,

Since the first oil wells were drilled in 1859 along the banks of Oil Creek in Pennsylvania, United States, oil has become an increasingly significant part of our lives, especially because it possesses the most concentrated energy of any fossil fuel. Today, crude oil has become the world's most important commodity and all nations use it. The products of crude oil such as gasoline and diesel fuel can be easily transported in the fuel tank of a car, and the energy they contain is released when we step on the accelerator pedal.

The fact that oil is a finite resource means that there was a date when oil use began and there will be a date when it ends. At some point between those two moments there is a period when we are "producing" oil (extracting it from the Earth) at a maximum rate. That period of maximum production is what we call "Peak Oil." In the first chapters of this book we discuss our dependence on oil and define the term *Peak Oil*.

Many experts have attempted to calculate how much oil exists in our planet Earth and where it can be found. Every previous estimate of when oil production will reach its maximum rate has been based on the limited knowledge available at that time. Obviously, incomplete knowledge about the abundance and distribution of the Earth's crude oil can lead to predictions of maximum oil production that are premature. These premature estimates have been mistakenly compared to Aesop's fable of *The Boy Who Cried Wolf*. In that fable, the boy shouted that the wolf was coming despite knowing that it was a lie. However, previous premature warnings of an approaching oil production maximum were not intentionally mistaken. They simply used the knowledge and prediction methodologies available at the time, which is how science always works.

Today we know that the world's oil companies made their largest discoveries of oil during the 1960s. The average size of oilfields discovered has decreased in every decade since then. This trend is now sufficiently clear (and irreversible) that we can estimate how much oil will be discovered in the future. We know also that there are limits to how much oil can be produced in any year. All this knowledge and the fact that the world's largest oilfields are showing declining production mean, as we show throughout the book, that the hopes of oil companies and other national and international agencies for increased oil production must be regarded as wishful thinking.

In the 150 years since we first drilled for oil our knowledge regarding all aspects of the production and exploitation of this resource has increased enormously. In *Peeking at Peak Oil*, we discuss this knowledge as well as the factors that will determine future oil production. As we progress through the chapters of this book we describe the *global oil factory*, including where on our planet sedimentary layers exist that provide a source of oil and what geological conditions are required for it to be produced.

Geologist Colin Campbell then shares his insight on how oilfields are discovered through geological fieldwork and geophysicist Jean Laherrère tells us how technology assists in this process. We then explain some of the nomenclature that the oil industry uses to describe the quantity of oil in an oilfield before examining the technological art of producing this oil. The crude oil in an oilfield is a dark finite resource, as dark and as finite as the fluid in a bottle of Coca-Cola. Drinking a Coke and tapping an oilfield are very different processes but this book shows you that a Coke bottle can teach us a great deal about oil production! This then leads us into a discussion of the physical laws and economic principles that determine oil production.

Previously, the databases used to calculate future oil production have been owned by the oil industry. In the past decade, however, some oil databases have been assembled beyond industry control. The largest academic database is now maintained at Uppsala University in Sweden by the research group Uppsala Global Energy Systems (UGES). So far, UGES has used this information to examine various factors (parameters) that are important for future oil production. This research has been published in more than 20 scientific papers. It is obvious from this work that the largest 1% of the world's oilfields—the "giants" or "elephants"—originally contained almost two thirds of all the crude oil that was ever formed. Production from the giants will be decisive for the world's future oil production. One chapter summarizes the research that UGES has published about the giants including an important parameter that can be used to estimate future oil production, namely depletion of remaining recoverable resources, DRRR. Then, in Chap. 11, "The Peak of the Oil Age," we use this information to examine the accuracy of the predictions published by the International Energy Agency (IEA) which was established by the Organization for Economic Co-Operation and Development (OECD) to advise it on energy. You will find the results surprising!

The oil from the giant oilfields is usually described as "conventional" but UGES has also analyzed production of unconventional oil such as that produced from Canada's oil sands. Unconventional oil will be an important part of oil production in the future so we have dedicated an entire chapter to it. Another important variety of oil production that we examine is from areas under deep water. The Deepwater Horizon disaster in the Gulf of Mexico showed how challenging this oil production can be.

Access to energy is vital for the development of national economies, and so oil has geopolitical importance. Four nations have played, and will continue to play, a central role in the history and future of oil. These are the two largest oil-exporting nations, Saudi Arabia and Russia, and the (now) two largest importing nations China and the United States. In four chapters, we look at the past and future of oil production and consumption in these nations including one chapter on how military and intelligence agencies have been interested in Peak Oil and, especially, in our research. We feel it is important that we discuss this although we realize that there are those who would rather we did not.

In our chapter on Saudi Arabia we can, for the first time, present a detailed prognosis for the future production from the supergiant Saudi oilfield Abqaiq. We then extend the results of this analysis to make a detailed prognosis for the future oil production of Saudi Arabia as a whole. Saudi Aramco's former vice-managing director Dr. Sadad al-Husseini has examined our chapter on Saudi Arabia and has not disagreed with any of our conclusions. We have also examined the rosy projections recently presented by the US Department of Energy and can only conclude that (put politely) our analysis raises many questions.

Transport uses more than 70% of all oil produced so naturally we discuss the future of this sector of the economy. We also discuss the significance of oil for future climate change. Peak Oil signifies that only a limited amount of oil can be consumed in the future so we conclude this book by examining how our future will be affected by it. To comprehend better how restricted future oil production will be, consider this simple fact: all the oil remaining to be produced in Iraq is estimated to be 115 billion barrels. Every year the world economy consumes 30 billion barrels. This means that all the oil in Iraq could only supply the world for 4 years.

We realize that our discussion of Peak Oil may be considered controversial inasmuch as it paints a picture of reality that is different from that portrayed by national and international agencies and by the oil companies themselves. However, our picture is based on scientific research and the data on oil resources that are currently available. Of course, we have also attempted to estimate how new discoveries of oil and technological advances will affect future oil production.

Certainly there will be those who attempt to portray this book as yet another example of crying wolf but we must emphasize again that our calculations regarding Peak Oil are based on unbiased research. Those who wish to dismiss our warnings should be reminded that, at the end of Aesop's fable, the wolf did actually appear. The chief economist of the IEA, Fatih Birol, said in 2008 that four new Saudi Arabias would be needed to maintain a constant rate of oil production in the future. By 2011 he had changed his story a little and is now saying that we need "two new Middle Easts" [1]. Although he did not use the words "Peak Oil," Dr. Birol delivered a warning about our future as stark as that we give.

Because the scientific research of the UGES group is the basis for the conclusions found in this book we finish this introduction by presenting a recent assessment of UGES by an international review panel reviewing all the research conducted at Uppsala University. In the spring of 2011 they wrote [2]:

> The panel identified the controversial nature of some of the findings of the UGES but they felt it is very likely that the UGES point of view is correct, and that a distinguished university of long standing (Uppsala University) is a perfect home for well-informed, academically sound researchers who occasionally annoy senior politicians and business people. As Lord Luce said in London in May, 2011, 'It was the job of an independent university not to be afraid to annoy people,' and similar views have been expressed down the centuries by Aristotle [3, 4], Bacon [5], and Newman [6].

References

1. WEO: World Energy Outlook 2011. International Energy Agency. http://www.worldenergyoutlook.org/2011.asp (2011)
2. Quality and Renewal: An overall evaluation of research at Uppsala University 2010/2011, p. 371. http://uu.diva-portal.org/smash/get/diva2:461235/FULLTEXT01 (2011)
3. Wikipedia: Aristotle. http://en.wikipedia.org/wiki/Aristotle (2012)
4. Diogenes Laertius: The Lives and Opinions of Eminent Philosophers, Life of Aristotle, Section XI (Translated by Yonge, C.D.). http://classicpersuasion.org/pw/diogenes/dlaristotle.htm (1853)
5. Wikipedia: Francis Bacon. http://en.wikipedia.org/wiki/Francis_Bacon (2012)
6. Wikipedia: John Henry Newman. http://en.wikipedia.org/wiki/John_Henry_Newman (2012)

Chapter 2

Peak Oil

At the start of the new millennium, the expression "Peak Oil" was unknown. Nevertheless, a discussion about when the world's rate of oil production would reach its maximum had already begun when the geologist M. King Hubbert presented his model for future oil production in the United States in the 1950s. At that time, Hubbert worked for the Shell Company and his model was discussed for the first time at a conference organized by the American Petroleum Institute (API) from the 7th to the 9th of March 1956 at the Plaza Hotel in San Antonio, Texas.

Hubbert's written conference presentation, *Nuclear Energy and the Fossil Fuels,* was catalogued in June 1956 at the Shell Development Company, Exploration and Production Research Division, Houston, Texas, as "Publication No. 95" [1]. Early in 1957, the API published their 1956 issue of *Production Practice,* thus making the Hubbert model available to all API members.

Half a year before his death, in the spring of 1989, Hubbert described how, on the day that he was to deliver his lecture in San Antonio, Shell tried to get him to tone down his assertion of an approaching production maximum in the United States, but he refused [2]. He also related that, for several years after his presentation, Shell held internal courses for its personnel, and that he presented his model on these occasions. This means that the issue that today goes under the name of "Peak Oil" is something that Shell has known about for more than 50 years. The fact that the API published "The Hubbert Model" in 1957 means that other oil companies have had access to the same information as Shell from that publication date. The question then is why they have swept their discussion of Peak Oil under the rug and have not discussed it more publicly (Fig. 2.1).

K. Aleklett *Peeking at Peak Oil,* DOI 10.1007/978-1-4614-3424-5_2,

Fig. 2.1 The fact that API published the Hubbert model in 1957 means that Shell and other oil companies since then have had access to the information that oil production will peak. The question is why they have swept their discussion of Peak Oil under the rug and not discussed it publicly

When Hubbert developed his model, it was not the rate of oil production that was of concern but, rather, reported oil discoveries. A review showed that discoveries of oil in the United States' Lower-48 (the 48 states south of Canada), reached a maximum during the 1930s and that the trend was downward in 1955. We know that oil is a finite resource formed under unusual conditions millions of years ago. Therefore, the year when humanity discovered the first barrel of oil will certainly be followed by a year when we discover the last. This discovery history can be approximated by a curve that has a maximum (a peak) when half of the oil resources have been found. The determining factor for the curve's form is the total amount of oil that can be found. Based on the estimates available in 1956, Hubbert used limiting values of 150 and 200 billion barrels of oil for the United States' Lower-48. He further assumed that the production rate curve would have the same form as the discovery curve and constrained the curve to fit the production data up to 1955. Using these assumptions he could predict a maximum rate of oil production sometime between 1965 and 1971. Today, we know that the upper limit was close to reality and that production in the United States' Lower-48 reached its maximum level in 1971.

When Hubbert attempted a similar analysis for the world's oil production, he calculated an estimate of the maximum production rate as occurring during the 1990s. We now know that this was an underestimation. The main reason is that the world's oil production cannot be fitted to a single Hubbert curve because there are many petroleum-producing regions in the world, each of which has its own maximal rate of production, and each must be studied individually. The fact that oil production from the Middle East was restricted for political reasons during the end of the 1970s and beginning of the 1980s is another important factor, and today we should be grateful for all the oil this disruption saved. Some have tried to model oil production rates by combining Hubbert curves for various regions and have, in this way, succeeded in describing broadly the course of history [3]. However, all these curves have a maximum when half of the oil has been produced and detailed analyses of what has really happened in the various regions gives a different picture. Uppsala Global Energy Systems (UGES) at Uppsala University, Sweden, have published detailed studies on this topic that are described later in this book.

Hubbert modeling is a method based in statistics rather than physics. In a Hubbert model, oil production data are fitted to a type of mathematical curve called a logistic curve. Hubbert modeling assumes that the rate of oil production will be maximal when half of the oil reserves have been produced. When the petroleum geologist Colin Campbell began to study future oil production he introduced two fundamental changes. He began fitting curves on a nation-by-nation basis and, more significantly, he based this curve-fitting on "depletion" analysis. According to Campbell, depletion is a measure of what fraction of the oil reserves remaining at the beginning of every year in an oilfield or region can be extracted. In contrast to the Hubbert model, when depletion is measured in this way it also reflects the physical characteristics of an oilfield: the pressure in the field, the porosity of the oil-bearing rock, and the viscosity of the oil. Campbell's method does not assume that the history of the rate of oil production will be symmetrical. The rapidity with which oil production increases before the peak does not need to match the rapidity at which it falls after the peak and the peak itself need not occur when half the oil has been produced.

Some have criticized the Hubbert model for underestimating the rate of production during the latter phase of production from an oilfield or region. However, this does not mean that the Hubbert model has not been useful. To make predictions of future oil production rates, both the Hubbert and Campbell models require estimates of total available oil reserves to be provided. Therefore, the total amount of oil that can be consumed under the two models is the same. The two models differ only in the future production trends that each foresees. When the Hubbert model was developed in the 1950s information on oil discoveries and production profiles was limited and under those conditions the Hubbert model was very useful for making crude estimates of future production rates. Today we know far more about the history and practice of oil discovery and production for various types of oilfields and regions and this has enabled us to improve our methods for predicting future production rates. We examine these refined methods later but for the moment we simply state that Hubbert and Campbell did pioneering work that led to our current ability to estimate future rates of oil production.

In December 2000 Campbell began to discuss the formation of an organization that would study oil production rate maxima. At first the name proposed was *The Association for the Study of the Oil Peak*, ASOP. During a discussion between Colin Campbell and me in the same month, Campbell suggested that we should invert "Oil Peak" to read instead "Peak Oil," and so ASOP became ASPO, the acronym still used today. In January 2001 Campbell wrote his first newsletter for ASPO, the Association for the Study of Peak Oil and Gas, and a total of 20 people received that newsletter. In May 2002, at a meeting in Uppsala, ASPO was formally established, and Bruce Stanley from Associated Press (AP) used the expression "Peak Oil" for the first time in the international press [4]. Today (December 2011) a "Peak Oil" search on Google results in over 7,500,000 hits. Campbell has also given us a definition of Peak Oil [5]. "The term Peak Oil refers to the maximum rate of the production of oil in any area under consideration, recognizing that it is a finite natural resource, subject to depletion."

The future of oil production is decisive for the future of oil companies, and it is to their advantage if the public has limited knowledge of this issue. National bodies, such as the Energy Information Administration (EIA) in the United States, and international bodies, such as the International Energy Agency (IEA) based in Paris, have, for many years, made prognoses of future production. However, only limited information is available on how these prognoses are produced. None of them satisfies the requirements of a scientific publication, and for many there are indications that a political agenda might be influencing the prognoses. The fact that governments around the world use these prognoses to plan our common future—and that Peak Oil will be decisive for that future—means that everyone should possess knowledge of this subject.

At the first-ever *Peak Oil Conference* in Uppsala in 2002 ASPO set the bar for global oil production in 2010 at 85 million barrels per day (Mb/d) [5] (for oil production as defined by BP [6]). Today we know that the oil industry could not clear that height as they only reached 82 Mb/d in 2010 [6]. In Fig. 2.2 this is illustrated by a high jumper knocking off the bar. An analysis of future oil demand published by the IEA in *World Energy Outlook 2010* showed that the world needs more oil production to allow for future economic

Fig. 2.2 The path up to the peak rate of oil production, Peak Oil, has been long and bumpy with many events along the way that must be explained. The route down begins at the peak and how it will affect us all is of vital importance. Peak Oil will determine our future, and we need to build a substantial "crash mat" of alternative fuel production to cushion us from the fall in conventionally produced oil and the natural gas liquids that are produced in association with conventional oil

growth [7]. This means that there will be a great need for production of alternative fuels in the future as symbolized by the crash mat in Fig. 2.2. We discuss possible alternative fuels later in this book but it is already worth noting that it will be difficult to produce even the volume suggested by the thickness of the crash mat shown. If we fail to provide a crash mat of sufficient thickness the high jumper will suffer a very hard landing!

Peak Oil and Energy Demand

Any geographical area producing oil, no matter how large or small, will experience a moment of maximal oil production that we term Peak Oil. This applies to individual oilfields inasmuch as each oilfield is finite. It must also apply to any oil-producing region made up of these oilfields. Our finite world is a collection of oil-producing regions so it too must reach a point of maximal oil production, Peak Oil, before the rate of global production inevitably declines.

Discussions of global oil production concern only those nations that possess oilfields and produce oil but discussions of oil supply and consumption concern every nation. Today there is not a nation on Earth that does not use oil and so Peak Oil will affect us all. Peak Oil is only one aspect of global energy use. Therefore, before we discuss Peak Oil in detail in this book we must look at the world's use of oil relative to other energy sources.

When the IEA [7] and BP [6] discuss our sources of energy they categorize them in the following way.

- Fossil energy: coal, oil, and natural gas
- Nuclear energy
- Renewable energy: hydro, wind, solar, biomass, and other sources

Our use of energy over the past four decades is shown in Fig. 2.3. It is obvious that fossil fuels dominate our energy supply [6]. Indeed, all the nuclear and renewable energy combined is still less than 60% of the energy we derive from the least-used fossil fuel, natural gas. Figure 2.3 also shows that we used less energy in total in 2009 (a year of economic recession) than we did in 2008. However, in 2010 total energy use returned to record levels [6]. A closer look at oil production in the past decade shows that this leveled off since 2005 and demand (and so price) continued to rise. In other words, since 2005 our use of oil has been limited by production, not demand. The crucial question now is what will happen to oil production during the coming 25 years. In the prognoses presented by the IEA it sees the rate of oil production continuing to rise until 2035 but in *Peeking at Peak Oil* we show that this is not possible.

Activity requires energy and so increased economic activity (economic growth) requires an increased rate of energy use. Historically, increased use of oil correlates best with increased economic activity. (This is discussed in the section "The Economy and Peak Oil", Chap. 19.) All nations use oil so the economy of every nation, and the world economy as a whole, will be affected by Peak Oil. If economic growth and increased oil use go hand in hand then so too must increased carbon dioxide production from burning oil and other fossil fuels. Climate researchers and politicians tell us that we must halve our fossil fuel use by 2050, so from that point of view Peak Oil should be their (and our) best friend. However, to economists, the concept of Peak Oil (and finite resources in general) is like a red rag to a bull. Many economists dismiss Peak Oil on theoretical grounds that have nothing to do with physical reality and the laws of nature. Unfortunately, our politicians

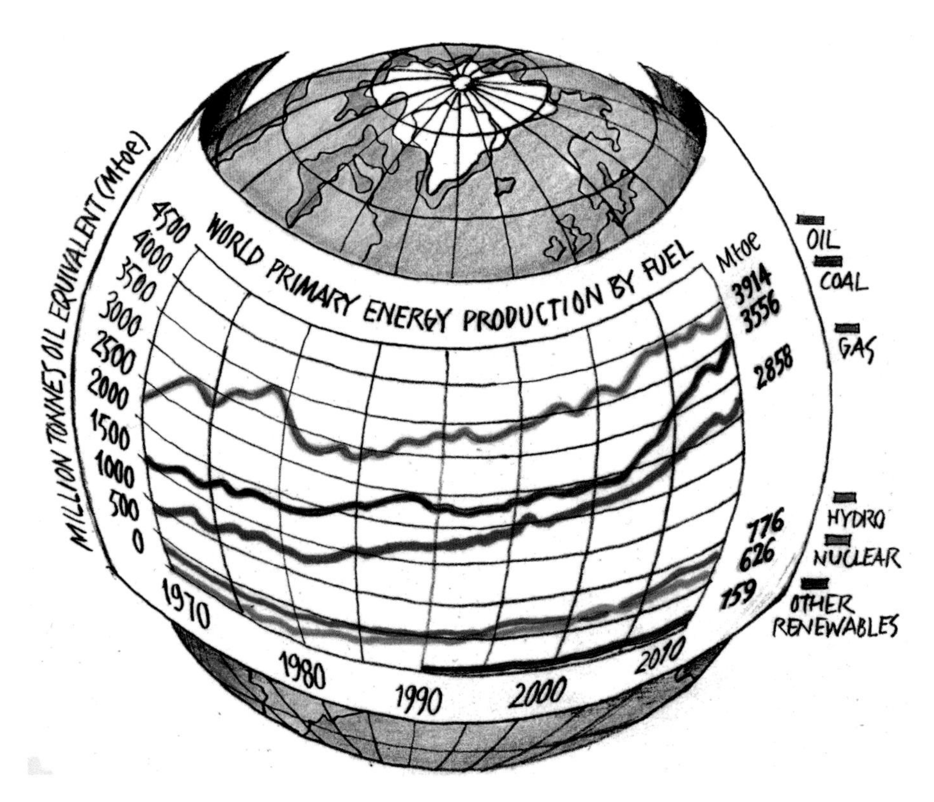

Fig. 2.3 The history of world energy production from 1970 to 2010 showing the contributions to the primary energy supply made by different energy sources. The numbers for "Other Renewables" are based on gross generation from wind, geothermal, solar, biomass, and waste [6]. To allow comparison of these different sources of energy to oil, the energy supplied by each is reported in terms of the heat it can provide (i.e., as thermal equivalence [8]) and is expressed in multiples of the energy in one million tonnes of oil (US: metric tons), that is, Mtoe (US: mt), million tonnes of oil equivalent

have listened to those economists and not the scientists who have been warning about Peak Oil for many years.

To maintain our current economy, a decline in oil use must be countered by an expansion in the use of renewable and/or nuclear energy. Wind and solar energy are popular with the general public and their use is growing dramatically. However, the contribution of wind and solar energy to total world energy use is still only minuscule. These energy sources produce mainly electricity. Most of the world's electricity is generated using coal and natural gas, so increased use of solar and wind energy will replace those fossil fuels but not oil.

Transport is that sector of the global economy requiring the most oil and Peak Oil will affect it severely. Local transport might use electricity stored in batteries but for transport over long distances liquid fuel is currently essential. Ethanol and biodiesel are liquid biofuels that can be used instead of oil but our analysis in this book shows that their potential to replace oil use is only marginal. Coal and natural gas might also be used to fuel transport so we examine their possible contribution.

The use of oil is deeply integrated into our global energy system so it is essential that we all become aware of the changes our society will need to make to cope with Peak Oil.

What Is Reported as Oil?

The standard unit for measuring volume is the liter (L) but when measuring oil volumes these are described in barrels. One barrel equals 159 L. The rate of production or consumption of oil can be stated as per day (commonly as millions of barrels per day, Mb/d) or per year (commonly as billions of barrels, or gigabarrels, per year, Gb/year, or per annum, Gb/a). Crude oil can vary widely in various qualities such as density, sulphur content, and so on. This is discussed in Chap. 10. Recently, various agencies have also begun to count other chemicals such as ethanol as part of the world's "oil" supply. However, changing the definition of what constitutes oil in this way can complicate comparisons of oil production and consumption between different eras.

In this book we use information primarily from the International Energy Agency, the US Energy Information Administration, and the *BP Statistical Review of World Energy*. The IEA usually discusses the oil the world needs, "demand," and what oil is available to meet that need, "supply." In their *Oil Market Report of March 2011* under the heading "World Oil Supply" they state that 87.4 Mb/d of supply existed for 2010 [9]. This volume includes 1.8 Mb/d of ethanol, 2.3 Mb/d of "processing gains" (which are increases in volume that can occur as oil passes through a refinery), and 0.2 Mb/d CTL (coal-to-liquid) and GTL (gas-to-liquid). From this we can calculate that the actual production of all varieties of oil is only 83.1 Mb/d.

Like the IEA, the EIA also includes processing gains when describing the oil supply but it does not include ethanol [10]. The *BP Statistical Review of World Energy* describes both production and consumption of oil [6]. For 2010 BP saw oil production as 82.1 Mb/d. In this they included crude oil, oil from oil shale, oil from oil sands, and natural gas liquids (NGL) that are produced in association with natural gas production. A barrel of NGL contains significantly less energy than a barrel of crude oil. This can lead to confusion if the NGL fraction of the "oil" supply is reported in simple barrels (as practiced by the IEA) rather than as barrel of oil equivalents (boe, as practiced by the EIA). Reported available volumes of liquid fuels can also include synthetic oil produced using coal or natural gas.

In this book we focus most of our attention on how much oil exists in oilfields and how much can be produced. When we discuss production of oil, as we do in Fig. 2.2, then we use the definition of this given by BP. For 2010 the oil produced (as defined by BP) was 82.1 Mb/d or 30.0 Gb/year. The most important component of oil production is conventional crude oil and in 2010 this represented 85% of total oil production.

References

1. King Hubbert, M.: Nuclear Energy and the Fossil Fuels, Shell Development Company, Exploration and Production Research Division, Houston, TX, Publication No. 95, 1956. http://www.energybulletin.net/node/13630 (1956)
2. Ronald, E.D.: Interview with Dr. M. King Hubbert, Jan 4–6, 1989. Niels Bohr Library and Archives. http://www.aip.org/history/ohilist/5031_1.html (1989)
3. Nashawi, I.S., Malallah, A., Al-Bisharah, M.: Forecasting world crude oil production using multicyclic Hubbert model. Energy Fuels **24**, 1788–1800 (2010)
4. Stanley, B.: Oil Experts Draw Fire for Warning. Associated Press, May 24, 2002. http://www.semissourian.com/story/75282.html (2002)
5. ASPO, The Association for the Study of Peak Oil and Gas. http://www.peakoil.net (2012)
6. BP: BP Statistical Review of World Energy, June 2011. http://bp.com/statisticalreview (historical data; http://www.bp.com/assets/bp_internet/globalbp/globalbp_uk_english/reports_and_publications/statistical) (2011)
7. WEO: World Energy Outlook 2010. International Energy Agency, November 2010. http://www.worldenergyoutlook.org/2010.asp (2010)
8. Since available data on the energy from oil, coal and gas are given in thermal energy units, the volumes of energy available from hydro, nuclear, and other renewables are converted on the basis of thermal equivalence assuming a 38% conversion efficiency as in a modern thermal power station (2010)
9. International Energy Agency: Oil Market Report 15 March 2011, April 11, 2011. http://omrpublic.iea.org/omrarchive/15mar11full.pdf (2011)
10. U.S. Energy Information Administration: Petroleum and Other Liquids, World Oil Balance. http://www.eia.gov/emeu/ipsr/t21.xls (2011)

Chapter 3

A World Addicted to Oil

A world addicted to oil was the title of my presentation on Capitol Hill on October 19, 2005. The Worldwatch Institute was host for the presentation. That was the day that the term "Peak Oil" was introduced to the corridors of power in Washington. There was standing room only in the seminar room packed with aides to US representatives, senators, and secretaries.

There is still debate over when we really will reach (or if we have passed) the maximum rate of oil production, Peak Oil. However, there is a consensus that we have passed that point in time when the rate of crude oil discovery was maximal. During the 1960s an average of 56 billion barrels was discovered each year. Of course, the attitude at that time was that we were literally swimming in oil because we were only consuming 10 billion barrels per year. Today, we consume over 30 billion barrels every year, far exceeding what the oil companies are finding. Figure 3.1 shows the average yearly amount of crude oil discovered for each decade up to 2009. The total amounts to some 1,900 billion barrels. Following the historic trend of discovery (see Fig. 6.4 for details), the rate of future discovery is also provided for the next 50 years, adding up to 200 billion barrels. Constant production at current levels for the next 50 years would require 1,500 billion barrels, which is 1,300 billion barrels more than the extrapolated discovery. To emphasize that we are engaged in a unique, once-only exploitation of a finite resource, we have shown these discoveries over a time period of 500 years.

Oil is the world's most important raw material. Every day we extract approximately 82.1 million barrels (Mb/d) from the Earth (see Chapter 1 for definition of oil). In the spring of 2011 the usual market price of a barrel of oil was around US$110 so the world was spending about US$9 billion every day to satisfy its need for oil. However, the market price has shown a tendency to be volatile and during July 2008 the price passed US$140 per barrel to

K. Aleklett *Peeking at Peak Oil*, DOI 10.1007/978-1-4614-3424-5_3,

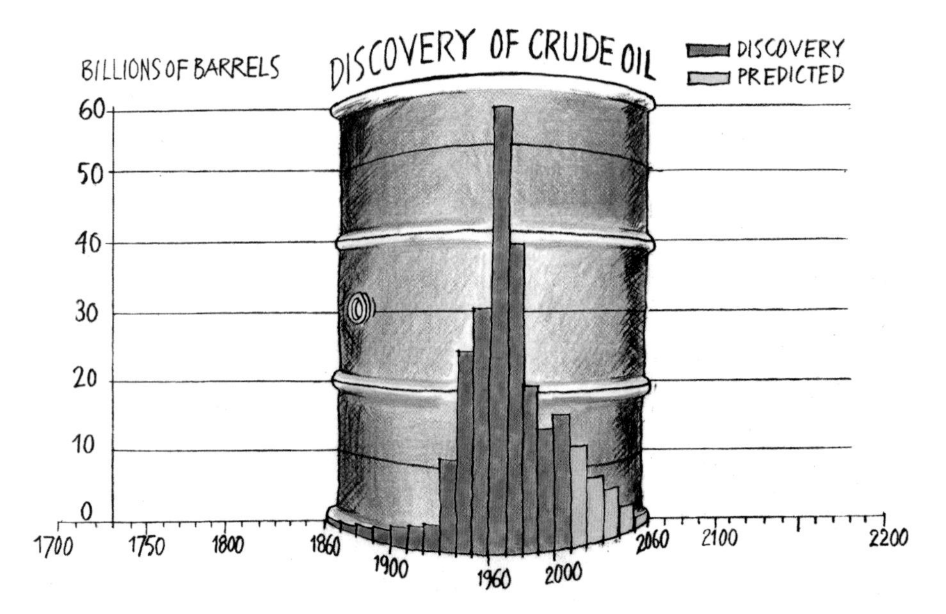

Fig. 3.1 Average yearly volumes of discovered crude oil per decade. The maximum average value of 56 billion barrels per year was reached in the 1960s during the period 1960–1969 [1]

reach a record of US$147. During that month the world's daily oil bill was around US$12 billion. This financial burden was so great that the world's economy was seriously affected.

Our everyday life is filled with hundreds of products, the making of which all require oil. This is why it is commonly asserted that we are addicted to oil. When oil is discussed (e.g., in the media) the unit of volume we use is the "barrel" which is about 159 L. However, the formal definition of a barrel of oil is US 42 gal (where 1 gal is 3.79 L). So our daily production of 82.1 Mb/d equals 3,448 million gallons or 13 billion liters, an almost incomprehensibly large amount.

Our dependency on oil is due to the energy that is released when we use it. Oil can perform a very large amount of work for us. The official unit for energy and work is the "joule" (J). However, one joule is a very small amount of energy so it is common practice instead to use units of measurement that describe how many joules one uses in an hour. The "watthour" (Wh) is one such unit. However, even this unit is often too small to be useful so it is more common to describe energy in thousands (kWh), millions (MWh), billions (GWh), or thousand billion (TWh) watthours (k, M, G, and T denote kilo-, mega-, giga-, and tera-, respectively). When considering the energy content of crude oil it is easiest to remember that 1 L of crude oil can release 10 kWh of energy.

Fig. 3.2 An illustration of work equivalent to 1 kWh

To begin to comprehend our dependence on oil we must first understand how much work 1 L of oil can do. Imagine that you park a small car weighing 1,200 kg at the base of the Eiffel Tower. You tie a rope in the car's tow-hold and then climb to the top of the tower (that is, 321 m high). Then, by hand (with the help of a pulley) you raise the car to the top of the tower. You have now done work equivalent to 1 kWh (see Fig. 3.2). The human capacity to do work varies between individuals but to perform a task equivalent to 1 kWh one would need to be strong and fit, and to work for at least 2 days. That means that the energy stored in 1 L of crude oil is equivalent to raising ten cars to the top of the Eiffel Tower or to the work that 20 people can perform during 1 day.

The gasoline tanks of many cars hold 50 L (13.2 gal). The next time you fill your car's tank take a moment to consider that it holds energy equivalent to the work of lifting 500 cars to the top of the Eiffel Tower. Alternatively,

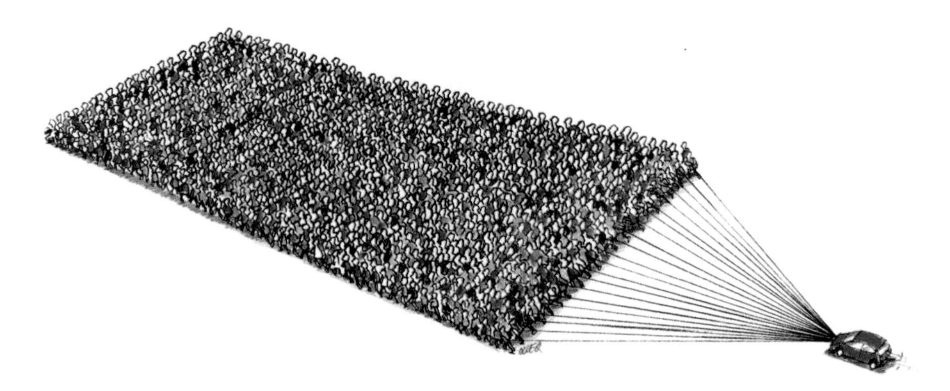

Fig. 3.3 The work of 1,000 people in the course of 1 day is equivalent to the energy content of 50 L of gasoline

imagine that you have 1,000 people in front of your car (see Fig. 3.3), and that, in the course of 1 day, they can pull your car about 500 km (300 miles). Once you understand this you can see how it may be justified to say that oil has helped abolish slavery.

During my presentation in Washington we also discussed the fact that the United States, with about 5% of the world's population, at that time consumed about 25% of the world's oil production, whereas China, then with about 21% of the world's population, only used 8% of the world's oil. When I asked the audience if they thought it was acceptable for China to have the ambition to consume 21% of the world's oil, the answer was yes. Presumably, they were not aware of what this would mean for the United States. If the world currently consumes 82.1 Mb/d, China's share would be over 17 Mb/d. At the moment Chinese consumption has risen to just over 10% of world oil production. If China is to increase this to 21% and total world oil production remains flat, then the additional 11% must come from a redistribution such that the United States, Europe, and Japan reduce their consumption by an equivalent amount.

My presentation on Capitol Hill in October 2005 sowed a seed of interest. Together with four others, I was invited back to Washington in December of that year to testify before the House of Representatives Subcommittee on Energy and Air Quality of the Committee on Energy and Commerce [2]. Some of those who testified presented the opinion that Peak Oil was not a concern. One of them was Robert Esser, senior consultant and director of IHS Cambridge Energy Research Associates (CERA), an organization that produces oil analyses for sale to industry. Even today CERA continues to advance the opinion that Peak Oil will not occur until after 2030 [3].

In 2005 we were engaged in intensive research on Peak Oil at Uppsala University and the testimony I gave to the subcommittee was based on our preliminary results. These results were summarized in an article in the *World Watch Magazine* with the title "A Bumpy Road Ahead" [4]. These preliminary results and other studies have now been published in more than 20 scientific articles [5] and form the backbone of the book *Peeking at Peak Oil*. The aim of this book is not to compare our approach to Peak Oil with what others have done. That comparison can be found in the report *Global Oil Depletion—An Assessment of the Evidence for a Near-Term Peak in Global Oil Production* [6].

The message I delivered on Capitol Hill (that the world is addicted to oil) was echoed 3 months later in an address to the nation by President George W. Bush in 2006. He stated, "Keeping America competitive requires affordable energy. And here we have a serious problem: America is addicted to oil, which is often imported from unstable parts of the world" [7].

References

1. Robelius, F.: Giant oil fields—the highway to oil: Giant oil fields and their importance for future oil production. Uppsala Dissertations from the Faculty of Science and Technology, Uppsala University Library, Uppsala, Sweden, ISSN 1104–2516, p. 69. http://uu.diva-portal.org/smash/record.jsf?pid=diva2:169774 (2007)
2. House of Representatives: Understanding the peak oil theory. Hearing before the Subcommittee on Energy and Air Quality of the Committee on Energy and Commerce (Aleklett page 32). http://www.peakoil.net/Aleklett/Hause_Peak_Oil_hearing_2005.pdf (also http://www.energybulletin.net/node/11621) (2005). Accessed 7 Dec 2005
3. Jackson, P.: The Future of Global Oil Supply: Understanding the Building Blocks. IHS Cambridge Energy Research Associates (CERA). http://www.cera.com/aspx/cda/client/report/report.aspx?KID=5&CID=10720 (2009)
4. Aleklett, K.: A bumpy road ahead. Word Watch Magazine, 19: 1. http://www.worldwatch.org/press/prerelease/191-peak-oil.pdf (2006)
5. Home page for Uppsala Global Energy Systems (UGES): Peer Reviewed Publications. Uppsala University, Sweden. http://www.physics.uu.se/ges/en/publications/peer-reviewed-articles (2011)
6. Sorrell, S., Speirs, J., Bentley, R., Brandt, A., Miller, R.: Global oil depletion—an assessment of the evidence for a near-term peak in global oil production. A report produced by the Technology and Policy Assessment function of the UK Energy Research Centre. http://www.ukerc.ac.uk/support/tiki-download_file.php?fileId=283 (2009)
7. Bush, G.W.: State of the Union Address, 31 Jan 2006. http://www.washingtonpost.com/wp-dyn/content/article/2006/01/31/AR2006013101468.html (2006)

Chapter 4

The Global Oil and Gas Factory

To discuss Peak Oil it is important that everyone have a similar understanding of what oil is and how the oil that is currently extracted from the Earth by large national oil companies (NOCs) and international oil companies (IOCs) was formed, but rarely, millions of years ago. In purely chemical terms, oil is a blend of molecules consisting of hydrogen and carbon: hydrocarbons. One can regard oil as the end product of a gigantic manufacturing process deep down in the Earth's crust. The hydrogen and carbon in oil come from water and carbon dioxide that were in circulation in the living world many millions of years ago. It was primarily algae, plankton, and tiny marine plants that bound the hydrogen and carbon into molecules in their bodies and then sank to the bottom of shallow seas and lakes where they accumulated in thick layers of biological sediment.

The oil extracted from the Earth and for which a price is set on the various oil markets around the world is called "crude oil." Some crude oils flow easily and are lighter than other oils that are more viscous and heavy. The difference between "light" and "heavy" oils is determined by the relative amounts of hydrogen and carbon in the oil. Because carbon atoms are heavier than hydrogen atoms, oils with more carbon are heavier.

Every oil field from which crude oil is produced contains its own individual blend of molecules. Therefore, the oils produced by particular oilfields have been chosen as standard varieties for which prices are set and against which other oils are compared. The price that is most commonly mentioned in news reports is the price of the standard oil type known as WTI, "West Texas Intermediate" crude. WTI is a light oil containing low amounts of sulphur. Oils with low sulphur content are described as "sweet," thus WTI is a "light, sweet" crude oil. When oil production began in the North Sea area in the 1980s they decided to adopt their own standard. For this they chose the

K. Aleklett *Peeking at Peak Oil*, DOI 10.1007/978-1-4614-3424-5_4,

Fig. 4.1 By blending together some easily available items one can recreate crude oil. A suitable blend is 4 L of gasoline, 3 L of diesel, 1 L of heating oil, 1 L of motor oil, and 1 L of tar. Sulphur can be added as a spice

oil extracted from the Brent oilfield in the British part of the North Sea. "Brent crude" is a little heavier and contains a little more sulphur than WTI but it is still described as "sweet." When crude oils contain a lot of sulphur they are described as "sour." The final price for the oil from a field or for a blend of oil from different fields is determined by how heavy and sour it is. Light, sweet crude oils are the most valuable. The heaviest oil that is extracted today comes from the oil sands of Canada, and it is as thick as tar.

Many people think of oil as something black and viscous. It is true that crude oil is black but its usual consistency is more like Coca-Cola than viscous tar. Oil refineries take crude oil and separate it into various "products" consisting of different types of hydrocarbon molecules. So, if you want to recreate crude oil you can do this by mixing the refined oil products back together. Here is a simple recipe for making something resembling crude oil: mix 4 L of gasoline (which can represent the combined products ethane, naphtha, gasoline, and aviation fuel) with 3 L of diesel, 1 L of heating oil, 1 L of motor oil, and 1 L of tar. To spice it up a little you can sprinkle in some sulphur (see Fig. 4.1). This blend flows quite freely as crude oil must if it is to flow through the porous rocks in which it is found deep underground.

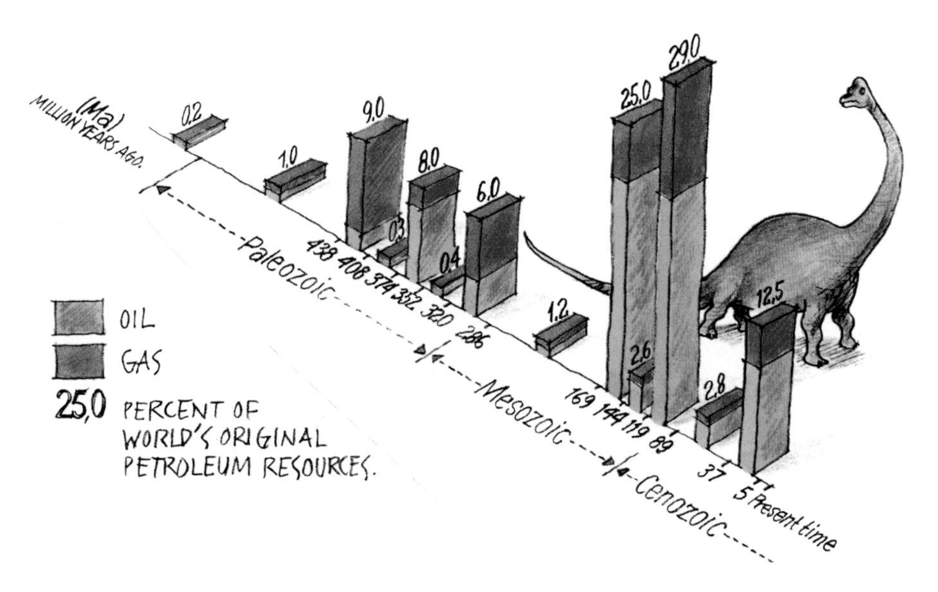

Fig. 4.2 The reserves of oil and natural gas that have formed during the last 600 million years. Most of the oil (about 60%) was formed during the Mesozoic era. During this era approximately equal amounts were formed during the Jurassic period of 169–144 million years ago and the Cretaceous period of 119–189 million years ago [1]. This information was published 20 years ago but the oil discoveries that have been made since then have not altered the situation significantly

Many people have heard of the geological eras spanning 600 million years: Cambrian, Ordovician, Silurian, Devonian, Carboniferous, Permian, Triassic, Jurassic, Cretaceous, Tertiary, and Quaternary. However, they may be unaware that it is only during these eras that our planet has taken on its current appearance as a blue and green oasis floating in the blackness of space. Only in the past 600 million years has complex multicellular life existed and our planet been able to form fossil fuels. As seen in Fig. 4.2 over 50% of the world's oil was formed during the Jurassic and Cretaceous periods approximately 150 and 100 million years ago, respectively [1]. During these periods the carbon dioxide content of the atmosphere was three to six times higher than today's 380 parts per million (ppm). These are levels far higher than those that have been discussed at various international climate change negotiations [2]. Therefore, it is not surprising that the average temperature during these periods may have been around 22°C [3], which is much higher than today's 15°C. The conditions for global warming were very favorable in those ancient times.

Algae, plankton, and marine plants thrived in the shallow waters of the late Jurassic and Cretaceous periods. Because the oxygen content of the

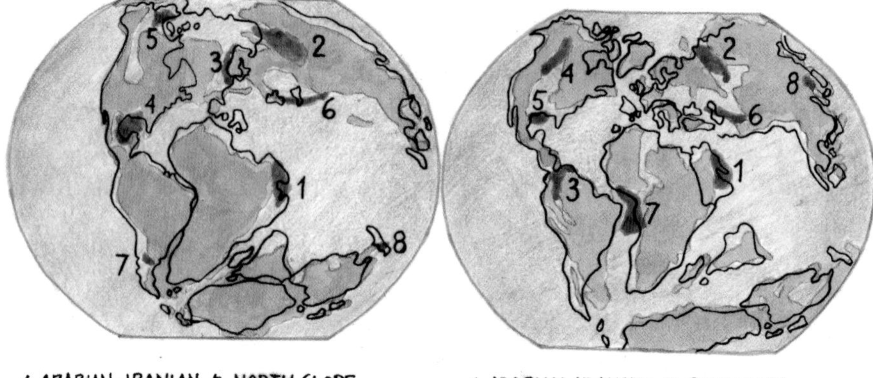

Fig. 4.3 During the Jurassic and Cretaceous periods our continents were arranged differently compared with today [1]. It is interesting to examine the conditions that prevailed at the times when today's large oil fields were formed. Note that several areas have oil-bearing sedimentary layers from both the Jurassic and Cretaceous periods. The fact that the Arabian Peninsula and Iran have continuously been located near the equator has meant large algal blooms and, ultimately, large oil fields

water was low this biological material could sink to the bottom of the seas and ocean basins and accumulate in thick layers of sediment without being lost through oxidation (conversion back into carbon dioxide). The biological components found in rocks from these periods are called "kerogen" and in this material is stored solar energy from millions of years ago. Kerogen type I is formed primarily from algae and kerogen type II is dominated by plankton. Some kerogen type II is contaminated with high levels of sulphur from the volcanic activity of those times.

Some of the fascinating geological history of the Upper part of the Jurassic and the Middle of the Cretaceous is illustrated in Fig. 4.3. There were two continents in the southern hemisphere 150 million years ago, one that later split into today's South America and Africa whereas the other was made up of today's India, Australia, and Antarctica. In the northern hemisphere North America and Greenland were joined and, like today, Europe and Asia were one contiguous land mass. Some of the oil discovered on the Arabian Peninsula, in west Siberia, the northern Caucasus, and in the Gulf of Mexico originates from the Upper (later) Jurassic period when some of

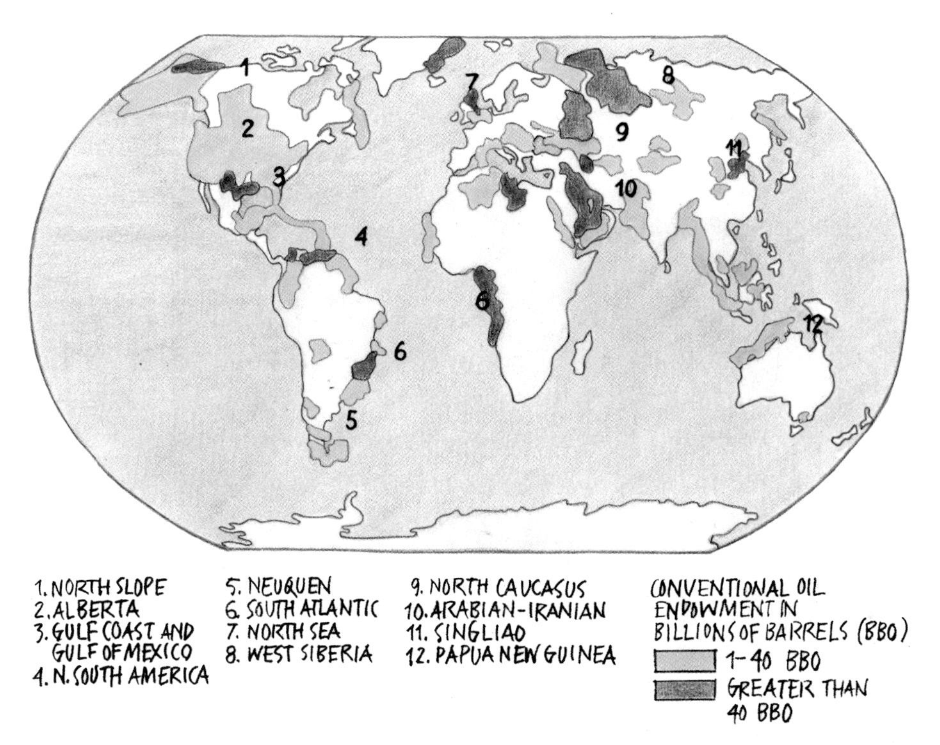

Fig. 4.4 Areas where sedimentary layers exist that may form oil are marked in shades of green. The sedimentary layers formed during the Cretaceous and Jurassic are indicated by numbers. The area east of Greenland was thought to be especially promising but no oil has yet been found there [4]

the largest dinosaurs lived. The oil from the North Sea and the North Slope of Alaska comes mainly from the Jurassic period [1].

During the Cretaceous period biological sediments continued to accumulate in the Middle East, west Siberia, northern Caucasus, and the Gulf of Mexico. There were also new areas of accumulation that became the oil-bearing sedimentary layers now found in northern South America (principally Colombia and Venezuela), Alberta in Canada, and in northeastern China around the Daqing oil field. The continents of Africa and South America had now separated and biological sediment accumulated under the waters between them. These became the oil fields now found off the coasts of West Africa and Brazil. The famed oilfields of Texas also originate from this time [1].

Today, the layers of biological sediment from the Jurassic and Cretaceous periods form a large part of the world's oil-producing sedimentary rocks. The most recent survey of the entire world's oil- and gas-producing sedimentary layers was published by the US Geological Survey [4]. A simplified summary of their analysis is presented in Fig. 4.4 in which we have also

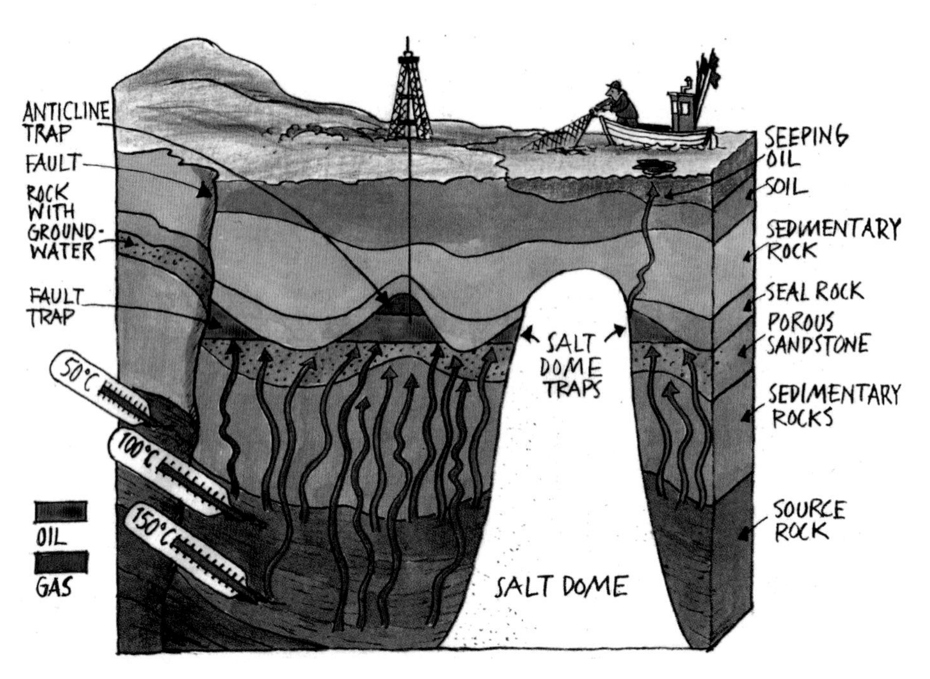

Fig. 4.5 Geological formations that can become oilfields. A layer must exist (*seal rock*) through which the oil cannot pass. Salt is a common form of seal rock. Under this are porous layers containing groundwater and beneath them is the source rock producing oil (*green*) and natural gas (*red*)

indicated the layers that were formed during the Jurassic and Cretaceous periods. The USGS regarded the area off the coast of Greenland as especially promising for the discovery of new oilfields. However, despite the drilling of several test wells no discoveries have yet been made.

To create a "factory" for oil that will form an oilfield a special set of conditions is needed. The layers of biological sediment that form kerogen-containing rocks must be covered by additional layers of sedimentary material that can form porous rocks such as sandstone and limestone. Above these layers, another layer of material is required that can become impermeable and seal the oil beneath it. These layers must be taken sufficiently deeply underground so that high pressures and temperature transform the kerogen-containing rock into a "source rock" for oil. In the many millions of years since these sedimentary layers were formed the continents have moved. This has usually disrupted the arrangement of the sedimentary layers. Only in very special situations have the correct conditions existed to produce an "oil factory."

Oil shale is a kerogen-containing rock that has not been processed in one of the Earth's oil factories. For oil to form, it is essential that the source rock be

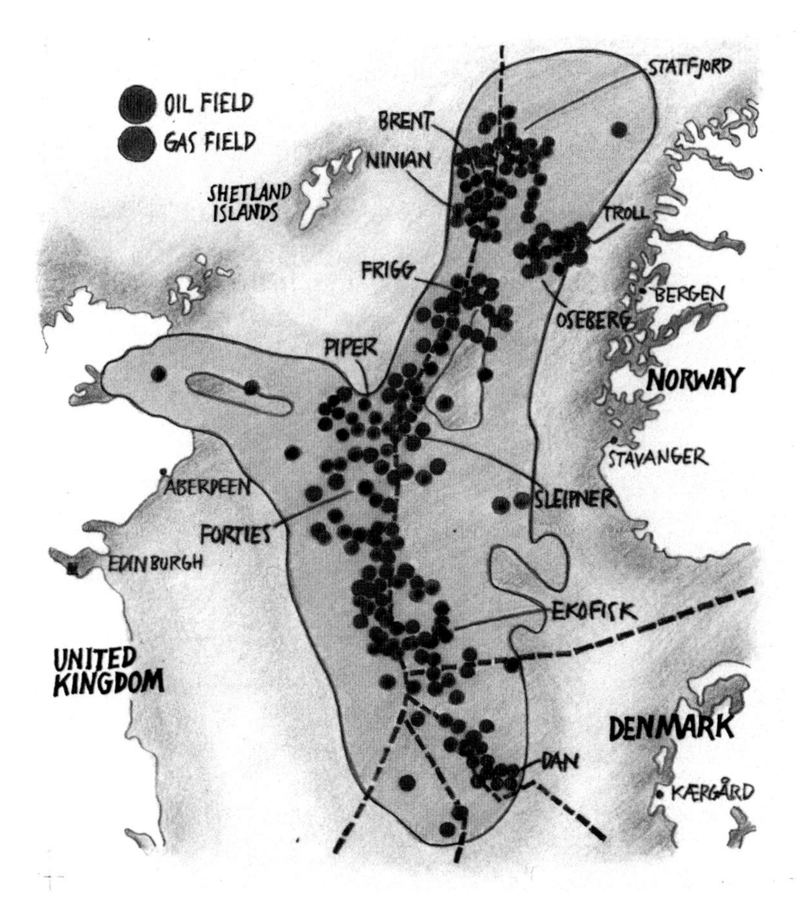

Fig. 4.6 Sedimentary layers from the late Jurassic period (light green area) and the locations of discovered oil fields under the North Sea

buried sufficiently deep underground to be exposed to a temperature of around 100°C. Under these conditions the kerogen breaks down into liquid oil and natural gas. The oil and gas are then forced up towards the Earth's surface because they are lighter than the groundwater that is also present. If there is nothing to stop the oil and gas on their way up to the surface then the lighter hydrocarbon molecules will evaporate and the heavier hydrocarbon molecules will form a tarlike residue like that found in Canada's oil sands.

When geologists search for new oil fields they look for geological formations that trap oil and natural gas on their way up to the surface. If the sedimentary layers have been forced together so that they form a dome shape called an "anticline," then the oil will move to the top of the dome. If an impermeable layer exists that can act as a seal, then the oil will replace the water in the porous rocks underneath the seal. The accumulating oil will form an oilfield inside the dome. An oil field can also be formed if there is a fault line through the sedimentary layers or if a salt dome forces its way up through the sedimentary layers (Fig. 4.5).

We have looked at some of the important conditions that must exist if one is to find oil in the Earth's crust. Now let's look more closely at the most recently discovered and exploited oil region, the North Sea. Geological mapping has shown us how the sedimentary layers from the Jurassic period are distributed under the North Sea. It is in these regions that we can expect to find oil. Figure 4.6 shows the locations of the layers from the Jurassic period and where the oil and gas fields have been found. There is no question that the theory of oil formation is very consistent with the distribution of oil fields that we see.

References

1. Klemme, H.D., Ulmishek, G.F.: Effective petroleum source rocks of the world: stratigraphic distribution and controlling depositional factors. AAPG Bull. **75**, 1809–1851 (1991). http://www.searchanddiscovery.net/documents/Animator/klemme2.htm
2. Berner, R.A.: Modeling atmospheric oxygen over phanerozoic time. Geochim. Cosmochim. Acta **65**, 685–694 (2001) (See figure at http://deforestation.geologist-1011.net/)
3. Scotese, C.R.: Paleomap Project. http://www.scotese.com/climate.htm (See figure at http://deforestation.geologist-1011.net/) (2001)
4. U.S. Department of the Interior, U.S. Geological Survey: USGS World Petroleum Assessment 2000. http://pubs.usgs.gov/fs/fs-062-03/FS-062-03.pdf (2000)

Chapter 5

The Art of Discovering an Oilfield

Many regard the 27th of August, 1859 as the day that the Oil Age began. It was on that day that Edwin Drake used a drilling technique previously used for mining salt to reach down to an oil field 21 m below the surface of a little island in the middle of a river named Oil Creek near Titusville, Pennsylvania (see Fig. 14.1). In 1859, cars and aircraft did not exist but there were steam engines, and it was a steam engine that powered Drake's drilling rig. The motivation for Drake's drilling attempt was the knowledge that the oil seeping out of the ground around Titusville could be used to produce paraffin (kerosene) that was then used in lamps for domestic lighting [2].

Historically, it is common that problems give rise to new innovations. In the mid-1800s the problem was that too many people wanted to use paraffin to light their homes. Before Drake's oil well, the best raw material for making paraffin was an oily substance found in the heads of sperm whales called "spermaceti." Up to 2,000 L of "oil" could be found in the enormous head of a sperm whale. The scale of whale hunting in this period even led to the literary creation that became the most famous sperm whale ever, Moby Dick.

Drake's oil well produced 3,000 L of oil per day, but the next year production in Pennsylvania was over 20,000 L per day, which is 10 times as much as from the head of a whale. By 1870, the oilfields of Pennsylvania were producing close to 15,000 L of oil per day and the hunt for sperm whales had come to an end. If sperm whales were more like humans they might have held a wonderful party to celebrate (Fig. 5.1).

K. Aleklett *Peeking at Peak Oil*, DOI 10.1007/978-1-4614-3424-5_5,

Fig. 5.1 A ruthless hunt for sperm whales was underway when the first oil well was drilled in Pennsylvania in 1859. Up to 2,000 L of "spermaceti" oil could be found in the enormous head of a sperm whale. This was bottled directly and sold as paraffin for lighting. Drake's oil well produced 3,000 L of oil per day undermining the profit in whale hunting. It was time for sperm whales to celebrate! [1]

The Discovery of Oil Seeping Out of the Ground

The discovery of oil in the United States and the subsequent development of oil-burning engines that could use petroleum products refined from crude oil was decisive for the development of the oil industry. However, long before this happened, oil from the earth was already used in the region between the Black Sea and the Caspian Sea known as the Caucasus. In particular, the area around Baku on the west coast of the Caspian Sea was known for its oil. This area was covered with oil-producing sediments during several periods when oil was formed (see Fig. 4.3) so it is not so surprising that oil seeped up out of the ground there like water from a spring to form natural oil wells (Fig. 5.2).

During his journey to China in 1270 Marco Polo traveled through this area and in his celebrated book, *The Travels of Marco Polo* he wrote [3]:

Fig. 5.2 In the Caucasus leaking oilfields formed oil "springs" and the inhabitants used the oil as fuel

> Near the Georgian border there is a spring from which gushes a stream of oil, in such abundance that a hundred ships may load there at once. This oil is not good to eat; but it is good for burning and as a salve for men and camels affected with itch or scab. Men come from a long distance to fetch this oil, and in all the neighborhood, no other oil is burned but this.

The first detailed description of an oil industry in Baku was made in 1683 by Engelbert Kempfer, a German working as a secretary to the Swedish ambassador in what was then called Persia. He was mainly fascinated by the natural gas that leaked from the ground and was lit. But it was 200 years later that two Swedes dominated the oil industry in Baku. In 1873, Robert Nobel (the eldest brother of Alfred Nobel who instituted the Nobel Prize) arrived in Baku and together with the second oldest brother, Ludwig, and some others he founded The Petroleum Production Company Nobel Brothers, Limited in 1876. In the telegraph communications of that time the name was shortened to Branobel. In 1879 the company was transformed into an investment company. Of the shares, 57% were owned by Ludwig Nobel, 3.8% by Alfred Nobel and 3.3% by Robert Nobel. Of particular interest is that Alfred Nobel bequeathed from these shares 12% of his fortune to the Nobel Foundation that now forms the economic base of the Nobel Prize.

The Nobel brothers were known for their innovations. Already in 1878 they had commissioned construction of the world's first oil tanker from the Motala shipyard in Sweden. At the end of the 1870s, they introduced oil pipelines in the Baku area, and they were the first to transport oil by train. Their company Branobel became the world leader in the oil industry but the Russian Revolution put an end to this when the company was nationalized in 1920.

The Discovery of Oil by Geological Fieldwork

During the 1960s, more oil was discovered than at any other time before or since. It was discovered through geological fieldwork. A detailed description of this epoch can be found in Daniel Yergin's famous book *The Prize: The Epic Quest for Oil, Money & Power*. To understand the hardship that such fieldwork involved I have asked Colin Campbell (the person who took the initiative to start ASPO) to describe his work in Colombia during the 1960s.

Colombia is a large nation of around one million square kilometers. The Andes mountain range has divided the nation into different geological provinces. In Fig. 4.3 we can see that the oil source rocks of Colombia were formed during the Cretaceous period. At the beginning of the twentieth century, oil was discovered in the Magdalena Valley between the central and eastern Andes and in a section of the Maracaibo basin of Venezuela that extends into Colombia.

The international oil companies Texaco, Shell, BP, and Esso dominated the search for oil in Colombia. Campbell participated in one of their expeditions to explore the eastern slopes of the Andes. Besides Campbell there were 13 Colombians, one of whom was a young geologist being taken along for training. The expedition left Bogotá, where it originally assembled, on May 31st, 1960, as Colin describes:

> At first, there were few outcrops to be seen, but then we entered a range of hills in which rather strongly deformed white porous sandstones with siliceous layers were exposed. These are known as the Guadalupe Formation and were formed during the Upper Cretaceous period. The sandstone strata were inclined. Our task was to measure the direction and angle of dip and to describe the composition of the exposed rock in detail, including taking samples. As we progressed through the landscape we could sketch in the surrounding geological features on a thick sheet of paper mounted on the planetable and we began to get a feel for the overall structural configuration. Our expedition was supported by a small truck and at first we were able to return home to Bogotá at the end of each working day.
>
> In due course, we came to spectacular cliffs that allowed us to measure the thickness of the geological formation in detail. Having good porosity, it was evident that the sandstone would form an excellent reservoir for oil if found

in suitable structures. We continued eastwards and found that these sandstones are underlain by black fossiliferous shales, from which we collected ammonites [fossil molluscs that died out at the end of the Cretaceous period]. Later we also found a thin indurated [hardened] sandstone, rich in fossil oysters (*Exogyra sqamata*), that we knew marked the Cenomanian Stage [rocks formed during a division of the Cretaceous period].

Eventually, the distances became too great to return to Bogotá, so we rented a house in the village of Caqueza as a base, sometimes renting mules to help us search for outcrops of rock in the adjoining country. Our expeditionary party included a cook who would produce simple meals for us. In due course, we came to another major sandstone scarp belonging to the Une Formation, which might form another, deeper reservoir if in suitable formations. We also noticed that the strata were becoming more disturbed and steeply dipping suggesting a subtle mid-Cretaceous epoch of deformation, which we were eventually able to map in detail. The survey continued eastwards for many weeks, crossing a Palaeozoic Massif (a block of the Earth's crust shifted to form a mountain range), to arrive at Villavicencio in the foothills, where some large gentle structures were observed.

After many weeks of work we completed the survey, and it was time to return to the office and write up the results, describing and computing the thickness of the various formations we had observed.

In the following year, another survey was mounted along a road to the north which ran roughly parallel to the one we had studied. We now had a good knowledge of the details of the rock sequence and could readily identify the various units, recognising their fossil content and noting how they had subtly changed in composition and thickness. The road ended at Gachala, which is known for its emerald deposits. We then needed to hire mules for a traverse following an ancient Inca trail across the wild Farallones de Medina, mountains which rise to an altitude of 4,000 m (Fig. 5.3). We had tarpaulins we used as tents, and sleeping bags. The massif was composed of highly deformed Palaeozoic rocks, from which we succeeded in collecting Devonian fossils. At length, we emerged into the foothills and rode, somewhat exhausted, into the village of Medina. It was a scene from a Wild West film, with single story simple houses, to which horses were tethered around a grass square. But as we rode in, we noticed that the inhabitants ran indoors and shut the shutters, some re-emerging with guns. Abel, our Colombian foreman, understood the situation and quickly ran over to the people to explain who we were. Evidently, a rough gang of unknown men riding into town usually meant an attack, but all was well when they discovered that we were *petroleros* not *bandoleros*. Field work was not without its colourful incidents!

Having completed the survey, it was time to return again to the office to evaluate the results. The fossil collections were photographed and taken to Dr. Bűrgl, a palaeontologist at the University, who identified them. We were thus able to identify all the classic Cretaceous stages in a sequence of rocks totalling some 10,000 m in thickness. It was evident that the mountainous

Fig. 5.3 "Petroleros" on mules following an ancient Inca trail across the wild Farallones de Medina

territory that we had surveyed was itself too deformed to be of interest for oil drilling, but the foothills offered great promise. We had identified two promising reservoir sequences as well as black shales in between, which we thought might be a good source of oil. We were further encouraged to have noted several natural oil seepages in the foothills, confirming that oil had indeed been generated.

The recommendation resulting from our work was that efforts should be made to secure concessions over the foothills and the territory immediately to the east of them, to map them in greater detail and also to conduct seismic surveys to define the detailed configuration at depth. While the prospects were obvious in geological terms, the area did not have economic merit because of its very remote location, being separated from an export terminal on the coast by high mountain ranges. However, twenty-five years later, a company did pluck up courage to drill in the right place despite its economic drawbacks. They found a major new oil province with several giant fields (Cañon Limon, Cusiana and Capiagua).

Today, astonishing techniques for discovering oil have been developed but the fact remains that the greater volume of crude oil has been found by geological fieldwork.

Geophysical Detection of Oil

When geologists leave the field, geochemists and geophysicists take over. Powerful methods have been developed that now make it possible to detect even the smallest of prospects lying thousands of meters deep underground. Breakthroughs in geochemistry made in the 1980s allow scientists to identify the original source rock of the oil in an oilfield and to determine how deep the oil has been buried during the millions of years since the source rock was formed. However, it is advanced geophysical tests that make it possible to find oil-bearing geological formations thousands of meters below ground. Such tests now include measurement of variations in the Earth's gravitational and magnetic fields. The most important method for detection of new oilfields has long been seismic measurement.

Jean Laherrère, is one of the founders of ASPO. He began work as a geophysicist in the 1950s. In 1956 he helped discover Africa's largest oilfield named Hassi Messaoud in Algeria. He did this using seismic methods. I asked him if he had an interesting story to tell from his years as head of exploration for the French oil company, Total. He does, but before we come to it we study the principle behind seismic detection of oil.

Seismic detection methods measure how rapidly sound travels through rock. The method was used first by the Germans during the First World War to detect the location of the Allies' artillery [4]. Today there are many sophisticated methods for generating sound waves but originally this was performed by causing explosions using Nobel's dynamite. The sound waves travel down through the various layers of rock and are reflected by these back up to measuring instruments at the surface. All the information on how long it takes for the sound waves to travel through the various sedimentary layers before they are reflected back to the measuring instruments is recorded digitally. It is then processed by the world's most powerful computers. The principle behind seismic detection of oil is illustrated in Fig. 5.4.

Colin Campbell concluded his experiences in Colombia with, "However, 25 years later, a company did pluck up courage to drill in the right place,

Fig. 5.4 Seismic detection of an oilfield. A sound wave generated by a suitable method travels down into the ground. When the sound wave meets significant boundaries between geological layers some of it is reflected back up to the surface. The remainder of the sound wave continues deeper into the ground and can be reflected back up when it meets deeper boundaries. Sensitive "geophones" that detect vibrations in the Earth are used to measure how strong the reflected sound waves are and how long it takes them to return to the Earth's surface. These collected data are processed by powerful computers and sophisticated programs to create an image of the geological layers within the Earth

despite its economic drawbacks, finding a major new oil province with several giant fields (Caňon Limon, Cusiana and Capiagua)." Jean Laherrère begins his story by describing the courageous company, BP, led by their then-chief for South America, Tony Hayward. He also notes that it was Total that initiated the project. As shown in the following story, the contractual relationships between companies that discover and exploit an oilfield can be quite complex. Here is Jean's tale:

> The right to drill the Cusiana prospect (the "lease") was owned by a small independent company, Triton. At the time I was Deputy Exploration Manager for Total. The Exploration Manager was on vacation so I proposed drilling the Cusiana prospect, under a "farmout" contractual agreement with Triton. My proposal was accepted but Total refused to be the operator [of the drilling project] because of the high risk that its personnel would be kidnapped by guerrillas of the "Revolutionary Armed Forces of Columbia" (FARC). Total proposed to find another partner to act as the operator. BP and Tony Hayward accepted to do so.
>
> Geologically, the prospect was a thrust structure above a big fault identified by seismic survey. The drilling took a very long time and the generator for the drilling rig was blown up by the FARC. After several months of drilling the operator decided to run a "seismic shot" [a sound pulse generated by an air-gun] with the geophone in the well [called "seismic coring"] to check the total depth of the well and compare it to the seismic profile through the location. After the seismic shot was processed, BP sent us a telex saying that the well had passed the big fault and that the expected reservoir was missing. At the end of the telex BP recommended abandoning the well.
>
> Since I was in charge of the exploration procedure I asked our best geophysicist to check BP's assessment. It took him only a few seconds to find that BP had confused the origin of the seismic profile (called the "datum plane") and the deck of the rig for the seismic shot. He told me, "we have still 200 milliseconds to drill." So we knew that the geological fault was still below the bottom of the well and we told BP to keep drilling. They did and sometime later we found the reservoir with oil.
>
> After taking another long time to drill though this thick reservoir, BP ran some tests and recommended again that we abandon the well because they believed the reservoir to be too tight [they thought the reservoir rock was insufficiently porous for good oil flow]. We at Total recommended that the flow rate be tested even if we had to do this without BP. Finally BP tested the well and it produced a good volume of oil. The giant field Cusiana had been discovered!

After the discovery Triton announced that Cusiana's reserves were 3 Gb, BP that it was 1.5 Gb, and Total's estimate was 1 Gb. Cusiana is now close to the end of its productive life and its original reserves are now estimated to have been 0.7 Gb, still a giant oilfield but not nearly as big as estimated when discovered (i.e., an example of negative reserve "growth"). Later, the giant field Capiagua was found close by.

Today, production of oil from the field is run by Colombia's national oil company Ecopetrol. BP was a part owner of Ecopetrol until the catastrophic oil spill in the Gulf of Mexico forced BP to sell its share in order to raise money for a fund that will pay future damages from the catastrophe. This sale must have been a bitter personal blow for Tony Hayward.

The Discovery of the Cantarell Oilfield

Oil seeping to the Earth's surface was important in the Caucasus and was decisive in the launch of the Oil Age in Pennsylvania. Yet another example of leaky oilfields playing an important role is the discovery of Mexico's largest oilfield. For many years, a fisherman named Rudesindo Cantarell had problems with oil contaminating his nets. Eventually he contacted Mexico's oil company PEMEX. At first they were not interested because the geological conditions did not appear suitable for oil. But when they finally decided to investigate the reason for the leaking oil they discovered one of the world's largest oilfields, subsequently named Cantarell after the fisherman who had alerted PEMEX in the first place. At the start of this millennium, Cantarell was the world's second most productive oilfield.

In geological terms Cantarell is a deviant inasmuch as its porous structure was formed by a gigantic asteroid impact 65 million years ago. This was not just any asteroid. It was the one that formed the Chicxulub crater on the Yucatan peninsula and many researchers think that this asteroid's collision with Earth changed the world's climate and led to the extinction of the dinosaurs. The energy released by the impact is thought to have been two million times greater than the most powerful thermonuclear bomb ever detonated.

References

1. The inspiration for this illustration comes from an illustration from 1861 in the magazine Vanity Fair (1861)
2. The Paleontological Research Institution, Ithaca, NY. http://www.priweb.org/ed/pgws/history/pennsylvania/pennsylvania2.html (2011)
3. Polo, M.: The Travels of Marco Polo. Penguin Classical, 1270, ISBN 0140440577.
4. Laherrère J.H.: "Memories and thoughts on 49 years of oil and gas geophysics" in book "What lies beneath: Tales of Petroleum Explorers", GeoPlanet Resources Co, Salt Lake City (2005)

Chapter 6

The Oil Industry's Vocabulary

Sometimes the language used in the oil industry and related organizations can seem confusing. However, when we discuss the future of oil it is important to understand this terminology. The International Energy Agency, IEA, is centrally important in discussions of future oil production. In the IEA's *World Energy Outlook* report for 2010, *WEO 2010*, it gives detailed definitions of various terms used in the oil industry [1]. We use these definitions when we discuss oil and the future:

- *Oil originally in place (OOIP)* refers to the total amount of oil or gas contained in a reservoir before production begins.
- *The recovery factor* is the share of the oil or gas originally in place that is ultimately recoverable (i.e., ultimately recoverable resources/original hydrocarbons in place).
- *A proven reserve* (or 1P reserve) is the volume of oil or gas that has been discovered and for which there is a 90% probability that it can be extracted profitably on the basis of prevailing assumptions about cost, geology, technology, marketability, and future prices.
- *A proven and probable reserve* (or 2P reserve) includes additional volumes that are thought to exist in accumulations that have been discovered and have a 50% probability that they can be produced profitably.
- *Reserves growth* refers to the typical increases in 2P reserves that occur as oil or gas fields that have already been discovered are developed and produced.
- *Ultimately recoverable resources (URR)* are latest estimates of the total volume of hydrocarbons that are judged likely to be ultimately producible commercially, including initial 1P reserves, reserves growth, and as yet undiscovered resources.
- *Remaining recoverable resources (RRR)* are ultimately recoverable resources less cumulative production to date.

K. Aleklett *Peeking at Peak Oil*, DOI 10.1007/978-1-4614-3424-5_6,

Oil Originally in Place (OOIP)

The natural place to begin discussing future oil production is with oil originally in place, OOIP. This can mean the total oil existing in an oilfield, in a region, or in the entire world before oil production begins.

When geologists and geophysicists have discovered an oilfield, the amount of oil that can be produced from the field must be calculated. First one needs to estimate how much oil exists underground in the field, the OOIP. Its volume is then used in subsequent calculations to estimate production volumes, rates, and so on. When one drills for oil, fragments of the reservoir rock are brought to the surface. These can be studied to see how the oil is held within the porous stone. In this case, to search for the black gold requires a microscope!

Oil reservoirs are formed when rising oil forces some of the water out of porous rock and then becomes trapped below an impervious rock layer. If we use a microscope to examine pieces of rock from a sandstone or limestone oil reservoir we can see clear differences in how oil is held within these materials (see Fig. 6.1). In sandstone, the water is found as a thin layer around the sand grains that make up the rock and the oil is in the remaining pore space. Limestone is made up of grains of carbonate to which oil adheres more readily than water. Therefore, in limestone, any water is found in the center of the pores. Oil originally in place is the volume of oil within the pores in the rock. If one knows the porosity of the rock and the proportion of water relative to oil in the pores then it is quite easy to calculate the volume of the oil relative to the total volume of the rock. (In Chap. 8, "The Size of the Tap—The Laws of Physics and Economics," we discuss the term "porosity" in more detail).

Normally a number of appraisal wells must be drilled before one can estimate the total volume of the reservoir itself. However, if we then also know the volume of oil that exists per cubic meter of reservoir rock we can calculate the total volume of oil in the field, the OOIP. This is at best an approximation. The OOIP for Saudi Arabia is estimated to be 700 billion barrels of oil and we discuss this later in Chap. 13.

Recovery Factor: The Amount of the OOIP that Can Be Produced

It is impossible to extract all the oil from an oilfield. If oil production from a field is to be profitable then a decisive issue is what percentage of the OOIP can, in fact, be extracted. In the section "Geophysical Detection of Oil",

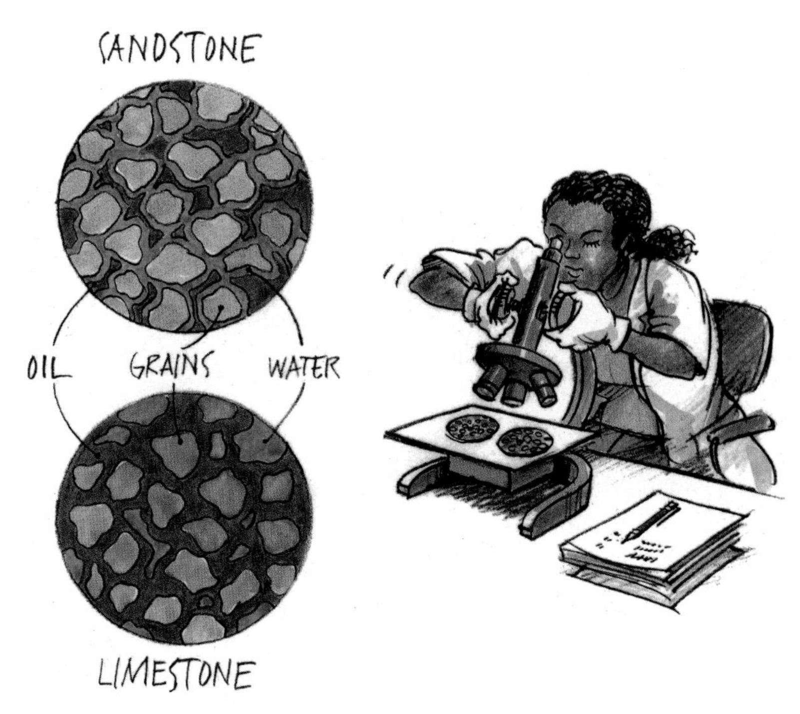

Fig. 6.1 If we use a microscope to examine a piece of sandstone or limestone from an oil reservoir we can clearly see both oil and water in the pores of the rock. This is because the oil has forced out most of the water in the rock as it rose into the reservoir from underneath. However, the distribution of the oil and water in the rock can vary. In sandstone the water surrounds the sand grains in the rock and the oil is found in the center of the pores. In limestone, it is the oil that adheres to the limestone matrix and any water is found in the center of the pores. Oil originally in place, OOIP, is the volume of oil within the pores in the rock. If one knows the porosity of the rock then it is quite easy to calculate the volume of the oil relative to the total volume of the rock. To extract oil from a reservoir, a difference in pressure must exist that forces the oil to move through the pores towards the well that has been drilled while most of the water remains

Chap. 5, Jean Laherrère said that, "After the discovery, Triton announced that Cusiana's reserves were 3 Gb, BP estimated that it was 1.5 Gb, and Total's estimate was 1 Gb. Cusiana is now close to the end of its productive life and its original reserves are now estimated to have been 0.7 Gb…". The proportion of the total OOIP that can be produced is called the "recovery factor." In the case of Cuisiana, the estimated recovery factor turned out to be too high. Normally we find that the recovery factor increases with time.

For many years, Leif Magne Meling from Norway's oil company Statoil has studied the recovery factors from thousands of oilfields. In 2005 The Royal Swedish Academy of Sciences' Standing Committee on Energy and

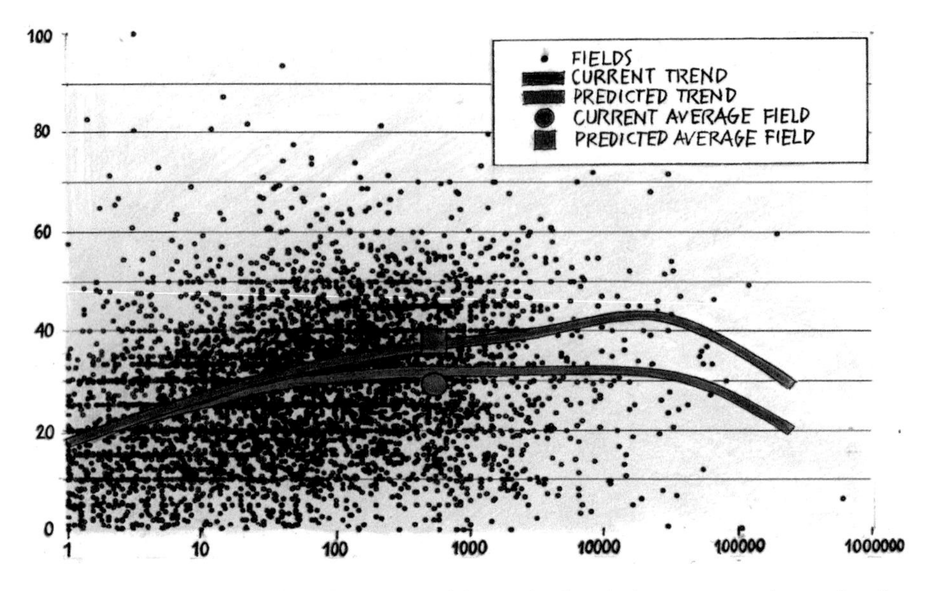

Fig. 6.2 Leif Magne Meling from Statoil has calculated the recovery factor for the thousands of oilfields for which this was possible in 2005. In this figure are shown the recovery factors for individual fields (*small dots*), the average for fields of a particular size (*purple line*), and, as a large purple dot, the average for all the field studies which was 29%. If one analyzes how the recovery factor has changed over time one can estimate the future recovery factor for a particular field size (*pink trend line*) and the average for all fields (*pink square*). Meling estimates that, in the future, the average recovery factor over all fields will be 39% [2]

the Royal Swedish Academy of Engineering Sciences' Standing Committee on Energy and Climate organized a symposium in Stockholm at which both Leif and I were invited to speak. It was then that I first saw his detailed research into recovery factors. The average recovery factor for the thousands of fields Leif had studied was 29%. For the largest fields, it was a little over 30%. The recovery factor could vary from 5% to 80% for fields of the same size. Leif Magne estimates that we will eventually be able to recover, on average, an additional 10% from oilfields (compared to today, see Fig. 6.2) [2]. Clearly this will increase by many billions of barrels the ultimate amount of oil that will be produced globally. New technology and the willingness of investors will be decisive for obtaining this increased production.

In the autumn of 2005 I was invited to a conference in Dubai where they discussed new methods of oil production. I remember one presentation especially well. It was delivered by the director for BP's activities in the Middle East at that time. He had studied the possibilities for increasing the recovery factors for giant oilfields, the oilfield class that originally held more than half of all global OOIP. With the new technologies discussed at the conference it was estimated that the giant oilfields could yield an additional 20 billion barrels of oil. That might sound like an enormous volume but not when the world consumes 30 billion barrels of oil every year.

Proven and Possible Reserves

Before an oil company decides to invest in production from an oilfield they want to know with a high degree of certainty that the field will produce a sufficient volume of oil. The volume of oil that they can recover from the field with 90% certainty is described as the "Proven" reserve. This is often abbreviated as "1P." They also calculate a volume that they believe will be produced with 50% probability and that is described as the "Proven plus Probable" (or "2P") reserves. The "Proven plus Probable plus Possible" or "3P" reserves represent the volume of oil that has only a 10% probability of being recovered from the field. When URRs for old oilfields are compared with the initial 1P and 2P values in general they are of the same order as 2P.

During the 1960s, approximately 56 billion barrels of oil were discovered per year and consumption was only 10 billion barrels per year. The statistics shown in Fig. 3.1 are what today would be called proven plus probable reserves (2P). The general public only has access to one database of oil reserves and that is the *BP Statistical Review of World Energy* (BPSR) [3]. Until the 1980s, the oil reserves reported in the *BPSR* were mainly proven reserves (1P). However, since then the *BPSR* has begun to report more of a mixture of 1P and 2P reserves. Normally the yearly update of the *BPSR* is released in May. It cites a number of different sources for its oil reserve numbers, but a comparison with the statistics reported by the *Oil and Gas Journal* (*OGJ*) shows that the *BPSR* mainly uses the numbers reported by the *OGJ*.

In our 2003 article, "The Peak and Decline of World Oil and Gas," Colin Campbell and I discussed the reporting of reserves in the *OGJ* and especially the increases in reported reserves by OPEC, the Organization of the Petroleum Exporting Countries [4]:

> The main OPEC countries expropriated the holdings of the foreign companies during the 1970s, following the precedent of Iran's action against BP in 1951. State enterprises were formed to produce the oil, inheriting the technical data and reserve estimates from the private companies. In 1984, Kuwait reported a 50% increase to its reserves overnight although nothing particular changed in the reservoir. It did so to increase its OPEC production quota, which was partly based on reserves. Then three years later, Venezuela doubled its reported reserves by the inclusion of large amounts of long-known heavy oil

> that had not previously been reported. This led Iraq, Iran, Dubai, Abu Dhabi and later Saudi Arabia to retaliate with huge increases to protect their quotas. Some revision was called for, as the earlier estimates were too low, having been inherited from the private companies before they were expropriated. But the revisions, whatever the right number might be, have to be backdated to the discovery of the fields concerned, which had been made up to fifty years before. In total about 300 billion barrels were added in this way during the late 1980s, greatly distorting the apparent discovery record. It is noteworthy too that in several cases the reported reserves remain implausibly unchanged for years on end despite production. It is staggering that such obviously flawed information is recorded in the public database, substantially without comment or qualification.

One can interpret these increased reserve numbers as due to the transition from reporting of 1P reserves to the reporting of 2P reserves. To provide a true picture of our progress in discovering oil reserves, it is important that, when reporting increased reserve numbers, we "backdate" these increases to the moment when the oilfield was discovered. Any increased estimate of reserve size in an oilfield should not be treated as a discovery of new reserves. The claimed increases in OPEC nation reserves during the 1980s that are described above have since been reported in the *BPSR* as new reserves. This distorts the picture of how oil discovery has progressed such that *BPSR* reports the 1980s to be the decade when most oil was discovered. However, the 2P reserve statistics in Fig. 3.1 show very clearly that the largest discoveries of oil were made during the 1960s.

In a recently published "View point" in the journal *Energy Policy* Swedish national economist Marian Radetzki asserts that oil reserves have increased markedly from the 1980s to today and this increase means that we do not need to worry about Peak Oil. This article is discussed in Chap. 8.

To be consumed, oil must first be discovered. Therefore, it is not surprising that, in any oil-producing region, a peak in oil discoveries comes before a peak in oil production. When any region is opened up for oil exploration and production we first see the volumes of oil discovered greatly exceeding the volumes of oil produced. Later we see oil production exceeding oil discoveries until, eventually, oil discoveries drop away to nothing before oil production finally ceases. The maturity of oil discovery and production in any region can be judged by calculating the growth in its oil reserves. Is more oil discovered in a year than is produced (positive reserve growth) or are total reserves declining because more oil is produced than is discovered (negative reserve "growth")? For the world as a whole, the growth in oil reserves is the difference between reported 2P discoveries (Fig. 3.1) and consumption. From the rates of oil consumption reported in the *BPSR* we can calculate an average consumption per decade. In Fig. 6.3 it can be seen that, since the 1980s, we have consumed more crude oil in each decade than

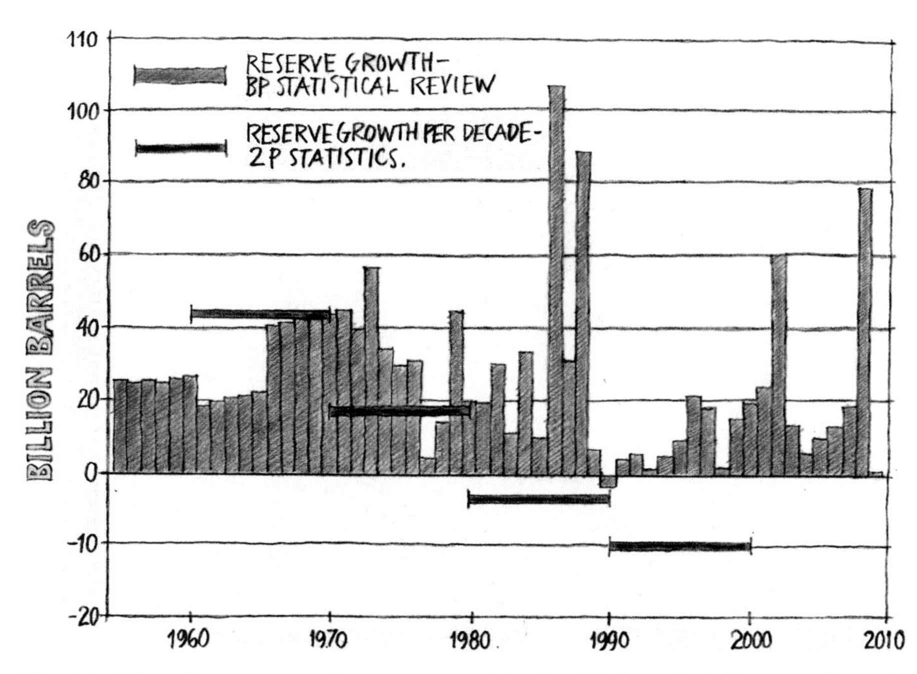

Fig. 6.3 Using the reserve volumes reported in the BP Statistical Review of World Energy, BPSR, one can calculate the growth in reserves for each year. The statistics for the period spanning 1980–2009 have been taken from the 2010 edition of BPSR whereas earlier numbers are taken from earlier editions. Average consumption per decade can be calculated from BPSR 2010. If this consumption is subtracted from the actual 2P reserve discoveries during the same decade (see Fig. 3.1) then we can determine the real growth in reserves. During the 1980s and later decades remaining reserves actually declined (reserve "growth" was negative)

we have discovered in new oilfields; reserve "growth" has been negative. The erroneous image of increasing world oil reserves reported by the *Oil & Gas Journal* (and that is then incorporated into the *BPSR*) may have very detrimental consequences for our future.

In the year 2000, the US Geological Survey (USGS) presented a detailed study of possible future discoveries of crude oil for the entire world from 1995 to 2025 [5]. The USGS study divided future oil reserves into two categories: as yet undiscovered oil and growth in already discovered reserves (reserves growth). They estimated that known remaining reserves had been 959 Gb in 1996 and that 717 Gb of reserves had been consumed before that year. When they summed together these categories they predicted that, by 2025, the world would have consumed and discovered (but not yet consumed) 3,345 Gb of oil (i.e., total ultimately recoverable resources). This was the average estimate and they thought there was a little less than a 50% probability that it would prove true. The USGS estimated that there was a

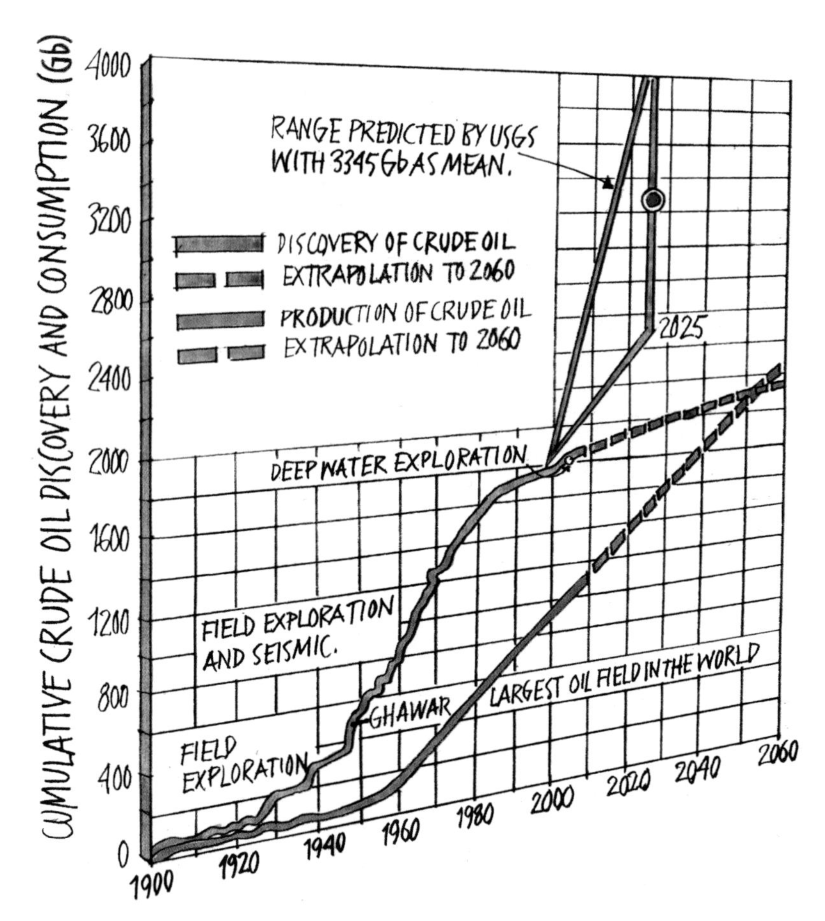

Fig. 6.4 The cumulative total of all oil discoveries over time (*solid green line* to 2010) appears to be leveling off and, if extrapolated accordingly, predicts that we will discover only an additional 200 Gb of oil by 2050. Different eras of oil exploration and the moment of discovery of the world's largest oilfield, Ghawar, are indicated on the green line. If we extrapolate the cumulative total of the oil consumed over time out to 2050 we see that total oil consumed equals total oil discovered at around that time. The discovery prognosis published by the USGS in 2000 is also shown in the figure (*solid green lines in the top right-hand quarter of the figure*) [5]. The lower of these lines shows the discovery trend they expected with 90% probability. The *upper green line* shows the trend with 10% probability. The *vertical line* shows the difference between these amounts predicted to be discovered in 2025 and the *red dot* indicates the mean (*average*) amount predicted

90% probability that total ultimately recoverable resources would be at least 2,500 Gb of oil by 2025 and only a 10% probability that it would be 4,495 Gb by that year.

The IEA appears to treat the USGS numbers as sacred. Since the Oil Age began, we have consumed 1,100 Gb and, if we compare this with the USGS's total ultimately recoverable resources number of 3,345 Gb above, this would imply that we will have 2,245 Gb to use and to find by 2025. In the IEA's *WEO 2010* report we can read that, "We estimate that around 2.5 trillion barrels of conventional oil remain to be produced worldwide as of the beginning of 2010." The reason that the IEA now cites this slightly higher value of 2,500 Gb is that, in 2008, the USGS presented a report on possible oil discoveries in the Arctic [6]. If the Arctic numbers are added to the 2,245 Gb value one gets the higher amount. *WEO 2010* also states that, as of 2010, 900 Gb of oil remain to be discovered without specifying the time frame. With the discovery rate currently at around 10 Gb per year and trending downwards (see Fig. 6.5) it would take more than 100 years to find this oil.

If you draw a diagram of the history of oil discovery where, for each year, you add that year's oil discoveries to the cumulative total of previous discoveries, you construct what is called a "creaming curve." Figure 6.4 is based on the 2P oil reserve statistics that are publicly available [7]. An extrapolation out to 2050 predicts that we can expect to discover an additional 200 Gb of oil. If the trend of previous and current oil consumption is extrapolated out to 2050 we see that we would have consumed all the crude oil that we had discovered by around 2050. Such an extrapolation will not represent reality as we discuss later. The discovery prognoses made in the USGS 2000 report are also included in Fig. 6.4. It is difficult to understand how the IEA can base its oil production prognoses on these USGS predictions.

Reserves Growth

In a doctoral thesis titled "Giant Oil Fields—The Highway to Oil," Frederick Robelius discusses the growth in oil reserves that occurs through the development of new technologies that provide oil producers with better knowledge about their oilfields. He compared the reserve data from 2005 with what was reported in 1994 (see Fig. 6.5). An interesting observation is that older fields from which oil was being produced in the 1960s and 1970s show much greater reserve growth than oilfields that were put into production during the 1980s [8]. During the 1980s it became increasingly common to use 3D seismic technology and other advanced methods from the start of oil production from new fields. This practice means that, in future, we cannot expect the same degree of reserve growth that we saw during the past 20 years as new

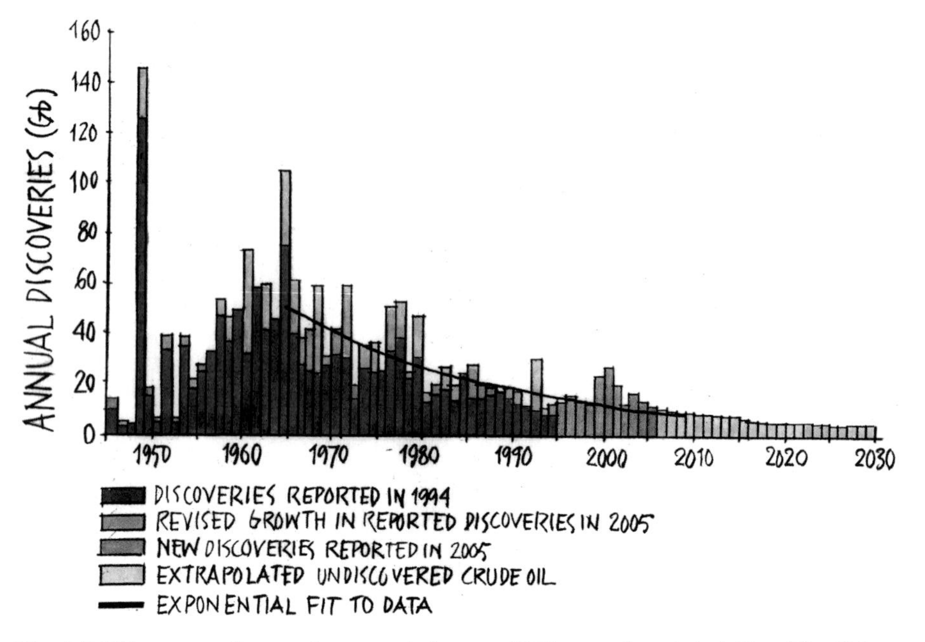

Fig. 6.5 Oil reserve discoveries reported up to 1994 are colored *dark blue*. The *lighter blue* color indicates the growth in reserves reported in 1994 and new discoveries from 1994 to 2005 are colored *orange*. By fitting a curve to the data reported in 2005 we can extrapolate to give an estimate of future oil discoveries [8]

technology was applied to the aging giant fields that began production in earlier decades. Most of the reserve growth that occurred was from such fields and there is now no further scope for increasing production from these mature giants. When making the prognoses described in their 2000 report the USGS used the same large degrees of reserve growth for future oil production as had been observed for older fields. Presumably, that is one of the reasons why their future prognoses are so excessively optimistic.

The Abbreviations URR and RRR

The oil industry terms that are possibly the easiest to understand are ultimately recoverable resources (URR) and remaining recoverable resources (RRR). URR is the total of all the oil that is produced from an oilfield or a region from the day production began until the day production is abandoned. RRR is the volume of oil remaining to be produced at a particular point in time. The volume of oil produced from the start of production up until a certain point in time is called the cumulative production (CP).

References

1. WEO: World Energy Outlook 2010. International Energy Agency. http://www.worldenergyoutlook.org/2010.asp (2010)
2. Meling, L.M.: Data available at: How and for how long it is possible to secure a sustainable growth of oil supply (PowerPoint presentation, 2004-02-27). http://www.bfe.admin.ch/php/modules/publikationen/stream.php?extlang=en&name=en_830170164.pdf (2004)
3. BP: BP Statistical Review of World Energy. http://bp.com/statisticalreview (historical data; http://www.bp.com/assets/bp_internet/globalbp/globalbp_uk_english/reports_and_publications/statistical) (2011)
4. Aleklett, K., Campbell, C.: The peak and decline of world oil and gas production. Miner. Energy Raw Mater. Rep. **18**, 5–20 (2003). http://www.peakoil.net/files/OilpeakMineralsEnergy.pdf, http://www.ingentaconnect.com/content/routledg/smin/2003/00000018/00000001/art00004
5. U.S. Department of the Interior, U.S. Geological Survey: U.S. Geological Survey World Petroleum Assessment 2000—Description and Results. http://pubs.usgs.gov/dds/dds-060/ (2000)
6. U.S. Geological Survey (USGS): Circum-arctic resource appraisal: estimates of undiscovered oil and gas north of the arctic circle. http://pubs.usgs.gov/fs/2008/3049/fs2008-3049.pdf (2008)
7. Laherrère, J.: Perspectives petrole et gaz a l'horizon 2030, Club de Nice VIIIe Forum energie et geopolitique 3–5 decembre 2009, Figure 20. http://aspofrance.viabloga.com/files/Nice09-long-part1.pdf (2009)
8. Robelius, F.: Giant oil fields—the highway to oil: giant oil fields and their importance for future oil production. Uppsala Dissertations from the Faculty of Science and Technology, ISSN 1104-2516, 69. http://uu.diva-portal.org/smash/record.jsf?pid=diva2:169774 (2007)

Chapter 7

The Art of Producing (Extracting) Oil

It is September 28, 2004 and the time has just passed noon. My mobile phone rings. Sweden's national TV Corporation (SVT) is calling and they want to visit me for an interview. The price of oil has just passed US$50 per barrel so they want to record my comments before their evening news broadcast. At that moment I am eating dates in the office of Abdulla M. Al-Malood who, in 2004, was head of production for the giant oilfield Bab in Abu Dhabi of the United Arab Emirates (UAE). Upon hearing this, SVT realizes it will be difficult for them to visit so they satisfy themselves with some comments via telephone. A few minutes later Sweden's national radio broadcaster rings me with the same agenda. The reason for my visit to the Bab oilfield is to learn more about oil production.

When I was invited to talk at the *Tenth Annual Energy Conference* of The Emirates Center for Strategic Studies and Research I asked from the start if it would be possible to visit any of Abu Dhabi's giant oilfields. I was very interested in making such a visit because, at that time, my research group was studying production from the world's "giant oilfields." An oilfield is considered "giant" if it initially has reserves greater than 500 million barrels of oil. This requirement is met by some of Abu Dhabi's oilfields. The immediate answer to my inquiry was no, but my host promised to check with the head of the Abu Dhabi Company for Onshore Oil Operation, ADCO.

It turned out that not even the head of ADCO had the power to grant me permission although he himself had nothing against me visiting the oilfields. Permission had to come from the Ministry of Internal Affairs. It would have been completely understandable if my host had then given up trying to fulfill my request. But he did not and, to my great happiness and other people's astonishment, I was granted permission to visit the Bab oilfield on September 28, the very day that the oil price passed US$50 per

K. Aleklett *Peeking at Peak Oil*, DOI 10.1007/978-1-4614-3424-5_7,

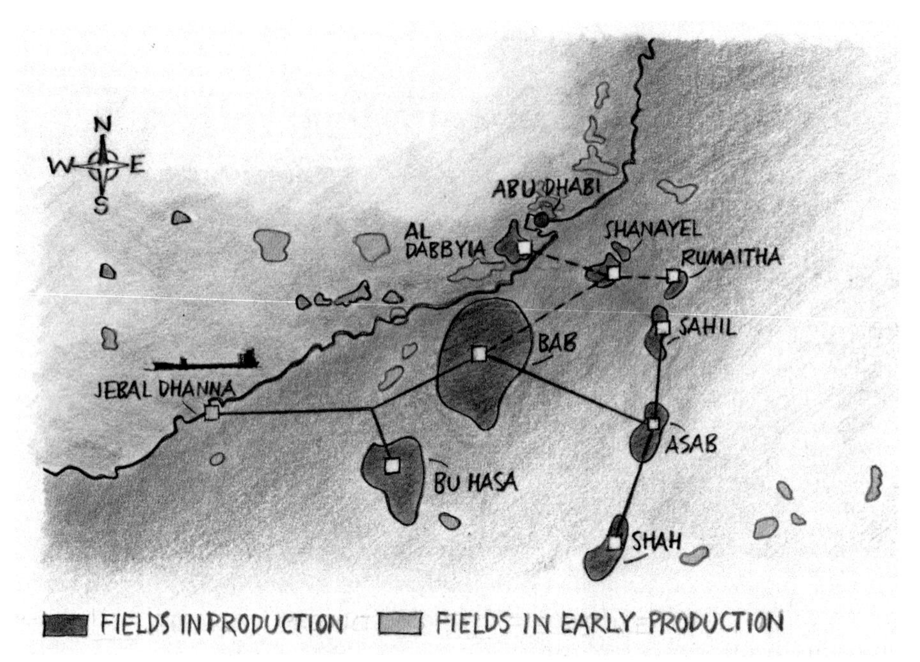

Fig. 7.1 The Emirate of Abu Dhabi lies on the southern shore of the Persian Gulf with Qatar as a neighbor to the northwest, Saudi Arabia to the west and south, Oman to the east, and the emirates of Fujairah and Dubai to the northeast. ADCO, the Abu Dhabi Company for Onshore Oil Operation, is responsible for the exploration and production of oil on land and in some nearby offshore areas. The Bab oilfield has the largest surface area of any of Abu Dhabi's fields on land although Bu Hasa is the most productive. The oilfields Al Dabbiya, Shanayel, and Rumaitha go under the combined name of North East Bab, NEB. During my visit in 2004 those oilfields were being rapidly developed. The oil produced from NEB will be transported by pipeline to the central processing plant at the Bab oilfield

barrel. As part of the opening ceremonies of the energy conference I met Sweden's ambassador to the UAE who told me that he had been trying unsuccessfully for 4 years to obtain permission for a similar visit!

When I received my written permit the only part of it I could understand was "Issued 26/9/2004" and "Expiry 30/9/2004." The rest was in Arabic. It was only when I showed the permit to the Head of Production Abdulla M. Al-Malood that I understood how special it was. I had been granted permission to visit all of Abu Dhabi's oil installations for 5 days without restriction!

During my visit to the Bab oilfield I was privileged to receive a lecture in the conference room of the amazing central processing plant on the art of producing oil. All the departmental heads for the Bab oilfield were in attendance as I received descriptions of current production, planned expansion, and so on. First I was shown a map on which the locations of the oilfields and gas fields for which ADCO was responsible were marked (Fig. 7.1).

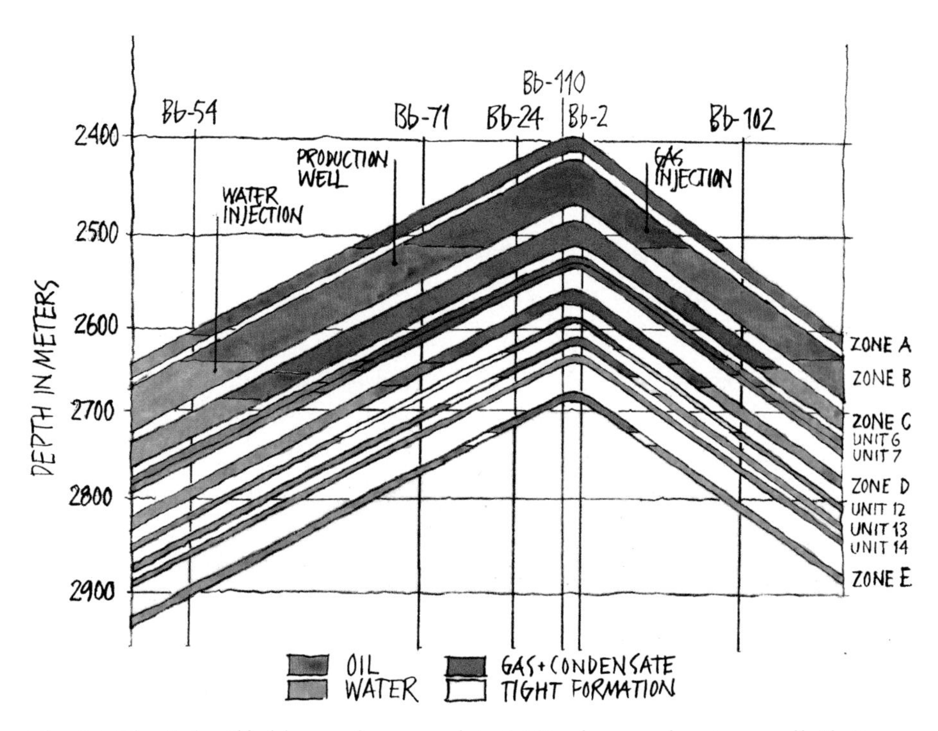

Fig. 7.2 The Bab oilfield was discovered in 1958 when exploratory well Bb 2 was drilled. Information can be combined from six exploratory wells lying on a line from southwest to northeast to show that 19 discrete oil and gas reservoirs have been discovered. The upper 10 of these are shown in the figure. The greatest amount of oil exists in *Zones A, B, and C. Units 12 and 13*, and *Zone E* are "tight" formations that do not allow passage of *oil, gas, and water*. This will be significant when they wish to pressurize the field to continue oil production as the field "matures"

In terms of surface area there is no doubt that Bab is the largest field. We had an interesting discussion about the smaller oilfields that had not yet been put into production. I was told that these fields were to be saved to provide for the future of the UAE's children and grandchildren.

Around halfway through the lecture I asked if there was any possibility of obtaining a copy of the presentation, and one of the departmental heads left the room. Some minutes later he returned and I realized that the answer was yes. No restrictions were placed on me regarding distribution of this information. That means that I can show you a detailed profile of the Bab oilfield (Fig. 7.2). The profile stretches from the southwest to the northeast. On it are marked six exploratory wells, one of which is "Bb 2." It was the drilling of this second exploratory well that disclosed the Bab field. The oilfield is in an anticline with the top of the dome at a depth of 2,400 m. If one follows well Bb 2 down to a depth of 2,700 m one passes through the first 10 gas- and oil-bearing layers of sedimentary rock. These are the layers shown in Fig. 7.2.

If one then follows exploratory well Bb 2 a further 300 m down one finds an additional nine layers with oil and gas but the volumes of these hydrocarbons in these layers are less. Because the Bab field contains both oil and gas this means that the source rock has been at a depth where it has been heated to a temperature of between 100°C and 150°C (see Fig. 4.5).

I was then taken for a tour out over the Bab oilfield. But before I describe that I discuss some of the fundamental principles behind the production of oil.

Build-It-Yourself Models of Oil Production

In Chap. 4 we learned how to re-create crude oil. In this chapter we construct two simple models that help explain how oil is (and is not) produced from oilfields.

Our first model, Bottle A, consists of a large bottle into which we put 1,333 mL of our re-created crude oil. Let us imagine that each little milliliter represents one billion barrels of oil. According to the 2010 edition of the *BP Statistical Review of World Energy* [1], "Global proved oil reserves rose by 0.7 billion barrels to 1,333 billion barrels, with an R/P ratio of 45.7 years." (The R/P ratio is the volume of reserves divided by the annual rate of production). So we can imagine that our 1,333 mL of crude oil in Bottle A represent all the world's oil reserves. Now let's imagine that a short second is the same as an entire year. Annual world oil production for 2009 is 29.2 billion barrels per year so this is the same as pouring out 29.2 mL from our bottle every second. If we pour all the oil out of our bottle at a constant rate of 29.2 mL/s the bottle will be empty in 45.7 s. Of course, this is not how oil production occurs. If oil production did occur in this way then the world would produce 29.2 billion barrels of oil in 2055 but absolutely nothing in 2056.

Unfortunately, journalists who report on oil often quote the R/P ratio of 45.7 years and this gives the public a false idea of what oil production will look like in the future.

The International Energy Agency, IEA, also makes projections of future oil use. However, inasmuch as economic growth requires growth in energy use, the IEA foresees rising oil production in the future. How is this possible if the world has finite reserves of oil? If we refill Bottle A with oil we can see that we can start to pour out the oil slowly and then steadily increase the rate of production until, once again, there is suddenly no more oil left. This is illustrated in Fig. 7.3. To avoid this nonsensical end to world oil production, the IEA speaks of discovering and producing new oilfields and of finding alternative sources of oil. We discuss this in detail in Chap. 11, "The Peak of the Oil Age."

Fig. 7.3 The world's finite reserves of oil mean that oil production cannot forever be constant or increasing. To show this using simple numbers we fill Bottle A with 1,333 mL of crude oil (representing the global reserves) and then pour it out at a constant rate of 29.2 mL/s (representing the yearly production of 29.2 Mb/d). The *black line* on the graph represents this production. After 45.7 s there is suddenly no more oil flow. Alternatively we can steadily increase "production" of oil from Bottle A by starting to pour at 20 mL/s and increasing this steadily to 38.4 mL/s at the moment before the bottle is empty (*red line* on graph). Once again oil production suddenly ends after 45.7 s

If world oil production cannot be constant or increase forever then what will actually occur? To understand this we must understand how oil is actually produced from an oilfield. It is time to build a model oilfield using Bottle B.

For an oilfield to contain large quantities of oil it must be composed of a porous form of rock. The two most common porous rock types are limestone and sandstone. Sandstone reservoirs have average porosities of around 10–20%. To represent this porous rock we need a large, fine-pored sponge. This is cut into pieces and then packed tightly into Bottle B. To represent the oil in the rock we then fill Bottle B with 1,000 mL of our re-created crude oil described in Chap. 4 (or with Coca-Cola).

Before we produce oil from our Bottle B oilfield we support the bottle upside down in a frame vaguely resembling a drill tower. For an added touch of realism and to honor the memory of the world's first drill tower (built 3,000 years ago in China) we construct this support using bamboo. The oil originally in place in an oilfield is under pressure from the water below it, the compressed gas above it, and the weight of the overlying rock. In Bottle B this pressure is represented by the force of gravity. In the initial phase of oil production, Phase I, the original pressure in the field drives the oil into (and out of) the production well (the mouth of the inverted bottle) so that it

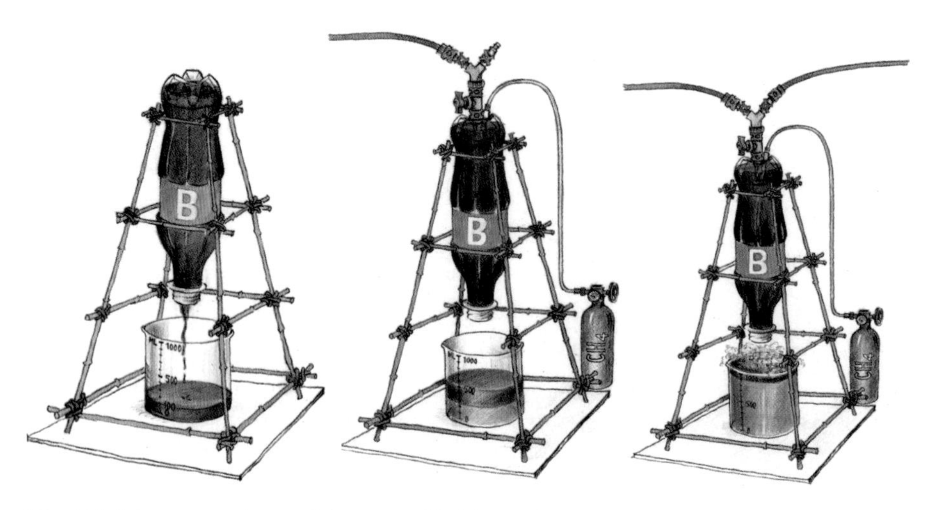

Fig. 7.4 A home-made model of an oilfield. Bottle B contains 1,000 mL of oil and is mounted in a frame. The three phases of oil production are explained in the text. The volumes in the beakers represent the volumes of liquids produced during the three phases. The setup on the left represents Phase I where production proceeds using the pressure that originally exists in the oilfield. A field with sufficient pressure may yield about 10% of its volume in this way (shown as 100 mL in the beaker). In the middle is Phase II where the pressure in the field is enhanced by pumping in water (or pumping back in natural gas if this is produced with the oil). In this phase both oil and water are produced from the well. The field may yield up to 25% of its oil (shown as 250 mL of oil accompanied by 250 mL of water). If half of the produced volume is water then the "water cut" is said to be 50%. The setup on the right represents Phase III where a chemical (e.g., a detergent) is injected. The flow of water increases and an additional 10% of the field's oil may be extracted (100 mL in the beaker and 900 mL of water). If one sums together the oil produced from our model oilfield during the three production phases it is 450 mL or a "recovery" factor of 45%, which is quite high. A working example of this model oilfield was built and was then filmed by the Australian popular science program Catalyst [2]

collects in a beaker below (see the arrangement on the left in Fig. 7.4). We can expect about 10% of the field's oil to be produced in this way.

After Phase I our technicians begin to work more intensively on our model "oilfield." They connect two pipes to the bottom of Bottle B. One is a water hose and one carries natural gas (methane, CH_4) from a cylinder. (See the middle diagram of Fig. 7.4.) A new beaker is placed below and then the water tap is carefully opened. This produces additional pressure in Bottle B that increases the flow of oil. Just as for a real oilfield, if we increase the water pressure too rapidly the fluid flowing out of our production well will consist of more water than oil. Fortunately our experienced technicians know that they must proceed cautiously. Slowly but surely the flow from the field is increased until the maximum oil production rate is reached.

However, the proportion of water in the production flow, the "water cut," steadily rises. In the beaker we can see the amounts of oil and water produced. When eventually most of what comes out of Bottle B is water we have produced most of what we can expect from Phase II, approximately 250 mL of oil (25% of the field's oil) and 250 mL of water. In reality we would expect more water than oil.

Sometimes it can be difficult to pressurize an oilfield with water. This occurs especially for oilfields that also contain natural gas. For such fields it can be necessary to pressurize them with natural gas. For example, in "Zone B" in Fig. 7.2 the positions of injection wells for water and for natural gas are indicated. By injecting both water and natural gas the oil is put under pressure from two directions, below and above, respectively. Usually, the natural gas that is used for injection is what is produced from the same field or a nearby field. When natural gas is produced from a field it usually consists of 15% "natural gas liquids" (NGL, a mix of hydrocarbons such as ethane, propane, and larger molecules) and the "dry natural gas" (primarily methane) is pumped back into the oilfield to maintain pressure. This is represented in Fig. 7.4 by the connection of Bottle B to a gas cylinder containing methane.

If we closed down production from the oilfield after our efforts in Phase II we would have achieved an oil recovery factor of 35%, that is, 350 mL from an original volume (OOIP) of 1,000 mL. However, we can continue with Phase III during which we attempt to "wash" the reservoir with various chemicals. In our model of an oilfield, we connect a new hose to our water injection hose, for example, one bringing in some form of detergent. We can also connect an additional gas injection hose. This new hose brings in a gas such as carbon dioxide, CO_2. Currently such methods are only applied to a small number of oilfields but their use is expected to grow as the world's oil reserves dwindle. If we place a new beaker under Bottle B and then open the valve controlling the water and detergent, the oil production will resume. After a while we might obtain 100 mL of oil floating on 900 mL of water with the detergent creating some froth on top.

In reality, the three phases of oil production from a large oilfield overlap somewhat as shown in Fig. 7.5. If it were possible to run our model of an oilfield over a period of only 150 s we would see the oil production history shown in this figure. In reality the three phases of oil production from a field span a much longer time. For example, oil production from the Statfjord field in Norway spanned decades. This is shown in Fig. 7.6.

Because of limits pertaining to oil transport through pipelines, oil producers normally try to maintain a steady rate (a "plateau") of production from a field. But eventually all oilfields reach a phase where production declines (despite that the world's thirst for oil is constantly increasing).

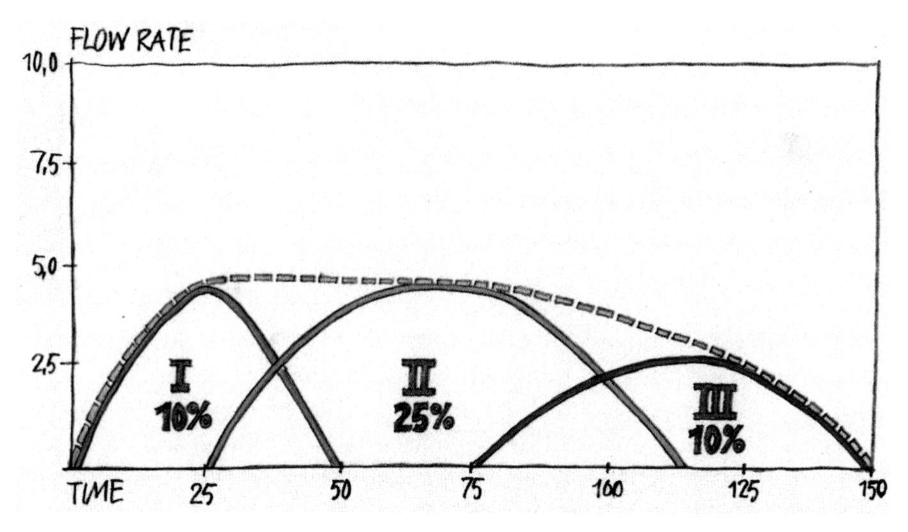

Fig. 7.5 The rate of oil production during the three phases described in Fig. 7.4. In this diagram the three phases of oil production partially overlap and the entire process spans 150 time units (e.g., seconds). If the time units are seconds and Bottle B holds 1,000 mL of oil then the flow rate would be measured in mL/s

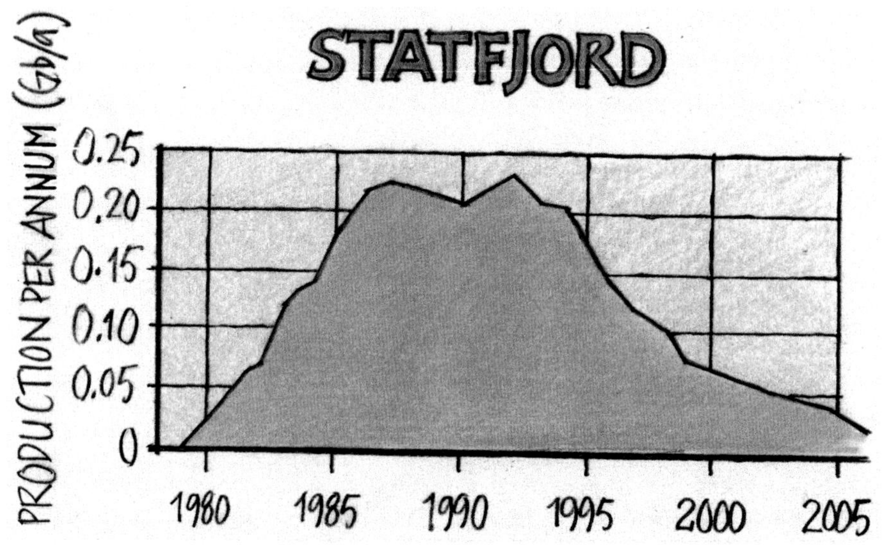

Fig. 7.6 The Statfjord oilfield is one of Norway's giant fields. Production began in 1979. By 1986 production had been expanded to the desired level and "plateau production" began. Factors that limit the rate of production are the ability of the production platform to handle the flow from the various production wells and to pressurize the field. In 1994 after 8 years of plateau production the field's maximum possible production rate was reached (when measured not as gigabarrels per annum, Gb/a, but as the proportion of the remaining recoverable oil) and so the decline of production (as Gb/a) began. This is discussed in detail in Chap. 9

In the case of our model oilfield the ultimate oil recovery factor is 45% which is somewhat higher than the value that Leif Magne Meling gave as the global average in Fig. 6.2.

Oil production, whether from a single oilfield or the world's entire oil reserves, never suddenly runs out. Instead, the oil becomes steadily more difficult to produce and so the flow of oil tapers off slowly.

A Tour by 4WD Out Over the Bab Oilfield

Let us conclude this chapter by returning to the Bab oilfield. It lies under the desert 160 km southwest of Abu Dhabi. It was found in 1952 after seismic studies identified a suitable geological structure to investigate. When they drilled exploratory well Bb 2 in 1958 they found oil and further investigation showed that the oilfield was a giant. The first oil was produced from the field in 1963. Initially the oil was produced by "natural decline" according to Phase I as illustrated above. However, by 1986 the remaining pressure in the field was so low that they were forced to shut down production. Since 1993 they have gradually instituted Phase II production using both gas and water injection. When I was there in 2004, they had just installed a new cluster of water-injection wells and new infrastructure for handling the gas released by processing the incoming flow of oil from the production wells. (Understandably I was not permitted to take photographs during my tour so my description of the oilfield is from memory only).

I was pleasantly surprised when it turned out that the guide for my tour was to be the Head of Production at Bab, Abdulla M. Al-Malood. We had an interesting journey among the sand dunes. The first stop was to look at their new wells for water injection. Technically speaking, water-injection wells are drilled down to the level of the water that lies underneath the oil in a reservoir/zone. Additional water is then pumped in which forces the overlying oil into the production wells. In Fig. 7.2 a water-injection well and a suitable production well are indicated for "Zone B".

The next stop was at one of the production wells. Most people think of an oilfield as covered in classic "pump jacks" (also known as "nodding donkeys") that rhythmically move up and down. So it was a bit of a disappointment to see instead only a "Christmas tree" with all its valves (Fig. 7.7). The pressure in the Bab oilfield is now sufficiently high that the oil simply flows up through the production well, through the Christmas tree (that controls the pressure in the flow) and then via a "production wing valve" to a "flow line." One could see a network of flow lines over the sand dunes all leading to the central processing plant's "oil train" (more on this below).

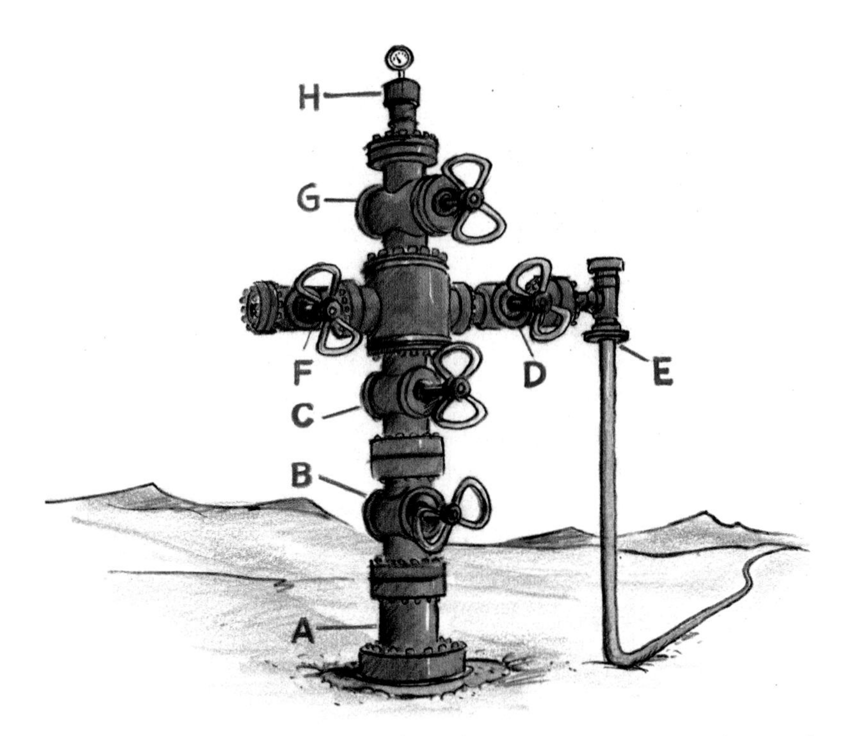

Fig. 7.7 A production well is lined with steel. At its top are mounted a number of valves pointing in different directions. All these "branches" decorated with valves have led to this equipment being called a "Christmas tree" by the oil industry. It is this equipment that is meant to prevent uncontrolled oil flow. The various valves of the Christmas tree are also used to restrict the flow of oil to a suitable rate. *A* – tubing-head adapter, *B* – lower master valve, *C* – upper master valve, *D* – production wing valve, *E* – flow line to production facility, *F* – kill wing valve, *G* – swab valve, and *H* – tree cap and gauge

On the way back to the central processing plant we chanced upon some of the field's resident camels. Once again I regretted not being able to take my camera with me. However, when we arrived back it was time for a wonderful lunch and all the oilfield's personnel had the opportunity to partake of it. These people work and live at the field for 2 weeks and they then have a number of weeks off. There is no doubt that working on the Bab oilfield is very desirable employment!

The "oil train" is not a train. Rather, it is an installation that separates water and gas and contaminants from the oil in a number of processing steps. During the Phase II upgrade of the Bab oilfield in the 1990s they installed five oil trains. When I was there in 2004 they were building an additional two trains to handle production from North East Bab, NEB. When discussing the field's production capacity the personnel did not refer to the flow rate from the oil production wells. Instead they discussed the flow capacity of the oil trains. Thus it became apparent that when people discuss Abu Dhabi's capacity to produce oil, they are actually referring to the flow rate through the oil trains. The nightmare of every oilfield operator is an increasing proportion of water in the flow from the production wells (the water cut) because this influences the volume of separated oil coming out of the oil train. In 2004 the Bab field had only 5% water in the combined flow from its various production wells. This is a uniquely low water cut. The world's largest oilfield, Ghawar in Saudi Arabia, has reported a water cut of 35% and some famous old fields in Texas have been said to have a water cut of 90%.

While at Bab we also discussed the costs of production and I asked what it costs to deliver a barrel of oil to the ships waiting at the terminal in Jebal Dhanna. On the same day as the price of a barrel of crude oil rose past US$50 I was told, "Currently the costs have risen to 98 cents per barrel."

My visit to Bab ended in a traditional Arabian manner. We removed our shoes and sat down on beautiful woven carpets. Traditional crystal glass from Sweden was exchanged for an inkstand of molded plastic in the shape of an oil drop and containing actual crude oil from the Bab oilfield. Instead of an inkstand's usual fountain pen, this one supported a ball-point pen with a casing made of polished marble.

References

1. BP: BP Statistical Review of World Energy. http://bp.com/statisticalreview (Data for 2009 in historical data; http://www.bp.com/assets/bp_internet/globalbp/globalbp_uk_english/reports_and_publications/statistical) (2011)
2. In November 2010 when this chapter was written the team (Michael Lardelli, Olle Qvennerstedt and I) was together in Adelaide in Australia. At that time Jonica Newby, a presenter on the Catalyst popular science program on Australia's ABC TV, was preparing a segment about Peak Oil titled The Oil Crunch. Jonica came to Adelaide to interview me for the segment. *The Oil Crunch* and extended interviews can be viewed online at http://www.abc.net.au/catalyst/oilcrunch/. There you can also find a video of Jonica demonstrating our model oilfield as shown in Fig. 7.4 (2010)

Chapter 8

The Size of the Tap: The Laws of Physics and Economics

Oil can be found underground in oilfields or above ground in storage tanks. When taking oil from a tank the rate of flow is limited by the diameter of the hole, pipe, or tap. However, when taking oil from an oilfield, it is not only the diameter of the oil well or the capacity of other infrastructure that is important. The rate of flow is also controlled by the physics of oil movement through rock.

When future rates of oil production are discussed by economists they mainly consider access to the resource and the demand for oil. Economists assume that, if the oil price is high, this should encourage producers to invest in exploration for new resources and in development of discoveries that they already control. For most economists, price regulates the flow of oil to the market in a way similar to how a tap regulates the flow of oil from a tank. For international oil companies registered on the world's stock markets it has become extremely important for them to show they have a sufficient flow of oil into their "reserves" (from discoveries, upward revisions of reserves they control, or purchases of reserves controlled by other oil companies) to compensate for the rate at which oil flows out of their reserves due to oil production.

Some years ago, Shell failed in its attempts to do this. Shell reported some discoveries as reserves that, according to the rules, should not have been classed as such. When Shell's reported reserves were subsequently revised down by 25%, its value on the stock markets fell by an equivalent amount and Shell was also forced to pay expensive fines. Since then rumors have circulated that factions were in conflict within Shell about how large the company's reserves actually were and that the optimistic faction had

K. Aleklett *Peeking at Peak Oil*, DOI 10.1007/978-1-4614-3424-5_8,

held the upper hand for too long. At that time, Shell's optimists believed that Peak Oil was just a fairy tale but today's leadership has a more realistic view of the future and regards Peak Oil as a possible truth in some distant future. However, Shell still has some distance to go before it accepts Peak Oil as reality.

The Flow Equation

Oil wells and other infrastructure are "above ground" factors determining the rate of flow of oil from a field. Economic factors can determine the capacity of this infrastructure. However, the ultimate limit on oil flow is determined by the rate at which oil flows through porous rock. As oil passes through this rock on its way to a production well, it is the laws of physics that matter, not economics.

Sandstone is a porous form of rock formed from sedimentary layers of sand. With a microscope it is possible to see the pores in the sandstone (see Fig. 6.1). The relationship between the volume of the pores and the volume of the rock is called the *porosity*. A rock with high porosity can contain more oil than one with low porosity. For rock to become an oil reservoir the pores within it must be connected to one another so that the rock is permeable. Oil moving upwards from the source rock can then force out the water that originally occupied these pores. Eventually the oil floats above the groundwater. When a production well is drilled down into a reservoir rock, oil is expected to move mainly sideways (horizontally) into the well. To make it easier for the oil to move into the well the rock is sometimes intentionally fractured. The rate at which oil moves from the rock into the well determines how many barrels of oil the well will produce per day, month, or year.

All porous forms of rock can become oil reservoirs if their pore size is sufficiently large to accommodate oil and if the channels between the pores are sufficiently large to allow oil to flow between them. In reality, most of the world's oilfields are formed of sandstone or limestone. When geologists search for potential oilfields they look for these types of rock. If a rock's porosity is 15% then it can form a very good reservoir and allow high rates of oil flow but such reservoir rocks are the exception. Most oil reservoirs exist deep underground where they are subjected to very high pressure and they commonly have poorer porosity.

A microscopic examination of a reservoir rock shows that there is competition for space within the pores. Even though oil displaced much of the water as it forced its way up into the rock, some water still remains. A reservoir's "oil saturation" describes how much of its pore space is occupied by oil and gas; for example, an oil saturation of 25% means that one quarter of the pore space is occupied by hydrocarbons. For oil to flow from a reservoir rock into a production well the oil saturation must exceed a critical level.

The pores within a rock can hold oil but these same pores also form a labyrinth through which oil must move if it is to enter a production well. The speed at which oil can move through this labyrinth determines the rate of flow and, ultimately, the rate of oil production from the well. In the mid-1800s (at about the same time as the first oil was produced in Romania) the French engineer Henry Darcy studied how fluids flow through layers of packed sand. He discovered that the rate at which a fluid could flow through this porous material (called *permeability* and usually given the symbol q) is determined by the pressure difference through the sand bed (P), the distance through the sand bed (L), the cross-sectional area of the sand bed through which the oil is flowing (A), and how hard the sand is packed (i.e., the porosity of the sand bed, k). The *viscosity* of a liquid is a measure of how much internal resistance it has against flowing and is usually represented by the Greek letter μ. Slow-flowing liquids (e.g., honey) have a higher viscosity than fast-flowing liquids (e.g., water). Darcy's observations could be summarized by an equation that is now called Darcy's law:

$$q = -\frac{kA}{\mu} \cdot \frac{\partial P}{\partial L}$$

Because P and L are the pressure difference through the sand bed and the distance through the sand bed, respectively, then the expression $\partial P/\partial L$ means the pressure gradient down which the liquid flows through the bed. Darcy's law tells us the ability of a liquid such as oil to flow through a porous material such as sandstone and is illustrated in cartoon form in Fig. 8.1. Note especially how differences in viscosity have been illustrated ("fat" versus "slim"). One can add that even though Darcy assembled his equation purely from experimental observations (i.e., empirically) this law can also be derived from the Navier–Stokes equations that now form the basis of our understanding of fluid motion [1].

The permeability of rock for horizontal fluid flow is usually less than its permeability for vertical flow. A reserve rock is filled by oil vertically but oil mainly moves horizontally into a production well, thus the lower horizontal permeability means that it is impossible to empty a field entirely of its oil. This is one of the reasons why recovery factors are less than 100% (see the section "Recovery Factor: The Amount of the OOIP that Can Be Produced", Chap. 6). The formation of cracks in the reservoir rock (which commonly occurs in limestone) can increase permeability. One modern technology for increasing oil recovery is to generate such cracks artificially. However, caution is required because cracks that are too large can be problematic when one later tries to pressurize a field with water.

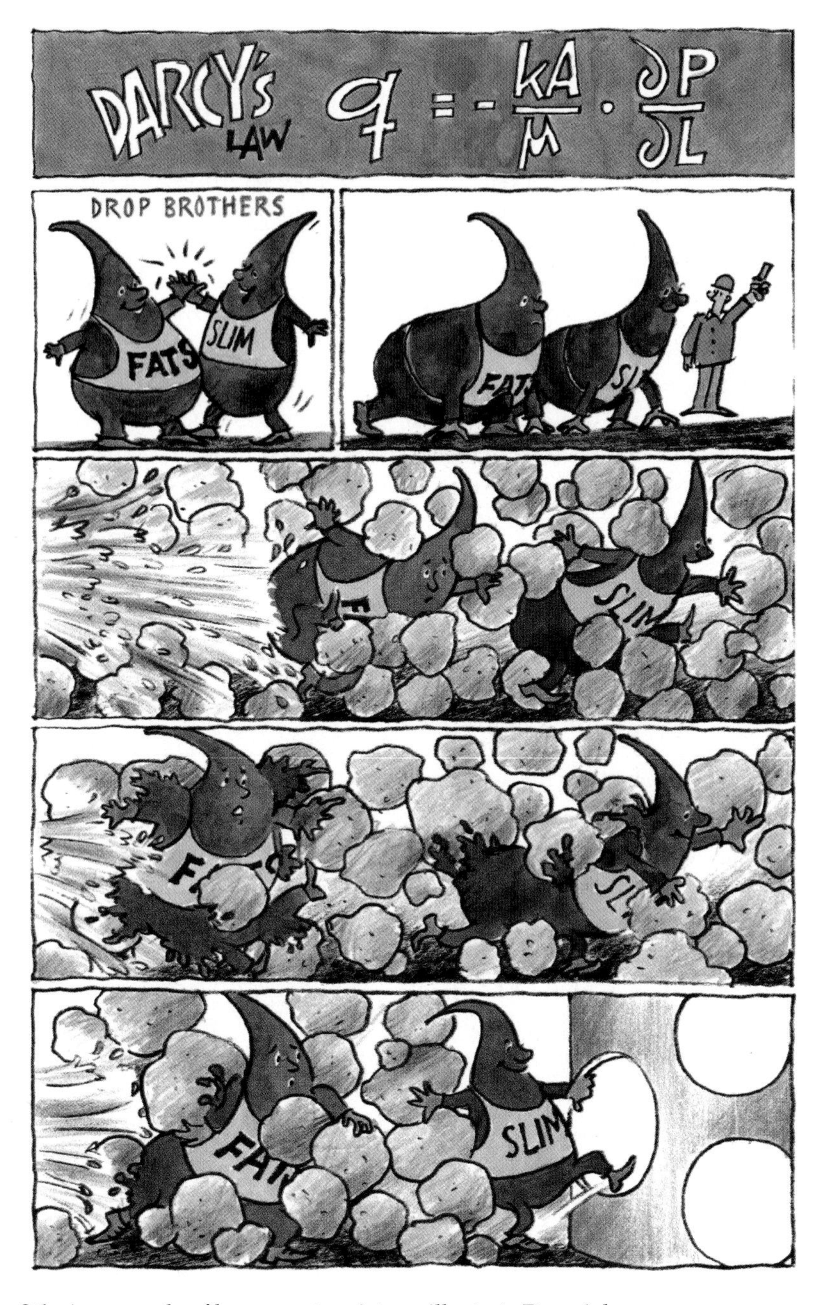

Fig. 8.1 An example of how a cartoonist can illustrate Darcy's law

The conclusion that we can draw is that a good oilfield should have high porosity and permeability. (A complete overview of reservoir rocks and their fluid flow properties can be found in "Petroleum Reservoir Rock and Fluid Properties" [2]).

The Market Economy and Peak Oil

In February 2001 I published an article on Peak Oil in one of Sweden's largest daily newspapers, *Svenska Dagbladet*. My article was subsequently challenged by an authority in national economics, Marian Radetzki, who asserted that I was completely wrong. One of his arguments was that then, in February 2001, there was no economic modeling by any prominent economic or energy agency or bank showing future oil prices exceeding US$27 per barrel and, if Peak Oil were true, then these prestigious institutions would have estimated future oil prices differently.

Another of my economic opponents is Michael Lynch, the director of Strategic Energy and Economic Consulting. We have debated Peak Oil publicly a number of times, most recently in Shanghai in November 2008. At that time he restated his belief that oil prices would be low and stable at around $40 per barrel for the next 10 years. In the presentation he delivered at that meeting he showed just how "flat" were the economic prognoses in 2004, including the one from the International Energy Agency (IEA). If we also consider the prognoses that the IEA has presented in more recent years then we can conclude that most economists' predictions for future oil prices are "flat" (Fig. 8.2). This flatness has led Colin Campbell to call economists such as Lynch and Radetzki "Flat-Earth Economists." Today we can see that these economic prognoses made between 2001 and 2007 were wrong, including the one that Lynch made in Shanghai in November 2008. The oil price today is much higher than they predicted. Our predictions made in 2008 see future shortages of oil and higher future oil prices as a possibility. Indeed, we can reverse the reasoning of Marian Radetzki to state that the recent history of oil prices indicates that Peak Oil is true.

"Peak Oil and Other Threatening Peaks – Chimeras Without Substance" is the title of a "Viewpoint" by Marian Radetzki published in the journal *Energy Policy* [3]. In the abstract (summary) of the article he echoes the attitude of many economists to peak oil:

> The Peak Oil movement has widely spread its message about an impending peak in global oil production, caused by an inadequate resource base. On closer scrutiny, the underlying analysis is inconsistent, void of a theoretical foundation and without support in empirical observations. Global oil resources are huge

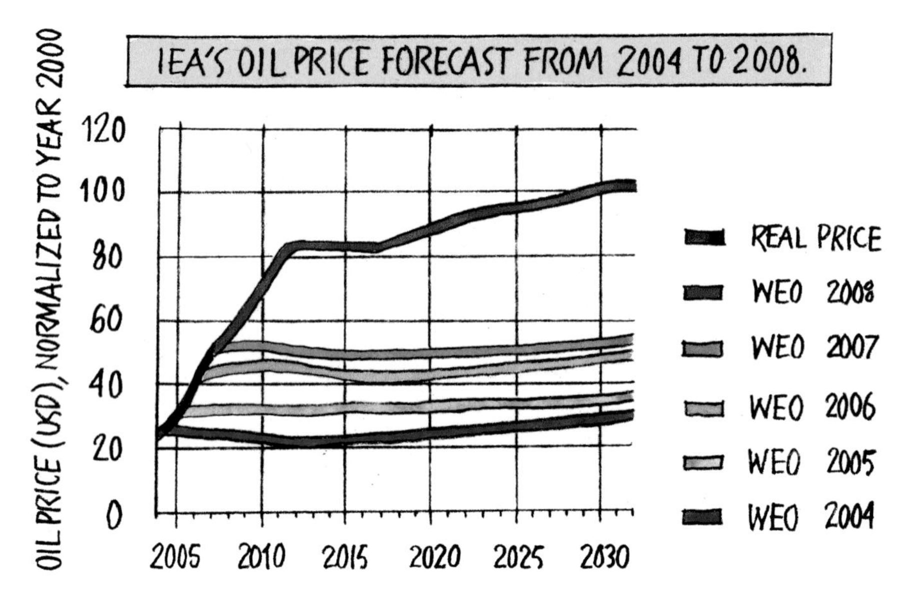

Fig. 8.2 The predictions of future oil price presented annually in the IEA's World Energy Outlook Report (IEA 2004–2008)

> and expanding, and pose no threat to continuing output growth within an extended time horizon. In contrast, temporary or prolonged supply crunches are indeed plausible, even likely, on account of growing resource nationalism denying access to efficient exploitation of the existing resource wealth.

Among the arguments that he presents against Peak Oil is the fact that Colin Campbell uses a much lower estimate for ultimate recoverable resources than that used by the United States Geological Survey (USGS; see the section "Proven and Possible Reserves", Chap. 6). If we study Fig. 6.4 we can see that this is not a strong argument. Radetzki asserts that the reserve growth figures given in BP's *Statistical Review of World Energy* are evidence that we do not have a problem despite the fact that, if we base our reserve growth estimates on 2P-statistics (i.e., proven and probable reserves; see the section "Proven and Possible Reserves"), we get a completely different picture. For estimates of future reserve growth from currently producing oilfields Radetzki refers to old US oilfields where production was begun using simpler technologies compared with what is available today. As Fredrik Robelius demonstrated in his PhD thesis, it is unrealistic to expect reserve growth in the future to resemble what was seen for those old US oilfields. Robelius' results are summarized in Fig. 6.5.

During the 1960s when actual oil discoveries averaged 56 billion barrels per annum (Gb/a) and oil consumption was only 10 Gb/a (Fig. 3.1), our reserves of oil were so great that economic models were generated that did not consider oil availability to be a limiting factor. The expectation was that, if more oil were needed, it could easily be produced from these huge reserves. Today the reality is different, and it is time for contemporary economists to put assumptions of unlimited oil behind them. They must familiarize themselves with the concept that future oil flows will be determined, in large part, by Darcy's law. They must reformulate their economic prognoses according to this.

References

1. Neumann, S.P.: Theoretical derivation of Darcy's law. Acta Mech. **25**(3–4), 153–170 (1977)
2. Dandekar, A.Y.: Petroleum Reservoir Rock and Fluid Properties, p. 488. CRC, Boca Raton, FL (2006)
3. Radetzki, M.: Peak oil and other threatening peaks – chimeras without substance. Energy Policy **38**, 6566–6569 (2010). http://www.sciencedirect.com/science/article/pii/S0301421510005793

Chapter 9

The Elephants: The Giant Oil Fields

"Study the past if you would define the future" is a famous saying of Confucius (551-479 BC) (The illustration in Fig. 9.1 is based on a photograph of a carving in Xi'an). The analysis of historical trends is a common form of research. Sometimes, extrapolation of trends is the only way we know to make predictions about the future. Nevertheless, we must be aware that such methods are not precise and divergence of the future from our expectations is possible. When making predictions of the future we must always clearly indicate the conditions and assumptions upon which the predictions are based. By doing so we will be prepared to change our predictions if, for example, new political decisions are made or new facts come to light (Fig. 9.1).

In Chap. 9 we discuss the research that we conduct in the Uppsala Global Energy Systems group (UGES) at Uppsala University in Sweden. All our peer-reviewed scientific articles, theses, special reports, and magazine and newspaper articles are available to download and read [1].

The Elephants

In Chap. 4 we learned that certain regions of the world possess thick sedimentary rock layers that can support the formation of oilfields whereas other regions lack the necessary rock types and geological structures. Oilfields that are estimated to contain an ultimately recoverable resource (URR) greater than 500 million barrels are defined as "giant oilfields" and the oil industry has nicknamed them "elephants." Historically, the giant oilfields have dominated the world's oil production and future production from these fields will be critical for our access to oil (Fig. 9.2).

K. Aleklett *Peeking at Peak Oil*, DOI 10.1007/978-1-4614-3424-5_9,

Fig. 9.1 Confucius was a Chinese philosopher who lived during the years 551–479 before Christ. The ancient values that he propagated were that one should respect one's superiors and show kindness to those of lower station. He also believed that we are all similar at birth and are then shaped through education and training

Cantarell in Mexico is one of the world's ten largest oilfields. The field has passed through all the phases of development described in Chap. 7 and the field's production is now declining dramatically. Technological interventions cannot hinder this decline and so Cantarell is approaching its final rest.

In December 2000 the term "Peak Oil" was coined (see Chap. 2) and in May 2002 the world's first conference on Peak Oil was organized in Uppsala, Sweden. Around that time Colin Campbell and I were invited to write an article on Peak Oil for the journal *Minerals and Energy* (see Chap. 11) and that inspired me to investigate the possibility of obtaining grant funding to research this topic. We were successful and in January 2003 I was able to appoint Fredrik Robelius as a Ph.D. student and Anders Sivertsson as a Diploma student. Colin Campbell, who has a Ph.D. in geology, became an affiliate of Uppsala University and acted as supervisor. In February 2003, we were able to hold our first group meeting in Ballydehob in southern Ireland.

At the meeting in Ireland we decided that Fredrik would investigate the world's giant oilfields and that Anders would work on a model of future oil production that Colin Campbell had been developing in recent years. Primarily Anders was to study whether the emissions scenarios presented

Fig. 9.2 The production story told by the elephant Cantarell is one of dramatic decline and soon it will be time for Cantarell to make its way to the elephants' graveyard. Some of the other gigantic elephants, Ghawar, Greater Burgan, and the younger Kashagan are also considering when it will be time for them to stand up and announce that they are going to retire

by the Intergovernmental Panel on Climate Change [2] in 2000 were reasonable in the light of Peak Oil. (The results of this study are discussed in Chap. 17). This established the research directions that the works of the Uppsala Global Energy Systems group were to follow. The discussion of giant oilfields that follows below is a summary of the Ph.D. thesis of Fredrik Robelius that is titled, "Giant Oil Fields—The Highway to Oil: Giant Oil Fields and Their Importance for Future Oil Production" [3].

Ghawar in Saudi Arabia is the world's largest conventional crude oilfield. It is 280 km long and 26 km wide and can be compared to a 26 km wide motorway between Brussels and Paris (see Fig. 9.3). In recent years, Ghawar has been reported as giving stable production of around five million barrels of oil per day which is equivalent to 7% of the world's total crude oil production. (We return to Ghawar in Chap. 13).

In addition to oil production from the giant oilfields, oil is also produced from a large number of small oilfields (and there are many small oilfields that have already been abandoned). According to the *Oil & Gas Journal* there were 34,969 productive oilfields in the United States [5] in 2006 and ISH Energy has stated that there were 12,465 productive oilfields outside the United States in 2005 [6]. In recent years new fields have also been discovered and/or put into production so an approximate total for all productive oilfields is 47,500. Of these, 507 are reported to be giant oilfields [3] which means that approximately 1% of the world's productive oilfields are giants.

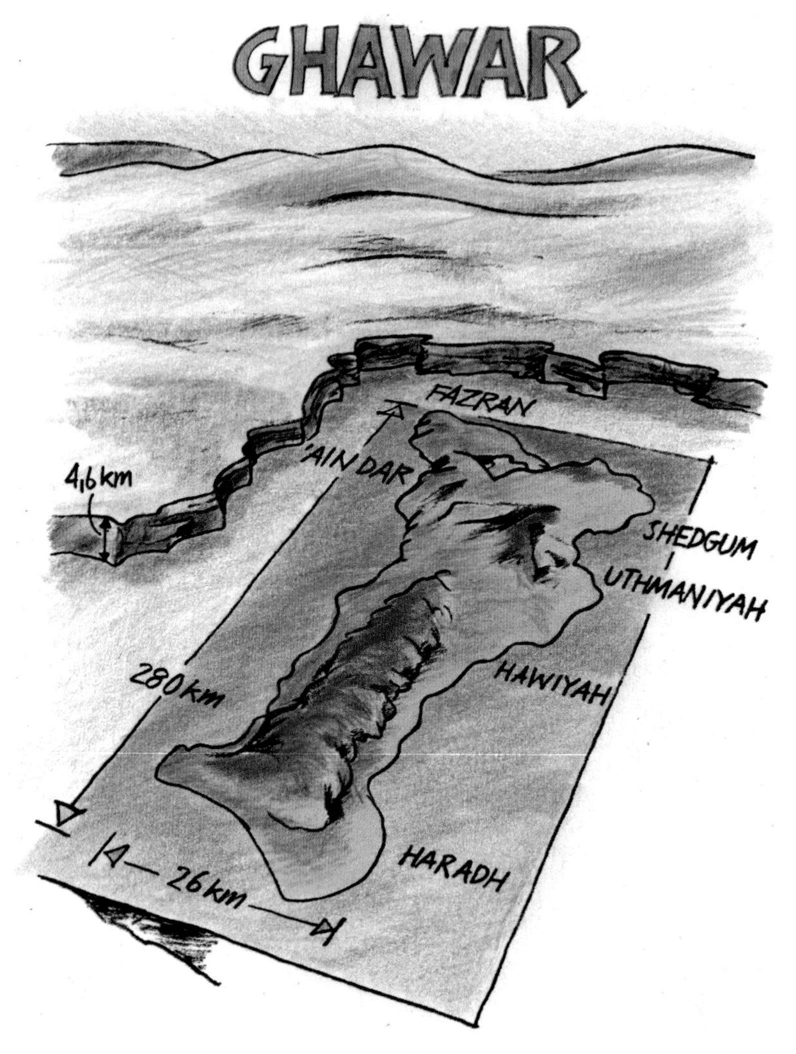

Fig. 9.3 Ghawar in Saudi Arabia is the world's largest conventional crude oilfield. It was discovered in 1948 and began producing oil in 1951. It is 280 km long and 26 km wide and can be compared to a 26 km wide motorway between Brussels and Paris. The field lies at a depth of approximately 4.6 km and contains both oil (*indicated in green*) and natural gas (*in red*). The Ghawar oilfield is divided into five production areas: Ain Dar, Shedgum, Uthmaniyah, Hawiyah, and Haradh. Note that the image is foreshortened on its longer axis [4]

If we add together the production from the world's 100 largest oilfields we find that they account for as much as 45% of the world's crude oil production. If we add together all the oil that has been produced and all the oil that we estimate will be produced (the ultimately recoverable resource,

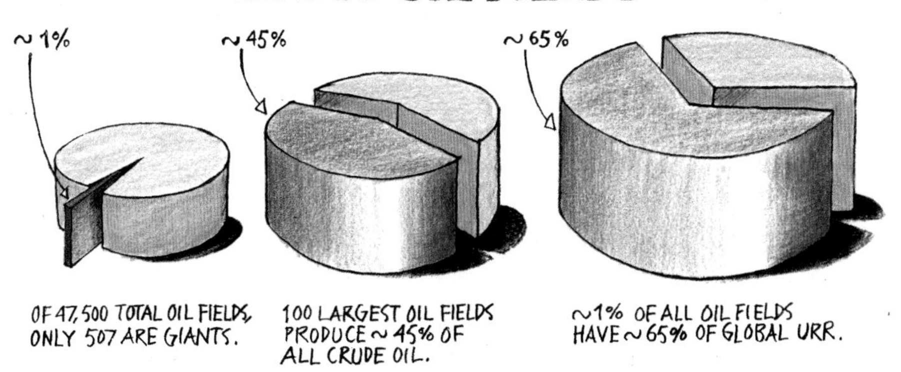

Fig. 9.4 Oilfields containing 500 million barrels or more of recoverable oil are called giant fields. In the database of the UGES we noted, in 2006, that there were 507 giant fields, which was approximately 1% of all the oilfields in production. Giant fields accounted for 65% of the oil that comprised the world's total URR and the world's 100 largest fields were responsible for 45% of the world's crude oil production [2]

URR) then the URR for the giant fields accounts for 65% of total world URR (see Fig. 9.4).

During the entire twentieth century production from the giant oilfields dominated global production. This is shown in Fig. 9.5 which is divided into three segments. The lowest segment is production from the 21 largest giant fields (see Table 9.1). The segment above that shows production from the remaining 291 giant fields found in our database. Finally, the uppermost segment is the production from the rest of the world's tens of thousands of oilfields [3]. Note that the number of fields in every new segment (i.e., from bottom to top) increases by approximately a factor of 10.

At the end of the 1970s production from a number of oilfields in the Middle East was restricted for political reasons. This decision created an artificial oil production peak (a "Peak Oil") for the collective production from the world's 21 largest oilfields. Oil production from the rest of the world could not compensate for this downturn (see Fig. 9.5). In this way, the Middle East showed that control of oil flow is a form of political and economic power. The fact that changes in production level from only some of the world's giant oilfields could affect global access to oil demonstrates that detailed studies of the production capacity of giant oil fields is critical for our ability to predict oil production. This observation has been one of the keystones of our research strategy.

It is difficult (bordering on impossible) to obtain precise information on URR for the giant fields. For example, if one gathers data on Ghawar from various articles and journals, one finds URR values ranging from

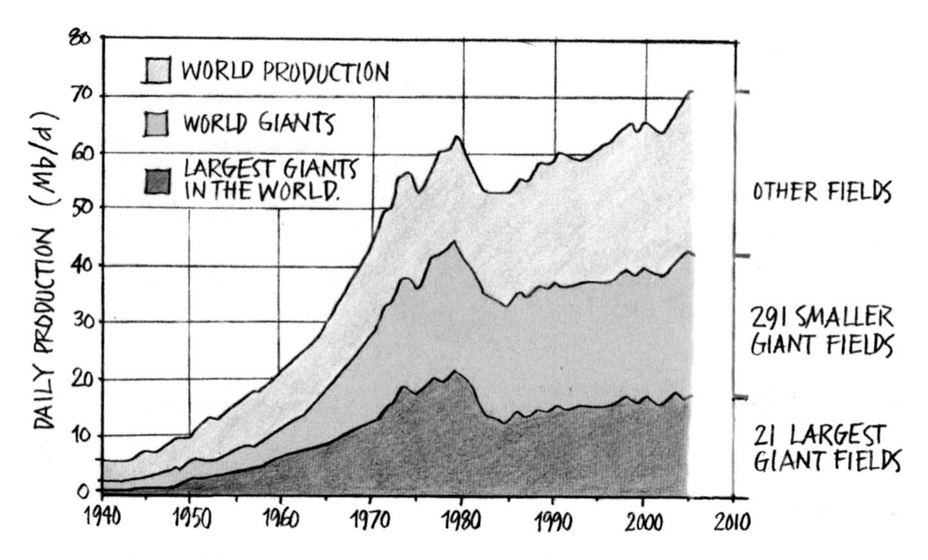

Fig. 9.5 The world's oil production can be divided up into three segments that differ from each other by approximately a factor of 10 in terms of the number of producing oilfields each represents. Politically motivated decreased production from some of the world's largest fields created an artificial production maximum at the end of the 1970s. When the largest fields begin to decline we can presume this means that the world has passed its production peak, Peak Oil (Data from Robelius' thesis [3])

66 to 150 billion barrels [3]. However, these URR values can be treated as lower and upper limits when making analyses of future oil production. In Table 9.1 is a summary of the lower and upper URR limits for the world's 21 largest oilfields obtained by analyzing scientific and other literature [3]. These values were used by Fredrik Robelius in formulating various scenarios of future oil production. The four scenarios he created were named "Best Case," "Standard Case High End," "Standard Case Low End," and "Worst Case." It is interesting to note that the large variation in total URR gives only a relatively small effect on the predicted total crude oil production from the giant fields under the four scenarios, which are presented in Fig. 9.6.

It is often asserted that we require much better data on the giant fields in the Middle East to be able to estimate future world oil production. However, our research has shown that quite large variations in total giant field URR only delay the predicted decline of world oil production by, at most, 10 years. If more reliable values for URR become available the time-point for the onset of decline can be estimated more precisely. However, considering that giant oilfields typically show dramatic rates of decline and that it takes a long time to construct alternative energy systems, our analysis shows that we should construct those systems without delay.

Table 9.1 The 21 largest oilfields in the world in terms of URR given in billions of barrels (Gb) [3]

Oilfield name	Nation	Year of discovery	Start year production	URR range (Gb)
Ghawar	Saudi Arabia	1948	1951	66–150
Greater Burgan	Kuwait	1938	1945	32–75
Safaniya	Saudi Arabia	1951	1957	21–55
N & S Rumaila	Iraq	1953	1955	19–30
Bolivar	Venezuela	1917	1917	14–30
Samotlor	Russia	1961	1964	28
Kirkuk	Iraq	1927	1934	15–25
Berri	Saudi Arabia	1964	1967	10–25
Manifa	Saudi Arabia	1957	1964	11–23
Shaybah	Saudi Arabia	1968	1998	7–22
Zakum	Abu Dhabi	1964	1967	17–21
Cantarell	Mexico	1976	1079	11–20
Zuluf	Saudi Arabia	1965	1973	11–20
Abqaiq	Saudi Arabia	1941	1946	13–19
East Baghdad	Iraq	1979	1989	11–19
Daqing	China	1959	1962	13–18[a]
Romashkino	Russia	1948	1949	17
Khurais	Saudi Arabia	1957	1963	13–19
Ahwaz	Iran	1958	1959	13–15
Gashsaran	Iran	1928	1939	13–14
Prudhoe Bay	USA	1968	1977	11–14

[a]In Chap. 15, "China and Peak Oil," we present data obtained in 2008 indicating Daqing's URR as 24 Gb

The preconditions for the "Best Case" scenario of Robelius were that a number of giant fields not then exploited maximally should have their production expanded to do so. The giant fields cited were Tengiz in Kazakhstan, "Northern fields" in Kuwait, "Zakum in Upper" in Abu Dhabi and Majnoon, West Qurnah, Halfayah, Nahr-Umr, Nasiryah, Ratawi, and Tuba all in Iraq. Note that seven of the giant fields for which production must be maximized are in Iraq.

Fredrik Robelius successfully defended his Ph.D. thesis on the world's giant oilfields on March 30, 2007. Some time after that a copy of the thesis was sent to Fatih Birol, the chief economist for the IEA. Birol replied with thanks for the thesis. In June of that year he was interviewed by the French newspaper, *Le Monde*. When asked about future oil production in Iraq he replied,

> If production does not increase exponentially in Iraq between now and 2015, we have a big problem, even if Saudi Arabia meets its commitments. The figures are very simple; one does not need to be an expert. It is enough to know how to do a subtraction. China will grow very quickly as will India and not even additional Saudi Arabian production of 3 Mb/d will be enough to meet the growing demand from China [7].

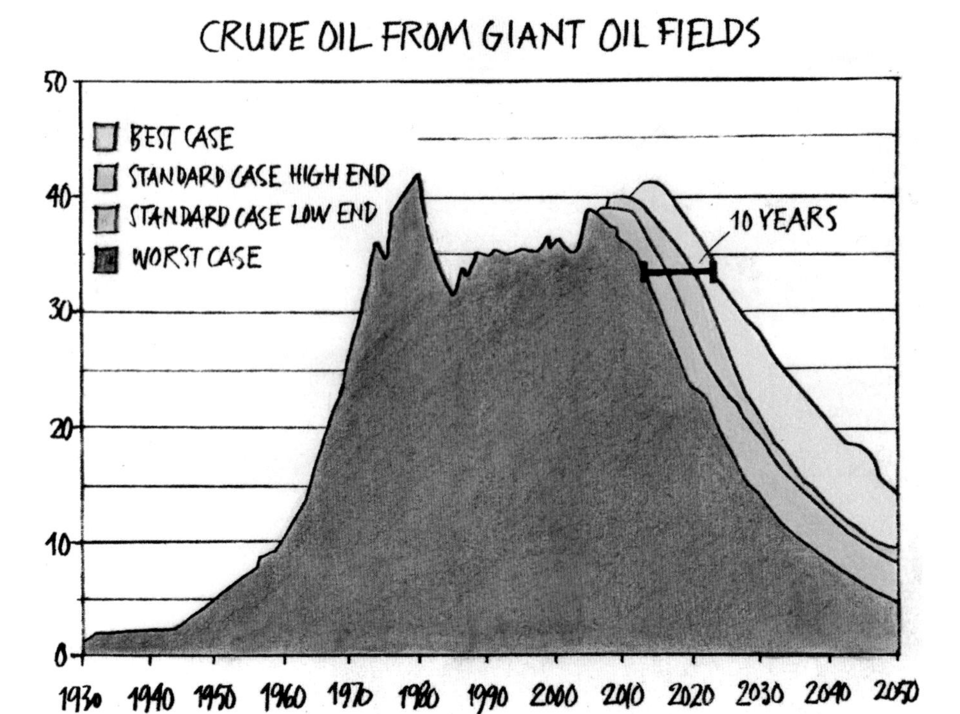

Fig. 9.6 Oil production from the world's giant fields expressed in millions of barrels per day (Mb/d) according to the four scenarios presented by Fredrik Robelius in his thesis [3]

It is interesting to note how the views presented by Fredrik Robelius in his thesis were thus confirmed by the chief economist of the International Energy Agency.

The main aim of Fredrik's thesis was to examine the importance of the giant fields to total world production. As shown in Fig. 9.5, their contribution was found to be very significant. The future projection of world oil production shown in Fig. 9.7 was made using the research data from the thesis. Of course, total world production as shown in that figure also includes other sources of oil such as from all the other, smaller oilfields, from oilfields in deep water, the unconventional oil from the oil sands of Canada and the Orinoco fields in Venezuela, and the "wet" portion of natural gas production called natural gas liquids. However, the projection shown in Fig. 9.7 (and Fig. 9.4 in the thesis [5]) does not include the production that may occur from oilfields we expect to find in the future: the discoveries marked in lighter green in Fig. 3.1 of Chap. 3. In our publication titled, *The Peak of the Oil Age* [8] we have shown that it is possible for

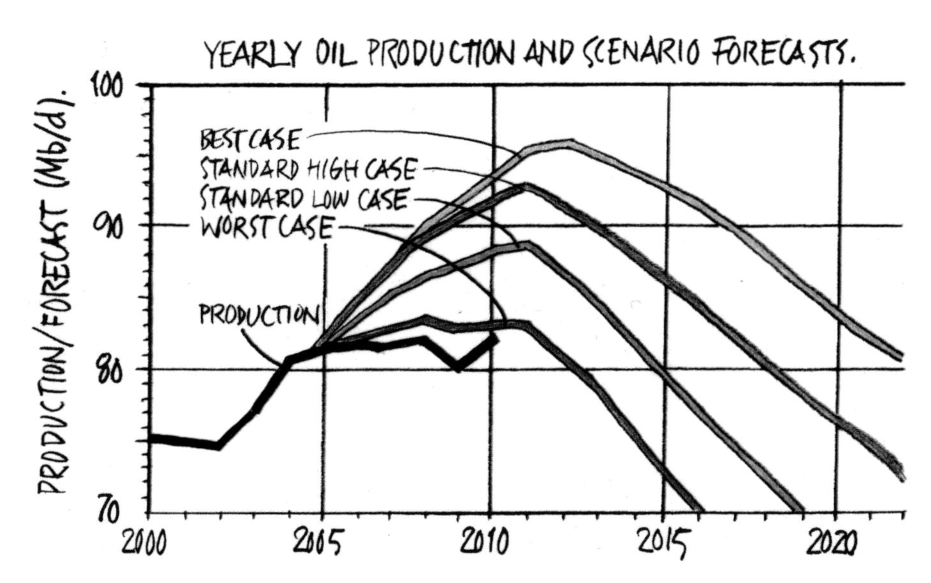

Fig. 9.7 Average monthly oil production in Mb/d from 2000 to 2010 as reported by BP (BP 2011) together with predictions of future production based on four different scenarios presented in Fredrik Robelius' thesis [3]

production from these as-yet-undiscovered oilfields to grow slowly to approximately 9 Mb/d by 2030. This production would not alter the overall situation shown in Fig. 9.7 but it would make the future decline in world oil production a little less steep.

Robelius' four scenarios for future production from giant fields were based on the information available in 2005. Now, 6 years later it is interesting to compare the scenarios with the oil production that actually occurred since then as reported in the *BP Statistical Review of World Energy 2011* [9]. In Fig. 9.7 we can see that oil production reached a plateau level in 2005 at 81.5 Mb/d which is slightly lower than the Worst Case scenario. If production had continued to follow the trend prevailing from 2002 to 2004 then it would have increased similarly to the Best Case or Standard High Case scenarios. The Worst Case and Best Case scenarios are based on the lowest and highest reserve values, respectively, for the giant fields (e.g., as seen for fields in Table 9.1). The production trend since 2005 indicates that production predicted according to the lower reserve values is more accurate and hence the lower reserve values may be, in general, more realistic. However, if the higher reserve values are valid then the current plateau of oil production may continue until 2015. The Best Case scenario required, among other things, that oil production in Iraq would be rapidly ramped up. Although oil production in Iraq now appears to be increasing it is too late for the Best

Case scenario to be achieved. However, this increase in Iraqi oil production may serve to extend the current production plateau.

If we take into account the fluctuations that have existed in reporting of URR values for the giant oilfields then in 2018 we can expect oil production to be somewhere between 70 and 90 Mb/d. This is a large degree of uncertainty but we can be sure that world oil production will not exceed 90 Mb/d. A possible update of Robelius' analysis that includes production information from 2005 to 2010 and new information on oil reserves will improve the certainty of our prognosis.

"The Elephants Retire": Production Decline in the Giant Oilfields

A large part of the research described in the thesis, "Giant Oil Fields—The Highway to Oil: Giant Oil Fields and Their Importance for Future Oil Production" was based on Robelius' construction of a database containing information on these fields. When Mikael Höök began as a Ph.D. student in our research group one of his tasks was to refine this database to study the nature of the production decline in fields that were already in decline.

"The Uppsala giant field database" includes 331 giant fields. Based on the estimates made in Robelius' thesis, these fields have a combined URR of 1,130 Gb. Land-based fields number 214 (about 65%) and the remaining 117 fields (35%) are offshore. To estimate the rate of decline in those fields that were already declining, Mikael Höök examined only those 261 fields that we classified as having passed their plateau phase of production and as showing actual production decline. Of these 261 fields, 170 were land-based and 91 were offshore. The results of this study of decline in giant fields were analyzed together with Robert Hirsch and published in the scientific journal *Energy Policy* under the title "Giant Oil Field Decline Rates and Their Influence on World Oil Production" [10]. In the summary (abstract) of the publication we state the following.

> The most important contributors to the world's total oil production are the giant oil fields. Using a comprehensive database of giant oil field production, the average decline rates of the world's giant oil fields are estimated. Separating subclasses was necessary, since there are large differences between land and offshore fields, as well as between non-OPEC and OPEC fields. The evolution of decline rates over past decades includes the impact of new technologies and production techniques and clearly shows that the average decline rate for individual giant fields is increasing with time. These factors have significant implications for the future, since the most important world

oil production base—giant fields—will decline more rapidly in the future, according to our findings. Our conclusion is that the world faces an increasing oil supply challenge, as the decline in existing production is not only high now but will be increasing in the future.

Production decline is something that the international oil companies do not normally wish to discuss. However, in 2002 Harry J. Longwell, the director and executive president for ExxonMobil Corporation published an article in *World Energy* in which he clearly noted that oil production from fields currently in production would decline dramatically by 2010 [11]. "The catch is that while demand increases, existing production declines. To put a number on it, we expect that by 2010 about half of the daily volume needed to meet projected [demand] is not on production today—and that's the challenge facing production." Longwell showed an estimated demand for oil in 2020 of around 110 Mb/d. He also showed a curve of "discovered oil volumes" that agrees well with what we show in Fig. 3.1 and Fig. 6.5. Incidentally, this means that ExxonMobil shares our understanding that all the oil in an oilfield (including later revisions of estimated reserves) should be dated as discovered on the day the field was found. This also implies that ExxonMobil does not share BP's attitude to oil discoveries that we discussed in Chap. 6.

In a February 2004 report to its shareholders titled, *A Report on Energy Trends, Greenhouse Gas Emissions and Alternative Energy*, ExxonMobil predicted a future as alarming as the red color in Fig. 9.8: "In other words, by 2015, we will need to find, develop and produce a volume of new oil and gas that is equal to eight out of every ten barrels being produced today" [11]. There is no doubt that this future worries them.

It is interesting to note that this report was released right after President George W. Bush presented his ten-point program to increase corporate responsibility and protect shareholders [13], a measure that became necessary after the Enron scandal.

A summary of our analysis of the decline rates in 261 giant fields is given in Fig. 9.9. In Table 9.2 this information is provided divided into various categories (land-based, offshore, OPEC, non-OPEC, etc.). Around 65% of the oilfields are land-based and 35% are offshore and significant differences in decline rate exist between these two groups. The average decline rate for offshore oilfields is around 10% whereas for land-based oilfields it is around 4%. The oilfields of the OPEC nations tend to decline more slowly than others.

In 2007, when we had decided to write our article on declining production in giant fields, Cambridge Energy Research Associates (CERA) had put together a report for its customers in which they estimated that production from existing oilfields of all sizes would decline by 4.5% per annum [14]. This was completely consistent with the projections presented in

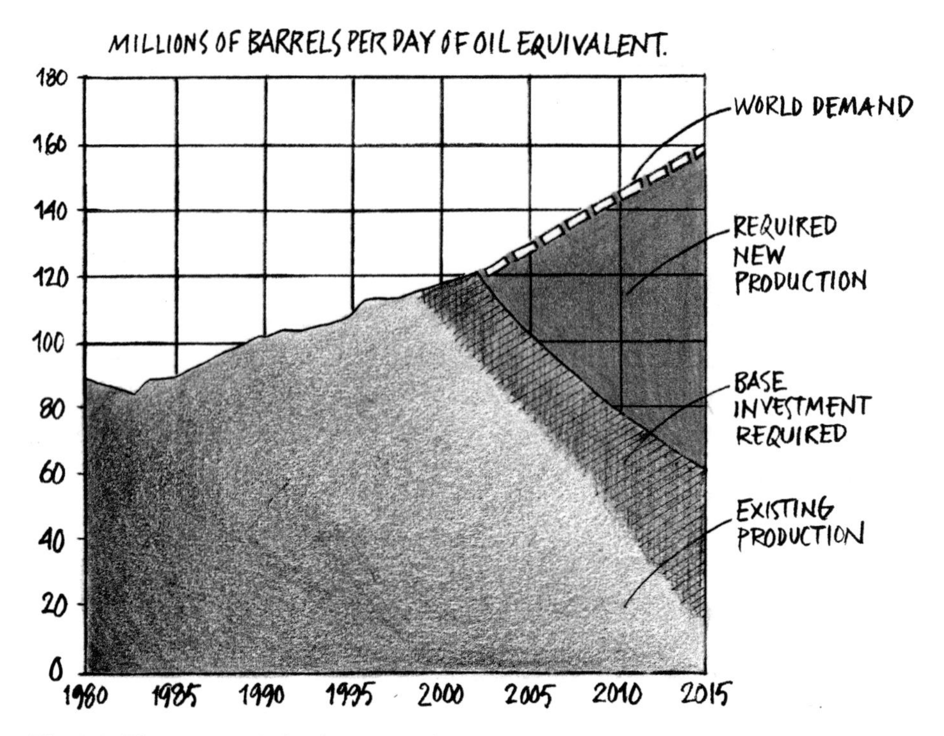

Fig. 9.8 The prognosis for future production presented by ExxonMobil to its shareholders in February 2004. The company noted that "supplying oil and gas demand will require major investments" [12]

ExxonMobil's 2004 report but less than what Andrew Gould, the chief executive officer (CEO) of Schlumberger, had asserted in 2005, "an accurate average decline rate is hard to estimate, but an overall figure of 8% is not an unreasonable assumption" [15]. It was apparent that there was growing interest in the decline rates of oilfields in production!

In the autumn of 2008 when we were ready to submit our article on decline rates in giant fields, the IEA announced that it was intending to present a similar analysis in their *World Energy Outlook* report to be released in November 2008 [16]. Therefore, we decided to postpone our submission so that our article could include discussion of the IEA study. It turned out that the IEA's estimates presented in *World Energy Outlook 2008* agreed well with our own analysis. However, the IEA study failed to point out that the future rate of production decline in those giant fields that had yet to begin their decline phase would be greater than for those fields already in decline. We return to this topic in the Chap. 11, "The Peak of the Oil Age" (Fig. 9.10).

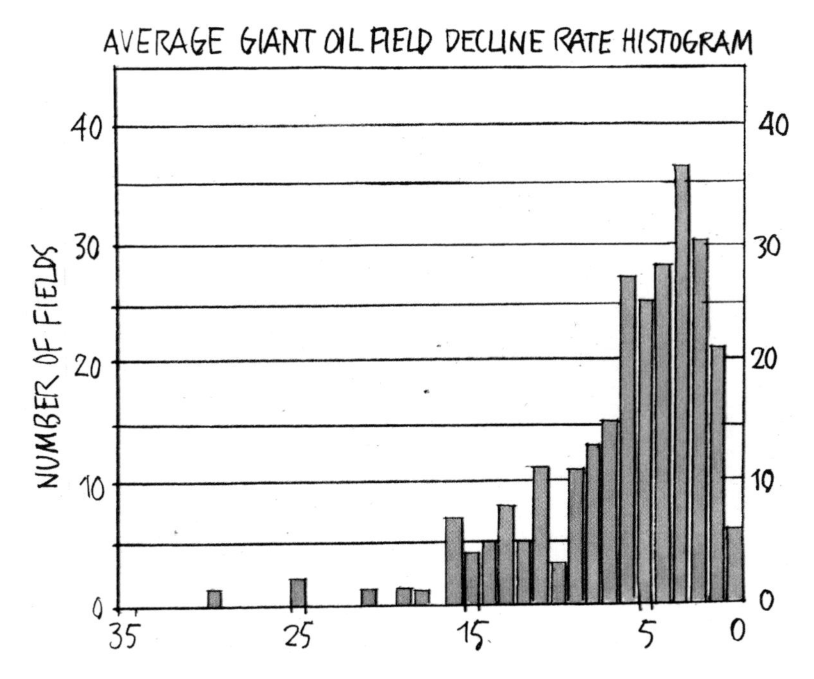

Fig. 9.9 A chart showing the different decline rates in percent seen for those 261 giant oilfields in decline at the end of 2005 [10]

Table 9.2 Characteristic production decline rates for various categories of giant oilfield as of 2005

# Fields	Category	Average (%)	Median (%)	Prod. Weight (%)
170	All land-based fields	−4.9	−4.4	−3.9
91	All offshore fields	−9.4	−9.0	−9.7
97	All OPEC fields	−4.8	−4.1	−3.4
73	OPEC onshore fields	−3.8	−3.8	−2.8
24	OPEC offshore fields	−7.7	−6.1	−7.5
164	All non-OPEC fields	−7.5	−6.3	−7.1
97	Non-OPEC onshore fields	−5.7	−4.7	−5.2
67	Non-OPEC offshore fields	−10.0	−9.4	−10.3

Fields that had not ended their plateau phase of production or were still in their buildup phase in 2005 were excluded. In total, 87% of all non-OPEC giant fields were classified as being in post-plateau phase in 2005. This meant that 83% of all URR in non-OPEC giant fields existed in oilfields showing declining production. A total of 67% of OPEC's giant fields can be classified as being in their post-plateau production phase representing 48% of all URR in OPEC's giant fields. "Prod. Weight" is the value weighted for the production volume for that oilfield class [10]

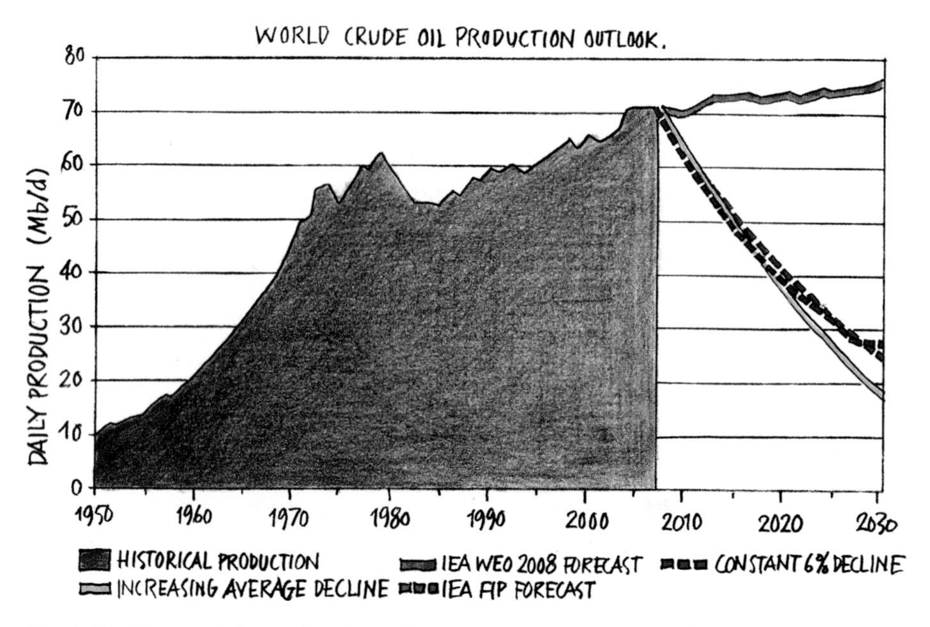

Fig. 9.10 The world's crude oil production up to the present day together with the reference scenario for future production provided by the IEA in its World Energy Outlook report of 2008 (*solid red line*) [16]. The *red dashed line* represents a constant decline rate of 6% of production from existing productive fields. The *blue dashed line* represents the decline rate reported by the IEA [16]. The *solid yellow line* is the decline prediction from our work [10]

The Elephants' Production Rate: Depletion of Remaining Recoverable Resources (DRRR)

Finding a new oilfield can be compared with being told you are to receive an inheritance from a deceased aunt. Expectations are high before the will is read and, when you hear that you will inherit a million dollars, you are fully justified in feeling like a millionaire! At that moment you do not care that your dear aunt has placed a "minor" restriction on how you can spend the money: that each year you can only use 6% of the balance remaining in the bank account at the beginning of that year. As a recently minted millionaire you want to travel the world, buy a new car, and much much more. You estimate that this will require $100,000 so you go to the bank to withdraw the money from the account holding the inheritance. To your amazement you find that you are not allowed to withdraw $100,000 but only $60,000, that is, 6% of your inherited million dollars according to your aunt's little restriction. You are also distressed to learn that the amount you can withdraw will decrease every year and when half the inheritance remains ($500,000) you will only be able to withdraw $30,000 that year. Your dreams of living like a millionaire have gone up in smoke!

In September 2008, I awoke one morning at 4 AM. I tried to get back to sleep but could not because my thoughts had turned to Peak Oil. It suddenly dawned on me that the crucial factor determining the rate of future oil production would be the ratio of annual production versus what remains to be produced, the depletion of remaining recoverable resources (DRRR). Internal discussions in our research group had, for some time, revolved around why our research produced predictions of future oil production rates that were so different from those of the IEA and others. I was convinced that it must be something fundamental because the difference was so large. The analysis of Peak Oil that Colin Campbell and I published in 2003 was based on a nation-by-nation analysis of production. For that analysis, when we determined future production it was based on a nation's current production rate in relation to its reported reserves and possible future discoveries. This value represented the nation's current DRRR and limited its production rate. Now, at 4 AM in the morning it suddenly struck me that our use of DRRR was probably the decisive factor producing the difference between our predictions and those of others.

We have now studied DRRR in two scientific articles in slightly different ways. The first examination was more theoretical and in it we discussed what we call "the maximum depletion rate model." As part of that article we also examined the future oil production scenarios presented by the IEA and the Energy Information Administration (EIA) of the US Department of Energy (DoE). In the second article we made a detailed analysis of DRRR for individual oilfields. Before looking more closely at those articles I must give you an exact definition of the ratio between annual production and future possible production, DRRR:

$$\text{DRRR} = \frac{\text{Amount produced in year x}}{(\text{URR} - \text{total amount produced before year x})}$$

Let us now look at the two articles. The first is titled, "How Reasonable Are Oil Production Scenarios from Public Agencies?" and the authors are Kristofer Jakobsson, Bengt Söderbergh, Mikael Höök, and Kjell Aleklett [17]. Below is the "abstract" (scientific summary) of its findings:

> According to the long term scenarios of the International Energy Agency (IEA) and the US Energy Information Administration (EIA), conventional oil production is expected to grow until at least 2030. The EIA has published results from a resource-constrained production model which ostensibly supports such a scenario. The model is here described and analyzed in detail. However, it is shown that the model, although sound in principle, has been misapplied due to a confusion of resource categories. A correction

of this methodological error reveals that the EIA's scenario requires rather extreme and implausible assumptions regarding future global decline rates. This result puts into question the basis for the conclusion that global "peak oil" would not occur before 2030. (Note: By allowing unreasonably high rates of DRRR the EIA was able to predict growth in oil production to excessively high levels followed by very dramatic declines).

The second article examines "The Evolution of Giant Oil Field Production Behavior" and its authors are Mikael Höök, Bengt Söderbergh, Kristofer Jakobsson, and Kjell Aleklett [18]. The abstract follows.

The giant oil fields of the world are only a small fraction of the total number of fields, but their importance is huge. Over 50% of the world's oil production came from giants by 2005 and more than half of the world's ultimately recoverable reserves are found in giants. Based on this it is reasonable to assume that the future development of the giant oil fields will have a significant impact on world oil supply.

In order to better understand the giant fields and their future behaviour one must first understand their history. This study has used a comprehensive database on giant oil fields in order to determine their typical parameters, such as the average decline rate and life-time of giants. The evolution of giant oil field behavior has been investigated to better understand future behavior. One conclusion is that new technology and production methods have generally resulted in high depletion rates and rapid decline. The historical trend points towards high future decline rates in fields currently on plateau production.

In giant oilfields, the peak production rate generally occurs before half the ultimately recoverable reserves have been produced. A strong correlation between depletion-at-peak and average decline rate is also found, verifying that high depletion rate leads to rapid decline. Our result also implies that depletion analysis can be used to rule out unrealistic production expectations from a known reserve, or to connect an estimated production level to a necessary reserve base.

In Chap. 7 we saw that oil production from a field usually occurs in three technological phases and in Fig. 7.5 we could see how these three technological phases combine to give an oilfield's typical lifetime production history/profile. The production histories of two actual giant fields are shown in Figs. 7.5 (Statfjord) and 9.11b (Prudhoe Bay). We can see three phases in the histories of these fields, the "buildup" phase when production is increasing, the "plateau" phase of relatively constant production, and then the "decline" phase. Of course, the remaining recoverable resources (RRR) constantly decrease during the productive life of a field. This means that, in order to maintain a constant rate of production during the plateau phase, the rate of depletion of the remaining recoverable resources must

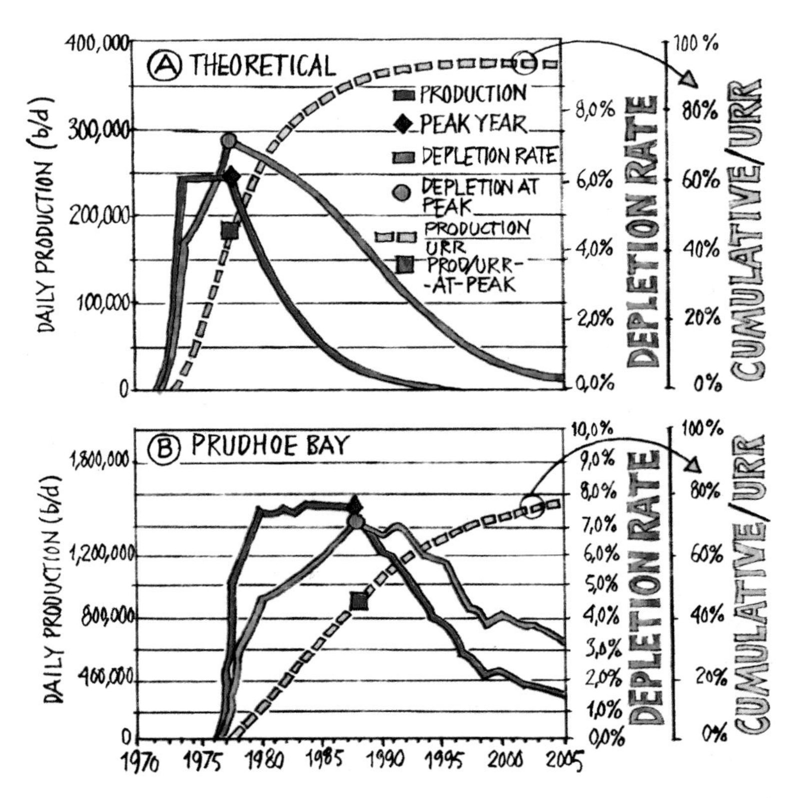

Fig. 9.11 (**a**) The theoretical production profile for a giant field is shown to illustrate how the rate of oil production ("Daily Production") and the fraction of remaining recoverable resources that can be produced in a year (the DRRR shown as "Depletion Rate") change as the field passes through its various production phases. The "peak" of production occurs when the plateau of production ends and that is also the moment when the rate of depletion is highest. The total (cumulative) oil produced as a fraction of the URR is also shown. (**b**) The famous "Prudhoe Bay" field in Alaska is given as an example of the behavior of a real field. The peak rate of depletion of 7.2% per annum was reached in 1988 when 46% of the URR had been produced [18]

actually be increasing. The DRRR reaches its highest point just at the moment that the rate of oil production begins to decline and then it too begins to fall. In Fig. 9.11b we can see how the DRRR changes over the life of a real field, the United States' largest oilfield, Prudhoe Bay in Alaska. The maximum DRRR of 7.2% occurs at the end of the plateau phase of production when 46% of the URR has been produced.

In total we have studied the rates of DRRR of 261 giant fields that have passed their production plateau phase. Of these, 65% are land-based and 35% are offshore fields. If we compare the rates of DRRRs of these two types of oilfield we see that the land-based fields have, in general, somewhat lower

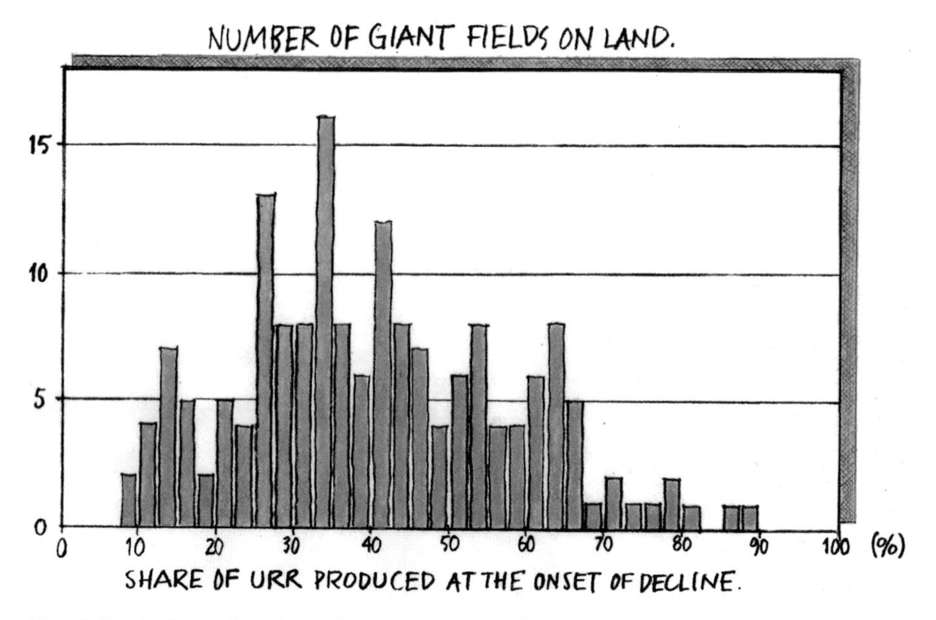

Fig. 9.12 A chart showing what proportion of the URR had been produced at the start of production decline of land-based giant oilfields. Those fields with over 55% of their URR produced when they reached peak production are nearly all in the United States and are those for which production was re-started during the oil crises of the 1970s and 1980s [18]

rates of DRRRs than the offshore fields. This is probably due to the fact that production from the offshore fields has generally started later than production from the land-based fields and so more modern technology (allowing higher rates of extraction) has been used from the start of their production. Another reason can be that it is much more difficult and expensive to institute "enhanced oil recovery" techniques (EOR) for offshore fields than for onshore fields so the decline phase for offshore fields would be shortened and this would reduce the total URR and increase the rate of DRRR.

Inasmuch as an oilfield only has a set finite volume of oil that it can yield, extending the plateau phase of its production must produce a faster subsequent decline phase. This means that the rate of DRRR during the decline phase will also be lower. In our theoretical work on DRRR [17] we showed that the maximum rate of DRRR is the same as the rate of production decline (i.e., the percent of DRRR at the end of the plateau phase equals the percentage rate of decline in annual production seen thereafter).

A remarkable observation that we can make is that oil production from many giant fields begins to decline long before half of the recoverable oil in the field has been extracted. For land-based fields (see Fig. 9.12) the average

Table 9.3 The characteristic average production fraction of URR of land-based and offshore giant oilfields[a]

	Average	Median	Prod. Weight	Standard Dev.[b]
Land-based oilfields				
DRRR at peak rate of production	6.8%	6.1%	5.8%	3.5%
Decline rate after peak	−4.9%	−4.4%	−3.9%	3.5%
Cumulative production/URR at peak	38.1%	36.2%	34.1%	17.5%
DRRR at peak rate of production	4.6	3.0	3.7	5.4
Years from first oil production until abandonment	21.4	16.0	21.0	17.9
Offshore oilfields				
DRRR at peak rate of production	10.4%	9.4%	11.0%	4.8%
Decline rate after peak	−9.4%	−9.0%	−9.7%	5.8%
Cumulative production/URR at peak	39.4%	40.3%	44.0%	15.7%
Years from discovery until first oil produced	6.3	5.0	5.3	6.0
Years from first oil production until abandonment	10.8	8.0	12.4	8.3

[a]The land-based group includes 170 post-plateau fields and the offshore group includes 91 post-plateau fields. Fields in 2005 that had not yet passed their plateau phase or were still in buildup phase were excluded. "Prod. Weight" is the value weighted for the production volume for that oilfield class (i.e., land-based or offshore) [18]

[b]In statistical analysis, "Standard Deviation" is used to describe how much variation there is in a set of data

percentage of URR produced when peak production occurs is 34% and for offshore fields it is 44%. Table 9.3 shows a summary of production information for land-based and offshore giant oilfields [18].

One of the main criticisms of the Hubbert model of resource production that is raised by some analysts is that the model assumes maximal oil production occurs when half of the URR has been produced. These analysts assert that maximal oil production can also be reached when more than 50% of the URR has been produced.

Our studies of DRRR support that Peak Oil production at 50% of the URR is not an accurate reflection of reality. However, we see the opposite of what the other critics of the Hubbert model expect. We see that Peak Oil production tends to occur when less than 50% of the URR has been extracted. This means that putting production data into a Hubbert model to make predictions of future production can lead to future production being underestimated.

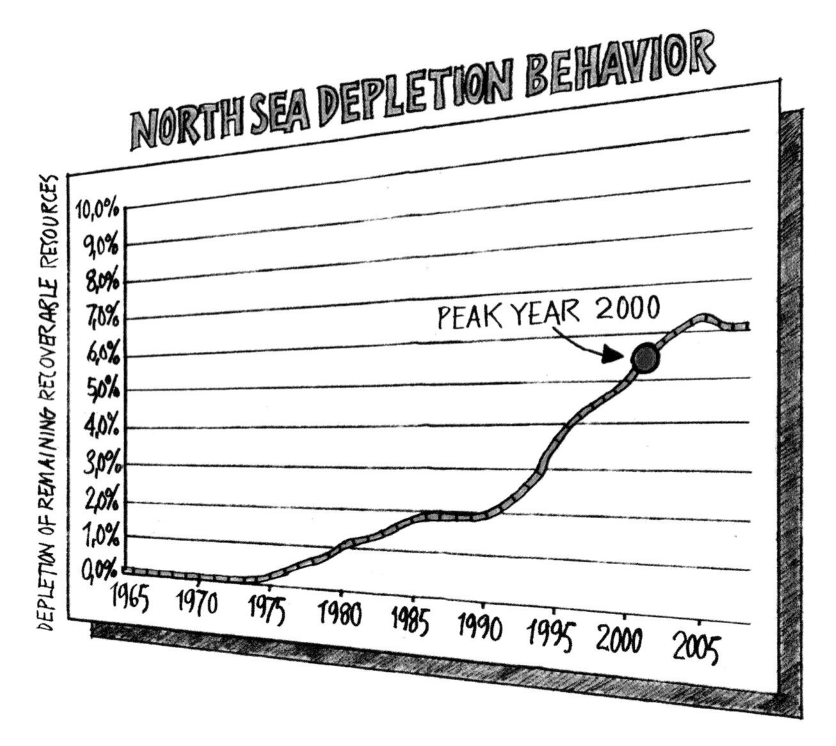

Fig. 9.13 Development of the DRRR parameter for the entire offshore North Sea region. The rate of depletion increased by 0.2–0.3% per year until it leveled off at the maximum rate at 6% per year for remaining recoverable resources [8]

We can now use the results of our research to investigate the factors determining oil production from fields that are in the plateau phase of production or have just begun to decline. In Table 9.1 the URR of the giant field Cantarell is stated to lie somewhere between 11 and 20 billion barrels of oil (Gb). Cantarell produced 0.73 billion barrels of oil in 2005 which was the last year before its production began to decline. At the end of that year, Cantarell had produced a cumulative total of 10 Gb. The upper and lower URR limits are stated as 11 and 20 Gb, respectively, and imply that remaining reserves are between 1 and 10 Gb of oil. These remaining reserve numbers give DRRR values of 66.2% and 7.5%. Therefore, we can be sure that Cantarell's URR is greater than 11 Gb. If we use the average DRRR number of 11% for offshore fields (see Table 9.3) then we can calculate a realistic URR number for Cantarell as being 16.6 Gb (URR = 10.0 + 0.73/0.11 = 16.6 Gb).

An oil-producing region consists of fields in various phases of production. The region that most recently reached its maximal production rate (Peak Oil) is the North Sea. If one analyzes the DRRR for that entire region one can see a depletion rate per year of 5.6% at the peak of production. It then grew to around 6% per year before leveling off (Fig. 9.13).

Other oil-producing regions in the world have lower DRRR values. In its *World Energy Outlook 2010* report, the IEA predicts a constant rate of conventional crude oil production of 25 billion barrels (Gb) per year until 2035. If we accept the North Sea's DRRR value of 6% as the maximum possible rate for the world's combined crude oil production, then the IEA's prediction would require the existence of conventional crude oil reserves in 2035 of more than 400 Gb. Constant world conventional crude oil production of 25 Gb for 25 years amounts to a total of 625 Gb and a realistic estimate of current conventional crude oil reserves is 800 Gb [19]. If reserves in 2035 must be at least 400 Gb then this means that additional discoveries of at least 200 Gb must be made in coming years for the IEA's prediction to be possible.

It is difficult to imagine that nations such as Saudi Arabia will allow a future DRRR as high as 6%. This means that maintaining world conventional crude oil production at its present level for the next 25 years will require even larger reserves than described above. We discuss this in Chap. 13 "Peeking at Saudi Arabia—Twilight in the Desert."

References

1. Uppsala Global Energy Systems. Uppsala University, Sweden, Publications: http://www.physics.uu.se/ges/en/publications (2011)
2. Intergovernmental Panel on Climate Change (IPCC): Special Report on Emissions Scenarios—SRES. IPCC. http://www.ipcc.ch/ipccreports/sres/emission/index.php?idp=0 (2000)
3. Robelius, F.: Giant Oil fields—the highway to oil: giant oil fields and their importance for future oil production. Uppsala Dissertations from the Faculty of Science and Technology, ISSN 1104–2516, p. 69. http://uu.diva-portal.org/smash/record.jsf?pid=diva2:169774 (2007)
4. Abdul Baqi, M.M., Saleri, N.G.: Fifty-year Crude Oil Supply Scenarios: Saudi Aramco's Perspective, Saudi Aramco. CSIS, Washington, DC (2004). http://csis.org/files/media/csis/events/040224_baqiandsaleri.pdf
5. Radler, M.: Special report: oil production, reserves increase slightly in 2006. Oil Gas J. **104**, 20 (2006)
6. Chew, K.: World oil and gas resource and production outlook. Presentation at OPEC-IFP Joint Seminar Paris. Downloaded from www.ifp.fr/IFP/en/events (2005)
7. Sans l'or noir irakien, le marché pétrolier fera face à un "mur" d'ici à 2015. Le Mond, 2007-06-27. http://www.lemonde.fr/economie/article/2007/06/27/sans-l-or-noir-irakien-le-marche-petrolier-fera-face-a-un-mur-d-ici-a-2015_928476_3234.html. Translation: Top IEA official: without Iraqi oil, we hit the wall in 2015, translation of article in Le Monde. http://www.eurotrib.com/story/2007/6/27/173221/933 (2007)

8. Aleklett, K., Höök, M., Jakobsson, K., Lardelli, M., Snowden, S., Söderbergh, B.: The peak of the oil age—analyzing the world oil production reference scenario in World Energy Outlook 2008. Energy Policy **38**(3), 1398–1414 (2010). (accepted 9 November 2009, available online Dec 1, 2009, http://www.tsl.uu.se/uhdsg/Publications/PeakOilAge.pdf)
9. BP: BP Statistical Review of World Energy. http://bp.com/statisticalreview (historical data; http://www.bp.com/assets/bp_internet/globalbp/globalbp_uk_english/reports_and_publications/statistical) (2011)
10. Höök, M., Hirsch, R., Aleklett, K.: Giant oil field decline rates and their influence on world oil production. Energy Policy **37**(6), 2262–2272 (2009). http://www.sciencedirect.com/science/article/pii/S0301421509001281; http://www.tsl.uu.se/uhdsg/Publications/GOF_decline_Article.pdf
11. Longwell, H.J.: The future of the oil and gas industry: past approaches, new challenges. World Energy **5**(3) (2002). http://www.localenergy.org/pdfs/Document%20Library/Exxon%20Future%20of%20Oil%20and%20Gas.pdf
12. ExxonMobil: A Report on Energy Trends, Greenhouse Gas Emissions and Alternative Energy, http://esd.lbl.gov/SECUREarth/presentations/Energy_Brochure.pdf (2004)
13. FT: This article discusses President Bush's ten point program for improved corporate responsibility and protection of shareholders (http://specials.ft.com/enron/FT3XIIYZIYC.html) (2002)
14. Jackson, P.M., Godfrey, K.M.: Finding the critical numbers: what are the real decline rates for global oil production. CERA. http://www.ihs.com/products/cera/energy-report.aspx?id=106592188 (2007)
15. Schlumberger Chairman and CEO Andrew Gould addressed the oil and gas investment community at the 33rd Annual Howard Weil Energy Conference on 4 April 2005 in New Orleans, LA, USA. Available from: http://www.slb.com/content/news/presentations/2005/20050404_agould_neworleans.asp (2005)
16. WEO: World Energy Outlook 2008. International Energy Agency. http://www.iea.org/textbase/nppdf/free/2008/weo2008.pdf (2008)
17. Jakobsson, K., Söderbergh, B., Höök, M., Aleklett, K.: How reasonable are oil production scenarios from public agencies? Energy Policy **37**(11), 4809–4818 (2009). http://www.sciencedirect.com/science/article/pii/S0301421509004558; http://www.tsl.uu.se/uhdsg/Publications/MDRM_article.pdf
18. Höök, M., Söderbergh, B., Jakobsson, K., Aleklett, K.: The evolution of giant oil field production behavior. Nat. Resour. Res. **18**(1), 39–56 (2009). http://www.springerlink.com/content/142004322x2885nk/; http://www.tsl.uu.se/uhdsg/publications/GOF_NRR.pdf
19. Laherrère, J.: Advice from an old geologist-geophysicist on how to understand Nature (Figure 51). Presentation at Statoil, Aug 14, 2008, Oslo, Norway. http://aspofrance.viabloga.com/files/JL_Statoil08_long.pdf (2008)

Chapter 10

Unconventional Oil, NGL, and the Mitigation Wedge

As I write this it is the spring of 2011 and if someone mentions "Black Swan" then most people would think of the recent film of the same name. Natalie Portman was awarded an Oscar for her role as the dancing black swan in that film. Those whose attention is more focused on the functioning of our society might think of Nassim Nicholas and his theory of "Black Swan Events." This refers to unexpected events of large magnitude that are significant from an historical perspective. When I hear the expression "Black Swan" I am reminded of November 22, 2005 when Bruce Robinson and his wife Sue took me and my wife Ann-Cathrine to the Swan River in Perth, Australia, to show us real black swans. Having seen them, I can understand why the first British sailors who found them early in the nineteenth century thought they had arrived in the lair of the Devil himself where everything is black (Fig. 10.1). Their discovery of black swans is just the sort of unexpected event that Nassim Nicholas describes in his theory. After visiting the Swan River I experienced my own personal "Black Swan Event" later that evening.

At 10:30 PM that evening my mobile phone rang. Naturally, I thought it was someone calling from Sweden. Instead, the call was from Maryam Sabbighian, Counsel to the House of Representatives Committee on Energy and Commerce in Washington, DC. She told me that the committee would hold a hearing on Peak Oil on December 7, 2005 (in only 2 weeks time) and they wanted me to come to Washington to testify. I had just begun my travels and from Perth I was due to visit Brisbane and Sydney in Australia and then Dunedin in New Zealand. I was then to return to Perth to fly home to Stockholm by December 4. To be in Washington on December 7 I would need to leave Sweden on December 6, and, on top of that, they wanted me

K. Aleklett *Peeking at Peak Oil*, DOI 10.1007/978-1-4614-3424-5_10,

Fig. 10.1 Black swans on the Swan River in Perth, Australia

to submit written testimony regarding Peak Oil. I felt very honored to be called to testify before a US House of Representatives Committee so, of course, I said that I would attend. Later I received the following email.

> Professor Aleklett
>
> I am sending this email per our phone discussion this morning. I am Counsel to the House of Representatives Committee on Energy and Commerce. The Chairman of my Committee is Mr. Joe Barton. The Committee is planning a hearing on the topic of peak-oil. The hearing is scheduled for December 7th at 9:30 am. Congressman Bartlett, who will also be testifying at the hearing, recommended that we contact you. We realize that you would have to travel some distance, but Chairman Barton and I would like to extend an invitation to you to testify at this hearing. Please let us know if you would be available. I can be reached at.......
>
> Best regards, Maryam Sabbaghian, Counsel US House of Representatives, Committee on Energy and Commerce.

Congressman Barlett is the House of Representative's greatest educator on Peak Oil so I felt doubly honored that the invitation came at his behest. An odd coincidence was that, some days later in Dunedin, New Zealand, I was scheduled to meet his brother Professor Al Bartlett who is world famous for his lectures on exponential growth [1].

What does all this have to do with "Unconventional Oil, NGL, and the Mitigation Wedge"? In addition to calling me to testify the committee had also summoned Dr. Robert Hirsch who, earlier that year, had submitted a report to the US Department of Energy (DoE) titled, *Peaking of World Oil Production: Impacts, Mitigation, & Risk Management* [2]. In the report, Dr. Hirsch and his co-authors explained that rather extreme measures (a "crash program") would be required to ameliorate the consequences of Peak Oil.

In their crash program, unconventional oil was seen as very important for keeping the wheels of civilization turning in the United States and the rest of the world.

The report by Hirsch and his colleagues is now widely known and is commonly referred to simply as the *Hirsch Report*. The report's authors have since written a book titled, *The Impending World Energy Mess*. Chapter 11 of that book is titled, "The Best That Physical Mitigation Can Provide," [2] and in it they estimate what a worldwide crash program might achieve. Based on common experience and their own judgment, they suggest there would be a delay in seeing the first ameliorative contributions from, for example, unconventional oil and far more efficient vehicles due to the time required to permit, plan, and construct the required facilities. They assumed that, after a 3–5-year delay, new vehicles and new sources of liquid fuels would increase linearly with time. Their studies assumed no limits on finances or raw materials, which Hirsch and his co-authors recognized was idealistic and not practical in reality.

Under such a "Crash Management Program" they suggested it is technically possible during a 20-year period to increase unconventional oil production by 25 million barrels per day (Mb/d). Increasing the energy efficiency of the vehicle fleet and applying new production technologies to existing conventional oilfields could reduce the necessary increase in unconventional oil production. In that case the crash program would only need to aim for an increase in unconventional oil production of 20 Mb/d over 15 years. In this chapter we examine the increases in unconventional oil production that can realistically be expected during the next 20 years under conditions as they exist today.

In this chapter we also discuss natural gas liquids (NGL) that, in international oil statistics, are counted as oil. Oil is made up of chemical compounds of hydrogen and carbon, "hydrocarbons." The lightest (simplest) hydrocarbon molecule is methane which is sold commercially as "natural gas" or "biogas." In natural gas fields there are also heavier (more complex) hydrocarbon molecules that form liquids under normal temperatures and pressures or that become liquid under higher pressure or moderate cooling. These liquid components are described as NGL and should not be confused with methane/natural gas that is transformed into liquid by extreme cooling. This liquid methane is known as liquid natural gas (LNG). NGL include molecules such as propane, butane, pentane, hexane, and heptane. NGL are a component of natural gas production and so NGL production can only be increased if production from natural gas fields is increased.

In some cases, production of unconventional oil involves severe environmental damage. The Uppsala Global Energy Systems group (UGES) considers that this cannot be ignored and in particular cases we discuss this issue.

Unconventional Oil

When the price of oil is discussed in news broadcasts it is the best (most useful) oil that they talk about such as Brent crude from the North Sea or West Texas Intermediate (WTI) crude oil. However, when these broadcasts mention that world oil production is currently 86 Mb/d (without processing gains 83 Mb/d) they are actually talking about a spectrum of many types/grades of oil. One can roughly divide this spectrum into conventional and unconventional oil. In the United States, the nation that founded the oil era, the American Petroleum Institute (API) has used the density of oil to define different grades of oil more precisely. They call this measurement "API gravity":

$$\text{API}^{\circ} = 141.5 / \text{Specific gravity} - 131.5$$

This measurement is made at the temperature of 15.6°C (60°Fahrenheit). Oil that has an API gravity of less than 10° is heavier than water and is classed as "extra heavy oil." Oil with an API gravity of between 10° and 20° is classed as "heavy oil." Between 20° and 30° oil is classed as "medium heavy" whereas above 30° oil is classed as "light" [3]. Bitumen is the heaviest naturally occurring hydrocarbon type and it is either solid or nearly so. When produced as the end product from a refinery these extra heavy hydrocarbons are described as "asphalt."

In Chap. 4 of its *World Energy Outlook 2010* report the IEA has defined the following oil types as "unconventional" [4].

- Bitumen and extra heavy oil from Canada's oil sands
- Extra heavy oil from Venezuela's Orinoco belt
- Oil produced from oil shale
- Oil produced from coal by CTL methods (coal-to-liquids)
- Oil produced from natural gas by GTL methods (gas-to-liquids)
- "Refinery additives" and the like

Using this definition of unconventional oil one finds that some oilfields with very heavy oil are treated as conventional, and the production from these fields is considered to be conventional oil. We accept this definition and discuss unconventional oil using the categories above.

The World Energy Outlook 2010 report discussed unconventional oil production for 2009. At that time the production of "shale oil" (explained later) from the Bakken geological formation in Montana and North Dakota was quite limited. However, since this form of unconventional oil production is increasing in importance we will examine it separately in the section "Shale oil production in the United States."

Canada's Oil Sands

When the debate on Peak Oil began in Sweden in 2001 the managing director of the Swedish Petroleum Institute (SPI) was one of those who repeatedly declared that I was wrong. He said that he and SPI did not care about Peak Oil and the message I was presenting. To support his argument he often cited Canada's oil sands as one of the sources of oil that would guarantee our future supply. Here is his statement from the Swedish newspaper, *Dagens Industri* (*Industry Today*) [5]: "There are large known, but undisturbed, resources in the form of oil sands that today are as large as all the oil previously discovered."

As CEO for SPI his authority was greater than mine. As a researcher, my natural response was to begin a scientific investigation of Canada's oil sands and how much oil they might produce in future.

My first step was to set up a project for a Master degree student in this area. Bengt Söderbergh was interested in the project and began work on it in 2005. As mentioned above, in that same year Hirsch and his colleagues published their report titled *Peaking of World Oil Production: Impacts, Mitigation and Risk Management* [2]. Their report included a projection of what Canada's oil sands might be able to produce assuming accelerated implementation of planned increases in production so that these occurred within 10 years rather than 25 years [2]. In contrast, Bengt made a very detailed study in which every planned oil sands project was analyzed. When Bengt subsequently began work in our group as a Ph.D. student in 2006 his first task was to rewrite his Diploma thesis in the format of a scientific article concerning the topic of the maximal possible production from Canada's oil sands. Like the *Hirsch Report*, our article also described a "crash management program" and was given the title, "A Crash Programme Scenario for the Canadian Oil Sands Industry." It was published in electronic format in August 2006 and became available in print in 2007 [6]. In his work we looked at production up until 2018 by assuming that all planned projects would be realized and by analyzing well-advanced projects under the assumption that there would be no delays. In this way we were able to make a projection of oil production from the oil sands up until 2050. As the abstract from the scientific article states:

> The implementation of a crash programme for the Canadian oil sands industry is associated with serious difficulties. There is not a large enough supply of natural gas to support a future Canadian oil sands industry with today's dependence on natural gas. It is possible to use bitumen as fuel and for upgrading, although it seems to be incompatible with Canada's obligations under the Kyoto treaty. For practical long-term high production, Canada must construct nuclear facilities to generate energy for the in situ projects. Even in

a very optimistic scenario Canada's oil sands will not prevent Peak Oil. A short-term crash programme from the Canadian oil sands industry achieves about 3.6 Mb/d by 2018. A long-term crash programme results in a production of approximately 5 Mb/d by 2030.

Some years ago "tar sands" was the most common name given to what is called "oil sands" today [7]. Indeed, the appearance and consistency of the oil that is extracted from these sands does remind one more of tar than crude oil. The industry term for the hydrocarbons found in the oil sands is "bitumen". Large deposits of bitumen exist in the province of Alberta in Canada. The total volume of this bitumen, the "Original Oil in Place" (OOIP), is 2,000 billion barrels (gigabarrels, Gb). Of this, 170 Gb are thought to be extractable. This would represent a recovery factor of around 9% which is similar to some of the lowest recovery factors of the world's oilfields (see Fig. 6.2). The bitumen exists at various depths. Approximately 20% of it (35 Gb) exists at depths shallower than 75 m and can be mined using conventional mining techniques. The total surface area for which this mining approach is viable is 4,700 km^2. The greater part of Canada's accessible bitumen must be extracted using other "in situ" methods. The most common in situ method is named Steam Assisted Gravity Drainage (SAGD).

Open-Pit Mining of Oil Sands

For open-pit mining of oil sands the forest above it must first be removed before excavators are used to remove the rest of the vegetation and the overlying layers of earth. This is all cleaned and stored for future restoration of the area. Then the layers of rock above the oil sands are removed. Some of the stone is used to build dams for the wastewater produced by the production process. The oil sands are mined using enormous scoops before loading into huge dump trucks which transport it to the crusher (see Fig. 10.2).

The crushed oil sands are mixed with hot water and form a slurry that is then pumped through pipes to an installation where the bitumen is separated from the sand. The slurry containing hot water, sand, mud, and bitumen is directed into large separation vats. By pumping air through the mixture in the vats a foam of bitumen forms at the surface and can be separated from the other components, the by-products or "tailings." The bitumen foam is transported to an installation for further processing and the liquid by-products are collected in large settling ponds. The solid particles collect as sediment at the bottom of the ponds and the surface fluid is pumped away for purification and reuse.

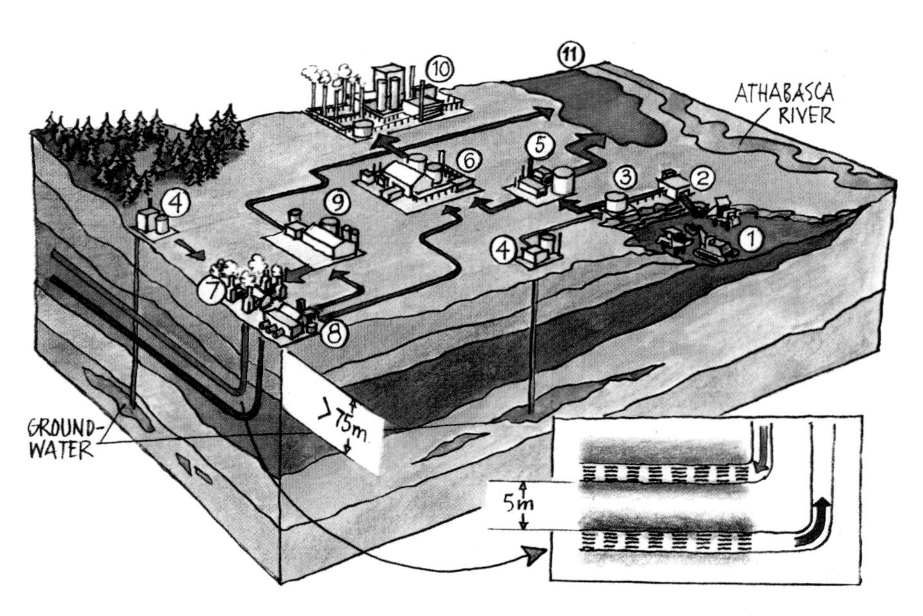

Fig. 10.2 A schematic illustration of the production of bitumen from Canada's oil sands. At the *lower right* (1) open-pit mining is occurring which is the technique used when the overburden does not exceed 75 m in depth. The oil sands are mined with enormous scoops and transported to a crusher (2). The crushed material is mixed (3) with water (4) to form a slurry of bitumen, mud, and sand. The slurry is transported to a facility that separates out the bitumen (5) and the remainder of the slurry is pumped to large sediment ponds (11). The bitumen is then pumped to an upgrading facility (6) where it is mixed with lighter hydrocarbons to produce synthetic crude oil. If the oil sands lie deeper than 75 m, in situ techniques such as SAGD are used. Two horizontal wells are drilled with one parallel and 5 m above the other. Steam is generated (7) from groundwater (4) and forced through the upper well and this makes the bitumen surrounding the well more fluid. Gravity then draws the fluid bitumen and condensed water down to the lower well from which the bitumen and water mixture is pumped to the surface. The bitumen is separated out (8) and then pumped to an upgrading facility (6). The rest of the fluid is passed through an installation (9) that removes around 90% of the water for recycling back to the steam plant (7) where it is reheated using natural gas. The synthetic crude can be processed in a nearby refinery (10) or transported in pipelines for processing farther afield

If the oil sands contain around 10% bitumen then 2 t of oil sands produce about one barrel of bitumen. About 85% of the bitumen in the original oil sands can be extracted and the remaining 15% collects in the settling ponds. This can be a sensitive issue because many people think that leakage from the settling ponds is very damaging to the environment.

The area of oil sands suitable for open-pit mining is divided up into mining concessions controlled by various companies who in turn propose a range of different projects for future mining. For our analysis of future oil production from Canada's oil sands we assumed that the mining projects would be completed according to the publicly presented plans. Using this assumption we saw increasing rates of oil production until 2020 at which time maximum production by open-pit mining would be achieved at 2.1 Mb/d. With a reserve of 35 Gb they could then maintain this level of production until 2040. In the analysis that the IEA presented in *WEO 2010* (page 148) [4] it predicts a significantly slower rate of expansion. From today's production at 0.7 Mb/d it predicts expansion to 1.5 Mb/d by 2030.

In Situ Production of Bitumen from Oil Sands

As explained above, the hydrocarbons in the Canadian oil sands are described as bitumen by the oil industry. Some of the oil producers in Alberta choose to describe bitumen as, "Petroleum that exists in the semi-solid or solid phase in natural deposits. Bitumen is best described as a thick, sticky form of crude oil, so heavy and viscous (thick) that it will not flow unless heated or diluted with lighter hydrocarbons. At room temperature, it is much like cold molasses" [8].

"In situ" production methods are used when oil sands lie at a depth of greater than 75–100 m. "In situ" is a Latin expression meaning "in place." For production of bitumen from oil sands, in situ means that the sand remains underground while the bitumen is removed. At normal temperatures the bitumen in oil sands is not fluid so if the sand is to be left behind the oil sands must be heated up to liquefy the bitumen.

Currently, the two main methods for in situ production of bitumen from oil sands are cyclic steam stimulation (CSS) and steam-assisted gravity drainage (SAGD). CSS is the older method and uses a vertical well to inject steam into a reservoir to heat up the bitumen after which the same well is used to pump the bitumen to the surface. For this reason, CSS is also called "Huff and Puff."

The development of new drilling technology has led to the current enthusiasm for SAGD. Two horizontal wells are drilled with one above, and parallel to, the other. The distance between the wells is about 5 m. The two parallel wells can be up to 1 km long (see Fig. 10.2). Steam is continuously injected into the upper well to liquefy the surrounding bitumen. Gravity then draws the bitumen and condensed water into the lower well. The bitumen/water mixture is first pumped to the surface and then onwards to a separation plant where the two substances are separated. The bitumen is mixed with lighter hydrocarbons to reduce its viscosity before it is sent to the next processing stage. The separated water is reheated into steam and recycled into the in situ production process.

Every barrel of bitumen produced by SAGD requires three barrels of water to be heated into steam but more than 90% of the water is reused. One SAGD installation can produce 100,000 barrels of bitumen per day and requires a separation plant that can receive around 300,000 barrels per day (b/d, ~12.6 million liters per day). In 2009 in situ production of bitumen was 670,000 b/d which meant that, every day, 80 million liters of water were purified. Currently natural gas is used to boil the water into superheated steam with an injection pressure of about 38 bars. (For comparison, the air pressure in a car tire is around 2 bars). Our analysis of oil production from oil sands showed that, if a crash program of production expansion were to be instituted for the Canadian oil sands, there would be insufficient natural gas in the region to produce the necessary steam. Instead, they would need to build new nuclear reactors to boil the water. In other words, they would need nuclear power to produce oil.

Bitumen can be diluted with lighter hydrocarbons (NGL) to facilitate transport to a refinery specially constructed to process extra-heavy oil. Another solution is for the bitumen to undergo preliminary processing to form synthetic oil that is then transported to a normal refinery [9].

Future Production of Bitumen

In our analysis, "A Crash Programme Scenario for the Canadian Oil Sands Industry," we presented a short-term scenario to 2018 and a long-term scenario to 2050. The short-term scenario was based on the 44 larger oil sands projects that were public knowledge in 2005: 18 open-pit and 26 in situ production projects. We assumed that the projects would be completed without delays. This led to a projection for production in 2018 of 2.3 Mb/d from open-pit mining and 1.3 Mb/d from in situ methods giving a total production of 3.6 Mb/d. In the long-term scenario, production from open-pit mining continued at 2.3 Mb/d until declining after 2040. For in situ production we projected a continuous increase until 2050 so that the combination of open-pit and in situ production reaches a maximum (a peak) in 2040 at 5.8 Mb/d after which it falls.

In its *WEO 2010* report the IEA shows an analysis of production from Canada's oil sands that resembles the one we presented in 2006. They studied oil production from the oil sands project by project but did not assume

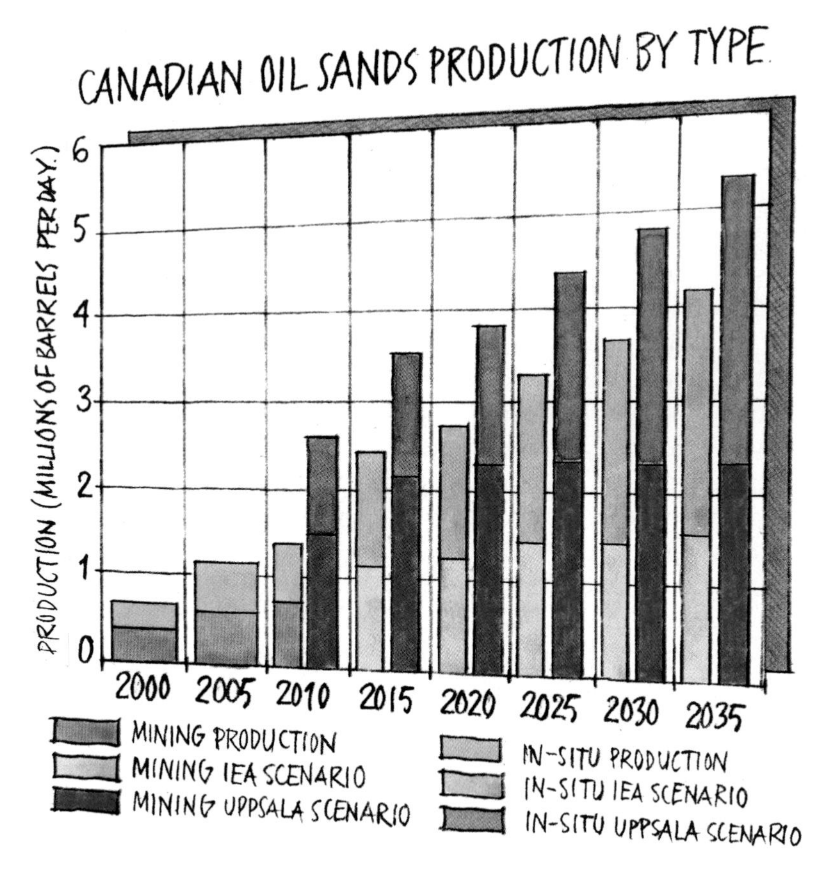

Fig. 10.3 A comparison of predictions for production of bitumen from oil sands until 2035 according to the IEA's New Policies Scenario [4] and the "crash program scenario" from UGES at Uppsala University [6]. The actual production that occurred in 2000, 2005, and 2010 is also shown

that development would be driven by a "crash program" scenario. In Fig. 10.3 we show oil production from oil sands by open-pit mining and by in situ methods up to 2010. From 2010 to 2035 the figure shows the levels of oil production that might have been achieved if the actions we had proposed in "A Crash Programme Scenario for the Canadian Oil Sands Industry" [6] had been instituted in 2006. For 2015–2035 this figure also compares our crash program predictions side by side with those of the IEA's "New Policies Scenario" in its *WEO 2010* report [4]. Interestingly, our analysis shows higher production in 2035 (5.3 Mb/d) than the IEA's scenario that is not a crash program scenario (4.2 Mb/d). This indicates that both scenarios are probably reasonable according to the premises upon which they are based.

Environmental Consequences of Bitumen Production

The total area that can be affected by open-pit mining of oil sands is 4,700 km^2. As reported by Total, the mining industry argues that this is not a great impact considering that it is only 0.1% of Canada's boreal forest area [9]. However, this is greater than Luxemburg's surface area of 2,600 km^2 and nearly as large as that of Trinidad and Tobago at 5,100 km^2. Because in situ production of bitumen also has environmental effects, the area that can be affected by production of bitumen from oil sands is actually much larger.

The indigenous people and others living in the area—mainly in villages along the Athabasca River—have complained for many years that contamination from the oil sands industry is poisoning the fish, other animals, and themselves. In many nations, including Sweden, people are beginning to question the morality of investing money in companies mining the oil sands. I have previously been invited to symposia where this question has been discussed [10].

The oil sands industry and its experts assert that they are not poisoning the area. However, in August 2010 a research report was published showing that 13 different forms of contamination resulting from bitumen production could be identified in the environment. The report was produced by researchers from Alberta University and is titled, *Oil Sands Development Contributes Elements Toxic at Low Concentrations to the Athabasca River and Its Tributaries* [11]. Below is a quote from the paper.

> We show that the oil sands industry releases the 13 elements considered priority pollutants (PPE) under the US Environmental Protection Agency's Clean Water Act, via air and water, to the Athabasca River and its watershed. In the 2008 snowpack, all PPE except selenium were greater near oil sands developments than at more remote sites. Bitumen upgraders and local oil sands development were sources of airborne emissions. Concentrations of mercury, nickel, and thallium in winter and all 13 PPE in summer were greater in tributaries with watersheds more disturbed by development than in less disturbed watersheds. In the Athabasca River during summer, concentrations of all PPE were greater near developed areas than upstream of development. At sites downstream of development and within the Athabasca Delta, concentrations of all PPE except beryllium and selenium remained greater than upstream of development. Concentrations of some PPE at one location in Lake Athabasca near Fort Chipewyan were also greater than concentration in the Athabasca River upstream of development. Canada's or Alberta's guidelines for the protection of aquatic life were exceeded for seven PPE—cadmium, copper, lead, mercury, nickel, silver, and zinc—in melted snow and/or water collected near or downstream of development.

The discussion of the environmental consequences of oil sands mining that is now underway will, no doubt, intensify. However, economic considerations mean that production of bitumen is certain to continue despite the following conclusion from the Alberta University researchers: "Contrary to claims made by industry and government in the popular press, the oil sands industry substantially increases loadings of toxic PPE to the Athabasca River and its tributaries via air and water pathways."

Let us return for a moment to the statement made by the managing director of the Swedish Petroleum Institute in 2005: "There are large known, but undisturbed, resources in the form of oil sands that today are as large as all the oil previously discovered." By 2005, the total of all the 2P reserves of conventional crude oil ever discovered was approximately 1,800 Gb. With a global recovery factor of 30% this implies the existence of around 6,000 Gb of OOIP compared with 2,000 Gb of OOIP for the oil sands. This means that the oil sands are equivalent to only one third of the crude oil that originally existed in the Earth. If one examines reserves (recoverable hydrocarbons) rather than OOIP then 170 Gb of recoverable hydrocarbons from oil sands can be compared with 1,800 Gb from oilfields. This means that the recoverable volume of hydrocarbon from the oil sands is only one tenth that of conventional crude oil. Only if one compares the total volume of recoverable conventional crude oil (1,800 Gb) with the total oil sands OOIP (2,000 Gb) can one speak of "… resources in the form of oil sands that today are as large as all the oil previously discovered." but such a comparison does not make sense. Replacing declining oil production of 4 Mb/d year on year with bitumen from Canada's oil sands is impossible according to both the IEA and UGES. The oil sands will not save us from Peak Oil and, considering the consequences for the environment, it is questionable if we should even try to use them for this.

Production of Heavy Oil from Venezuela's Orinoco Belt

The Orinoco Heavy Oil Belt lies north of the Orinoco River. It occupies an area of around 50,000 km^2 (see Fig. 10.4). The field is divided geographically into four areas; Boyaca (Machete), Junin (Zuata), Ayacucho (Hamaca), and Carabobo (Cerro Negro). (The names in parentheses are the older names for these areas.) These names are often included in the names of projects that are underway or are planned for the Orinoco Belt.

The oil reservoirs of this belt have a similar character to those of Canada's oil sands but lie significantly deeper underground at approximately 500–1,000 m depth. This means that the temperature of the heavy oil is around 55°C which is sufficient to permit the oil to be extracted without heating.

In 2009 the US Geological Survey (USGS) released a detailed study that summarized existing information on the Orinoco Belt. They asserted that

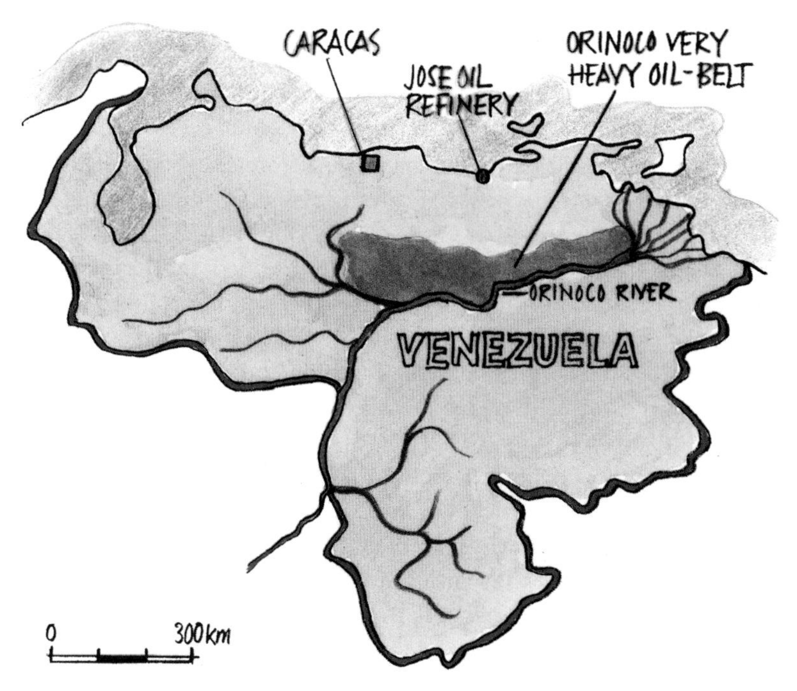

Fig. 10.4 The Orinoco Heavy Oil Belt north of the Orinoco River occupies 50,000 km^2 and holds at least 380 Gb of producible heavy oil. The Belt is divided geographically into four areas, Boyaca, Junin, Ayacucho, and Carabobo. These names are often included in the names of projects underway or planned in the Orinoco Belt

the OOIP of the fields was, with 95% probability, 900 Gb but there was a 5% chance that it was as much as 1,400 Gb. The median value of the estimates was given as 1,300 Gb which is the same value accepted officially in Venezuela [12]. For comparison, note that the total OOIP for all of Saudi Arabia is 700 Gb. Estimated recovery factors for the Orinoco Belt vary from 15% to 70% and it is not clear how these numbers have been used to arrive at the given minimum recoverable oil volume of 380 Gb, the maximum recoverable oil volume of 652 Gb, or the median value of 513 Gb. However, it is clear that, even at the minimum stated volume of recoverable oil, the Orinoco Belt comprises the world's largest oilfield. The frequently heard statement that Ghawar is the world's largest oilfield should be qualified: it is the world's largest field of conventional crude oil.

The Orinoco field was discovered as early as 1935 but at that time the very heavy oil it contains (API gravity 7°) was not of interest to extract. Venezuela's national oil company Petróleos de Venezuela SA (PDVSA) made a first attempt to extract oil from the field in the 1980s but it was only in the 1990s when international oil companies such as ExxonMobil, Statoil, and Total became involved that production actually began. A characteristic of production from the Orinoco field is that the production volume is determined by the size of the project (i.e., the size of the production infrastructure, etc.) and, once that is in place, the production volume will be roughly constant for at least 30 years. The heavy oil from these projects is transported via pipeline to refineries that can process it into lighter grades in the city of José, 200 km north of the Orinoco Belt (Fig. 10.4).

In his Ph.D. thesis Fredrik Robelius summarized the known Orinoco Belt projects in 2006. The total planned production from these projects in 2012 could amount to 1.2 Mb/d. Then he made the assumption that higher oil prices would stimulate new projects so that production in 2020 would have increased by an additional 1.0 Mb/d and by 2030 total production would be 2.4 Mb/d [13].

In its *WEO 2010* report the IEA states that oil production in the Orinoco Belt has expanded to 0.7 Mb/d (although production in 2009 was only 0.4 Mb/d) and that plans now exist to expand production by an additional 2.3 Mb/d [4]. However, at the same time, they show in a diagram of future anticipated production in Venezuela that production from the Orinoco Belt will be around 1.3 Mb/d in 2020 and that production only exceeds 2.0 Mb/d in 2035. We can state that the calculations that Fredrik Robelius made in his thesis agree quite well with the IEA's future production scenario for the Orinoco Belt.

Oil from Oil Shale

Oil shale contains kerogen and is the source rock for conventional oilfields. This means that oil shale can be found in all the world's oil provinces. Oil shale usually exists at great depths where it is heated so that the kerogen is transformed into oil (see Fig. 4.5). However, there are also large oil shale resources found today at shallow depths. These can form the raw material for industrial production of kerogen which, in turn, can form the raw material for synthetic oil.

The world's largest oil shale resources exist in the area around the Green River in the United States. Collectively the world's oil shale resources are estimated to be equivalent to 3,500 Gb of oil. Approximately 85% of these resources exist in the United States. Despite this, the United States has no commercial production of kerogen or kerogen-based synthetic oil. Instead, the world's largest mining of oil shale occurs in Estonia where it is mainly burned for electricity generation. In the United States coal is used for electricity generation and, for the foreseeable future, they will not need kerogen for that purpose.

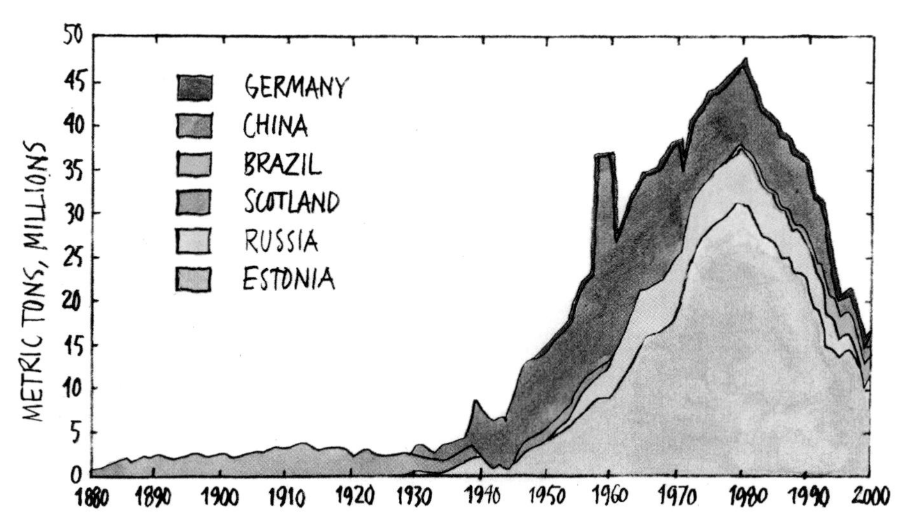

Fig. 10.5 World production of kerogen up to 2000. The largest production existed in the former Soviet Union (Estonia and Russia) but production fell with the dissolution of that state. At the end of the 1950s production increased dramatically in China due to an oil shortage but when the Daqing oilfield was discovered the need for kerogen for oil production declined

Global synthetic oil production from oil shale is 15,000 b/d which is 5.5 Mb/year [4]. This is equal to the world's total oil production in 1870 and is irrelevant in global statistical terms (Fig. 10.5). By 2035 the IEA estimates that, in addition to production in Estonia, there will also be limited synthetic oil production from oil shale in the United States, China, Jordan, and Brazil. However, the combined world production will level off at 300,000 b/d which is 110 Mb/year. World conventional crude oil production was 110 million barrels during 1896. Thus, over a period of 25 years oil production from oil shale is estimated to increase by the same amount that world conventional crude oil production increased during the 26 years from 1870 to 1896. So, when we examine oil shale production from the perspective of global Peak Oil we see that oil production from oil shale is insignificant.

Shale Oil Production in the United States

When shale containing kerogen is buried sufficiently deeply the kerogen can begin to be transformed into oil. The porosity of shale itself is very low compared to limestone or sandstone. However, it is possible to extract "shale oil" from these "tight" oil-bearing shales at relatively low flow rates using techniques such as horizontal drilling and hydraulic fracturing ("fracking").

In September 2011 the US Energy Information Administration (EIA) released a detailed report on the historical development of shale oil production from the Bakken formation in Montana and North Dakota. Production from the Bakken formation constitutes 90% of current shale oil production. The EIA notes that it is primarily technological progress in production of shale gas that has made it possible to increase shale oil production. During 2010 production of shale oil averaged 260,000 b/d compared to only 5-6 b/d in 2000! At the start of 2011 shale oil production had reached 344,000 b/d and by August 2011 it had grown to 445,000 b/d [24].

Another geological formation from which shale oil production has begun to increase is the Eagle Ford Shale in Texas. The EIA estimates that shale oil production from this area may increase by up to 500,000 b/d. Limited production of shale oil is also seen from the Barnett, Woodford and Marcellus formations.

Production of oil from shale is similar to production of gas from these formations. In an online article from the EIA titled *Bakken formation oil and gas drilling activity mirrors development in the Barnett* [24] there is an animation showing the progress of drilling activity in the Bakken shale from 1985 until the end of 2010. The animation shows when and where every well was drilled into this formation. Production of oil and gas from the Bakken shale has required the drilling of thousands of wells and one can see how rapidly production from the individual wells declines. This is clearly apparent if one examines the Middle Bakken area in Montana. In the animation the green dots representing oil-producing wells in the Middle Bakken begin to increase rapidly in number and production rate in 2002 and reach a maximum in 2005 before oil production rapidly declines. By 2010 mainly low rates of gas production remain. The animation is a beautiful illustration of the fact that Peak Oil also applies for areas producing shale oil but that the timeframe for the peaking and decline of shale oil production is much shorter than for conventional oil.

Production of shale oil requires the drilling of many more wells and much larger energy investments than for conventional oil. This means that the net production of energy (the energy profit) from shale oil is less than for conventional oil. It is net energy production that supports economic development and this means that shale oil gives less economic benefit to our society than conventional crude oil.

Public expectations are currently very high regarding future production of shale oil but it is important that we do not assume too much. In the EIA's report *Annual Energy Outlook 2012 (AEO 2012) Early Release Reference Case* [25] we can read that the EIA estimates production of "tight oil" (shale oil) will only grow by 1 Mb/d by 2035. There is nothing to indicate that the United States will become self-sufficient in oil production or even become an oil exporter despite enthusiastic reportage in the news media. There is a real need for more critical analysis by the media in future when they report the exaggerated dreams of some members of the oil industry. For our analysis here we will accept the increase in production of shale oil predicted by the EIA. However, we also urge caution since the history of production from the Middle Bakken shows such a clear Peak Oil profile.

Oil Production from Coal by CTL (Coal-to-Liquids)

In the debate regarding Peak Oil it is often argued that endless amounts of coal exist and that the coal can be transformed into oil in such large quantities that we do not need to worry about declining oil production. UGES regarded this argument as a new challenge for investigation. When Mikael Höök was appointed as a Ph.D. student his main task became the investigation of the world's coal reserves. We have now published a research report that shows that global coal production will reach a maximum around 2040. This means that conservative use of the remaining coal reserves must be seen as a priority. We have also produced a special report focusing on oil production from coal titled, *A Review on Coal to Liquid Fuels and Its Coal Consumption* [14]. Note this quote from the abstract of the report:

> Conversion ratios for CTL are generally estimated at between 1 and 2 barrels/t coal. This puts a strict limitation on future CTL capacity imposed by future coal production volumes, regardless of other factors such as economics, emissions or environmental concerns. Assuming that 10% of world coal production can be diverted to CTL, the contribution to liquid fuels supply will be limited to only a few Mb/d. This prevents CTL from becoming a viable mitigation plan for liquid fuel shortage on a global scale.

Historically, CTL technologies have been useful when nations have been cut off from deliveries of oil. During World War II Germany produced liquid fuel from coal, primarily for its aircraft and in the 1950s South Africa was already using this technology when a fuel blockade was imposed upon it by the rest of the world. The method most commonly used to transform coal into liquid hydrocarbon fuel is the Fischer–Tropsch process in which iron or cobalt is used as catalysts. The advantage with this technique is that the end product is high-quality diesel-type fuel that can be used directly for aviation. Today, the world's largest producer of liquid fuel using CTL technology is the South African company Sasol Ltd (in Afrikaans, "Suid Afrikaanse Steenkool en Olie" which means South African Coal and Oil). Their production capacity is approximately 150,000 b/d.

The fact that CTL technology is used on an industrial scale in South Africa makes it possible to make an empirical estimate of its coal requirements. South Africa has stated that 24% of its coal consumption in 2003 was used for making synthetic oil and that its total coal production in that year was 238 million metric tons. This means that 55 million barrels of synthetic oil per year (0.15 Mb/d) requires 57 million metric tons of coal. Thus almost one metric ton of coal is required to produce one barrel of synthetic oil. All of South Africa's coal is classed as bituminous which means that it has a carbon content of between 50% and 80%. It is this type of coal, or coal of poorer quality, that is available globally for CTL.

It should be a requirement that all future predictions of CTL production also include a calculation of the volume of coal needed. However, by referring to the data on CTL from South Africa we ourselves can determine what is required for such future projections to be realized. For example, the National Petroleum Council (NPC) in the United States has presented a number of future production estimates for CTL arguing that Peak Oil can be partially mitigated in that nation by large-scale implementation of CTL technology.

In the *Hirsch Report* [2] the authors estimated the oil production that might be possible from a crash program to expand use of coal-to-liquids. They argue that it should be possible to build five new plants of 100,000 b/d capacity every year which is 0.5 Mb/d or 180 Mb/year of new CTL production capacity. The experience of South Africa indicates that these five plants will require annual coal production of 180 million metric tons. In 2009 total US coal production was 973 million metric tons which means that, if the CTL plants were to be built in the United States, the yearly growth in CTL capacity would consume 18% of the United States' current coal production. Alternatively, coal production would need to be increased by this amount. We have made a detailed study of coal production in the United States and have concluded that the potential exists to increase annual production by 40% if production is expanded in Montana [15]. However, this increase would only allow for 2 years of CTL expansion.

The NPC report on future CTL production states that Peak Oil can be mitigated and predicts that production of synthetic oil by CTL will be 5.5 Mb/d in 2030 [16]. 5.5 Mb/d equates to 2,000 Mb/year and would require 2,000 million metric tons of coal per year. Thus, the NPC's prediction requires that coal mining in the United States must more than double by 2030, and that all coal should be used for CTL. We can state with certainty that this prediction is completely unrealistic and so concerns over Peak Oil are, in fact, justified.

We are aware that more efficient methods of CTL than are used in South Africa may be developed in the future. Using the experience of Sasol and

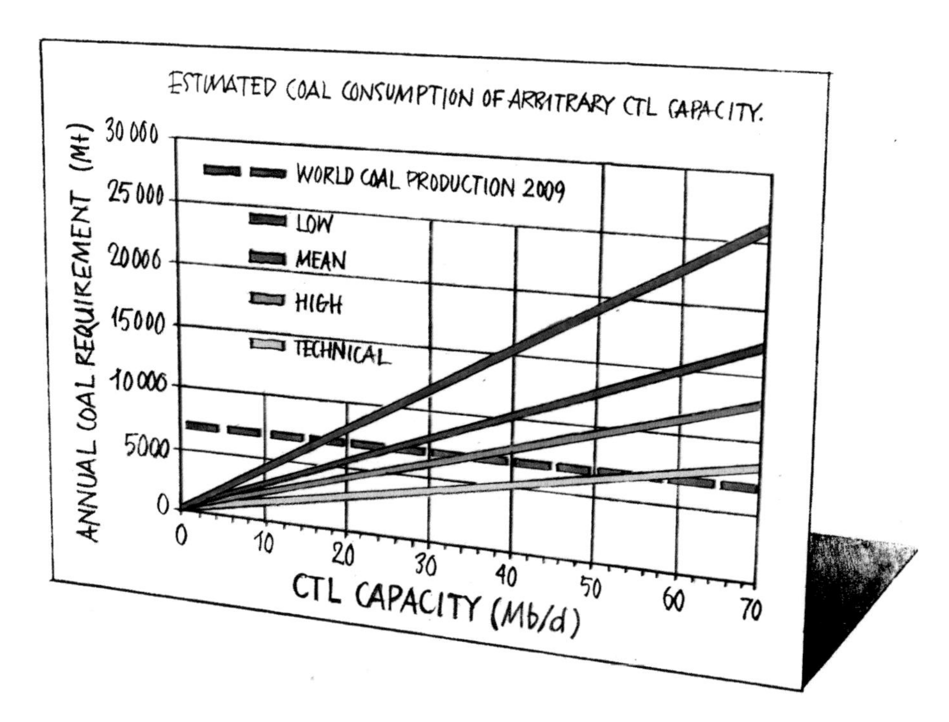

Fig. 10.6 Estimated coal use in millions of metric tons (Mt) as a function of CTL capacity for four different conversion factors of between 1 and 4 barrels of oil per metric ton of coal. Current global coal production could provide 18–55 Mb/d of synthetic oil depending on the conversion factor. Current CTL practice in South Africa represents the lowest conversion factor [14]

estimates found in the literature, it is possible to suggest four different conversion ratios (see Fig. 10.6). We have chosen to set a "low" conversion ratio at 1 barrel/t, a "mean" value at 1.5 barrels/t, and a "high" estimate at 2 barrels/t. We can also imagine that it might be technically possible to achieve a very high CTL-conversion ratio of 3 barrels/t coal [14] ("technical" in Fig. 10.6). However, empirical experience from Sasol indicates that realistic conversion ratios are in the range of 1–1.5 barrels/t coal.

Figure 10.6 shows how much coal is required by CTL in order to produce various volumes of oil under these different conversion ratios. The figure also shows the level of global coal production in 2009 which was 6,940 million metric tons. The analysis shows that current world coal production could provide at least 18 Mb/d of synthetic oil (if the coal were used for nothing else) and that more efficient CTL technologies might raise production to, at most, 55 Mb/d. The technology currently used in South Africa would be classed as "low." It is nonsensical to imagine that all the world's coal production would be diverted from electricity generation into production of synthetic oil by CTL and, in any case, research by UGES has demonstrated that we are approaching a future peak of coal production. In the IEA's *WEO 2010* report they project that production of synthetic oil by CTL will be 1 Mb/d in 2035. This would require 365 million metric tons of coal or about 5% of current global coal production. Expansion of CTL to this level within 25 years is feasible if the necessary investments are made. The IEA also notes that the price of oil must be between $60 and $100 per barrel for CTL production to be profitable.

Our conclusion is that CTL can provide a meaningful supplement to the oil supply of some nations but that it will not compensate in any significant way for declining production of conventional crude oil after the global peak.

There is considerable interest in CTL technology worldwide and especially in China. Originally, the Chinese planned to increase oil production from CTL so that, by 2020, it would equal 16% of the volume of their 2008 conventional oil production. They have now decided to postpone all their CTL projects other than two that are almost ready to produce oil. Research in one of our scientific publications shows that China will soon see a peak in its coal production, "Peak Coal." That may be one of the reasons why the Chinese have now reconsidered how they will use their remaining coal reserves [17].

Oil from Natural Gas by GTL (Gas-to-Liquids)

Total global oil production from GTL was approximately 50,000 b/d in 2009. In the *WEO 2010* report (page 175) the IEA estimates that this will grow to 200,000 b/d in 2015 and nearly 750,000 b/d in 2035. Qatar is expected to become the dominant producer with production of 160,000 b/d [4]. However, in that report we can also read that there are problems in Qatar, "Some technical problems with the commissioning of a new plant in Qatar and a sharp rise in construction costs, together with increased interest in LNG, which competes with GTL for gas feedstock, have led to many planned GTL projects being shelved in the last few years." So there is every reason to lower our future expectations of GTL production in Qatar.

"Exxon to Abandon a Big Investment in Qatar" is the headline of an article in the *New York Times* of February 21, 2007. The appearance of this article is connected with the discussion of GTL production in Qatar in *WEO 2010* [18]. In the *New York Times* article we can read the following,

> Exxon Mobil announced on Tuesday that it would abandon one of its biggest investments ever, a project with Qatar's state-run oil and gas company to produce clean-burning diesel from natural gas. Instead, Exxon Mobil said that

> it could concentrate on a new gas drilling project in the emirate's rich Barzan field, which is close to the site of the gas-to-liquid project. The Barzan project will initially produce 1.5 billion cubic feet of gas a day and eventually much more for the fast-growing Qatar domestic market in 2012.

The article implies that the reason for the decision is increased costs but ExxonMobil will not confirm this. We can also read in the article that,

> Energy companies have shelved or delayed several projects in Canada over the last year or so because of cost overruns in oil sands and conventional fields, and Qatar itself announced a moratorium on new gas projects last year. But Tuesday's announcement came as a surprise since the chief executive of Exxon, Rex W. Tillerson, had said as recently as September that the company was moving forward with the project, albeit with efforts to control costs.

So why would ExxonMobil abandon the world's largest GTL project when they had only signed the contract with Qatar Petroleum to build the installation less than 3 years earlier?

North Field (in Qatar) and the South Pars field (in Iran) are, in reality, parts of the world's largest single natural gas field that lies under the Persian Gulf. However, the size of this combined field has been questioned [19]. That the world's largest GTL plant should be connected to the world's largest natural gas field seems fairly logical especially when we know that there is no infrastructure to enable sale of the gas to other nations via pipeline. On July 14, 2004 ExxonMobil and Qatar Petroleum signed the contract to produce 154,000 b/d of ultra-pure diesel. The question is whether there was any reason other than cost for the world's largest GTL project to be abandoned so soon afterwards.

In a report from Simmons and Company International (SCI), *Simmons Oil Monthly* [20], we can read the following quotation from an interview with Matt Simmons, "Now, the North Field has basically two producing platforms, Alpha and Bravo, and while ConocoPhillips last summer was drilling the wells for the Charlie platform, they hit dry holes." It was intended that the Charlie platform would supply gas to the world's largest GTL project but the gas simply was not there. The veracity of Matt's statement has been confirmed for me in private conversations I have had with people who know about the issue. The gas that was not there had already been booked as reserves in ExxonMobil's annual accounts [20] and ExxonMobil's participation in the Barzan project can be regarded as compensation for the fact that the expected gas did not exist.

Although Exxon Mobil may have abandoned plans to build a GTL plant in Qatar another oil company has completed one. Shell is responsible for the world's largest GTL plant in Qatar. The plant is named Pearl GTL and it has a capacity of 140,000 b/d [23]. Nevertheless, the IEA's projections for future GTL production must be reduced and GTL will definitely not compensate for Peak Oil.

Additives as a Form of Unconventional Oil

When reading detailed accounts of the world's unconventional oil production one often finds a line with the title "Additives" and the volume stated is not negligible. Additives can be chemicals that are blended into crude oil before it is pumped to a refinery or blended into the products of the refinery itself. These chemicals are produced by the petrochemical industry from various raw materials such as oil, natural gas, coal, and biomass. Because they contribute to both the volume and energy content of oil products these additives can rightly be included as part of unconventional oil production if they are derived from natural gas or coal. In those cases, additives can be seen as variants of GTL or CTL production. In contrast, contributions made to production of additives by other materials such as conventional crude oil and biomass should not be included in this class of unconventional oil. However, it is often difficult to separate the contributions made by these various types of raw material so it is difficult to determine if all the barrels included as "Additives" in unconventional oil statistics are correctly attributed. Nevertheless, for the purpose of our discussion of Peak Oil we accept these numbers as correct inasmuch as any accounting errors in this class are insignificant relative to the contributions made to world oil supply by other classes of oil. In *WEO 2010* the additives are included under *processing gains* and for 2009 this volume was reported to be 0.9 Mb/d.

NGL, Natural Gas Liquids

Part of the volume of hydrocarbons that is commonly reported as oil production is, in fact, natural gas liquids (NGL). The IEA chooses to report NGL volumes in simple barrels (where 1 barrel is 159 L). The US Energy Information Agency (EIA) chooses instead to report NGL volumes in terms of the number of barrels of crude oil that would contain the same amount of energy, "barrels of oil equivalent" (boe). The energy content of the five main NGL components (ethane, propane, butane, isobutane, and pentanes) varies from 3.25 to 4.56 gigajoules per barrel (GJ/b), which is significantly less than the 6.1 GJ contained in a barrel of crude oil. In reality, a barrel of NGL only contains energy equal to 0.7 barrels of oil. In 2008, we note that the IEA reported

10.5 Mb/d of NGL production in 2007 and the EIA reported this as 7.92 Mb/d. The ratio of the two volumes is 0.75 which is approximately the ratio of the energy content of a barrel of NGL relative to a barrel of crude oil [21].

The lightest NGL components become liquid when put under pressure and the required pressure is so low that they can be mixed together and stored in liquid form in pressure vessels. They are commonly used for cooking or as LPG, liquid petroleum gas, in motor vehicles. In some nations that produce large volumes of natural gas (such as The Netherlands and Australia) many vehicles are powered by LPG.

The heaviest NGL components are liquid at normal temperature and pressure and are also described as "condensate." In Canada, condensate is mixed with bitumen to produce a liquid that can be transported through pipelines.

The amount of NGL that can be produced is determined by the volume of natural gas production. Statistics from natural gas production over the last 30 years show that the proportion of production that is NGL varies between 14% and 16%. The higher percentage value is approached as the gas field becomes depleted. Because the proportion varies so little, we can use these percentages to predict the proportion of NGL that will be produced as part of natural gas production in coming years. If we accept the IEA's prediction of natural gas production in 2030 of 4,424 billion cubic meters [11] and we say that 15% of this will be NGL, then this equals 15.5 Mb/d of NGL production or 11.5 million barrels of oil equivalent per day (Mboe/d) [22].

Final Comment

The volumes of future oil production that will be contributed by unconventional sources are significant, but are they enough to significantly delay Peak Oil? Economists often make statements in which they assert that unconventional oil will replace crude oil but the broad analysis we have made in this chapter shows that these statements are not founded in fact. The overview given in Chap. 5 of the *WEO 2010* report supports this [4]. The Hirsch report from 2005 stated that a "Crash Management Program" should start 20 years before Peak Oil to ameliorate the effects of Peak Oil [2]. However, in their book *The Impending World Energy Mess* [2] from 2010 the same authors now conclude that Peak Oil is not 20 years away and so their Crash Management Program can no longer fully ameliorate its effects. Our final conclusion is that the total volumes of future unconventional oil production will not allow us to avoid Peak Oil or compensate for the subsequent decline in conventional crude oil production.

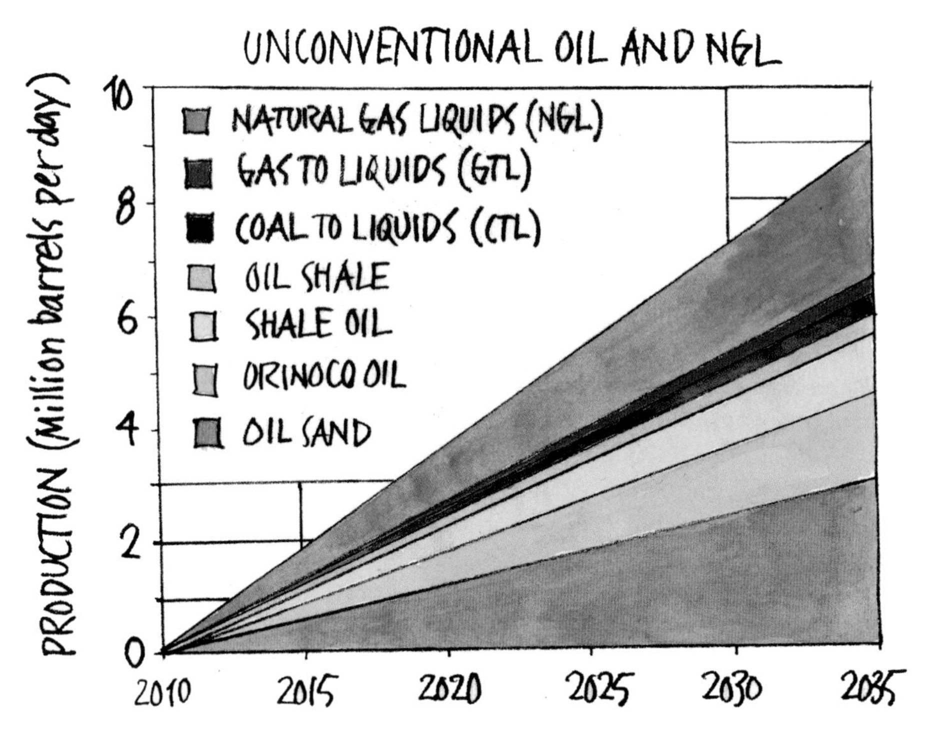

Fig. 10.7 Mitigation curves for unconventional oil production and NGL

The detailed analysis that we have done began with today's reality and considered a realistic rate of expansion (except for Canada's oil sands where we also did a crash program analysis). In addition, we have looked at some other unconventional sources of oil that are commonly discussed. The realistic contributions to oil production that can be made by the various sources of unconventional oil and NGL are shown in Fig. 10.7. In total we can see an increase in production of 8 to 9 Mb/d during the next 25 years. As described in Chap. 9, the total flow of oil from fields currently in production is decreasing by 4 Mb/d every year. This means that the total increase in unconventional oil and NGL production during the next 25 years can only compensate for the decline in conventional crude oil production occurring in the next 2–3 years. Even if we have underestimated the possible increase in unconventional oil and NGL production by 100% this would only allow compensation for less than 5 years of decline in conventional crude oil production rather than 2–3 years.

As for my testimony before the US House of Representatives Committee on Energy and Commerce, I return to that in a later chapter.

References

1. Bartlett, A.: The Most Important Video You'll Ever. See YouTube (more than 3,000,000 have seen it). http://www.youtube.com/watch?v=F-QA2rkpBSY (2007)
2. Hirsch, R.L., Bezdek, R., Wendling, R.: Peaking of world oil production: impacts, mitigation, and risk management. Report to US Department of Energy. http://www.netl.doe.gov/publications/others/pdf/Oil_Peaking_NETL.pdf (2005) and Hirsch, R.L., Bezdek, R.H., Wendling, R.M.: The Impending World Energy Mess. Apogee Prime, ISBN 978-1926837-11-6 (2010)
3. Corbett, P., Couples, G., Gardiner, A.: Petroleum geosciences. Technical Report. Department of Petroleum Engineering, Heriot-Watt University, Edinburgh, Scotland (2004)
4. WEO: World Energy Outlook 2010. International Energy Agency. http://www.worldenergyoutlook.org/2010.asp (2010)
5. Dagens Industri: Analytiker: Oljepriset kan mångdubblas. http://di.se/Default.aspx?pid=67936__ArticlePageProvider (2005)
6. Söderbergh, B., Robelius, F., Aleklett, K.: A crash programme scenario for the Canadian oil sands industry. Energy Policy **35**(3), 1931–1947 (2007). http://www.sciencedirect.com/science/article/pii/S0301421506002618
7. WEO: World Energy Outlook 2004. International Energy Agency. http://www.iea.org/textbase/nppdf/free/2004/weo2004.pdf (2004)
8. Canada's Oil Sands. http://www.canadasoilsands.ca/en/glossary.aspx (2011)
9. Total: Oil sands, production techniques. http://www.total.com/en/special reports/oil-sands/canada-s-oil-sands/production-techniques-200931.html (2011)
10. Aleklett, K.: Oil sands and our future pensions. Aleklett's Energy Mix. http://aleklett.wordpress.com/2009/05/15/oljesand-och-framtidens-pensioner/ (2009)
11. WEO: World Energy Outlook 2010. International Energy Agency. http://www.iea.org/textbase/nppdf/free/2010/weo2010.pdf (2010)
12. Kelly, E.N., Schindler, D.W., Hodson, P.V., Short, J.W., Radmanovich, R., Nielsen, C.C.: Oil sands development contributes elements toxic at low concentrations to the Athabasca River and its tributaries. PNAS **107**, 16178 (2010). http://www.pnas.org/content/107/37/16178.full.pdf+html
13. USGS: An estimate of recoverable heavy oil resources of the Orinoco Oil Belt, Venezuela. Fact Sheet 2009–3028. http://pubs.usgs.gov/fs/2009/3028/pdf/FS09-3028.pdf (2009)
14. Robelius, F.: Giant oil fields—the highway to oil: giant oil fields and their importance for future oil production. Uppsala University, Dissertations from the Faculty of Science and Technology. ISSN 1104–2516, p. 69. http://uu.diva-portal.org/smash/record.jsf?pid=diva2:169774
15. Höök, M., Aleklett, K.: A review on coal to liquid fuels and its coal consumption. Int. J. Energy Res. **34**, 848–864 (2010). http://dx.doi.org/10.1002/er.1596, http://www.tsl.uu.se/uhdsg/Publications/CTL_Article.pdf
16. Höök, M., Aleklett, K.: Historical trends in American coal production and a possible future outlook. Int. J. Coal Geol. **78**(3), 201–216 (2009). http://www.science-

direct.com/science/article/pii/S0166516209000317; http://www.tsl.uu.se/uhdsg/Publications/USA_Coal.pdf
17. National Petroleum Council: Facing the hard truths about energy and supplementary material. Topic Paper #18—Coal to Liquids and Gas. See also http://www.npchardtruthsreport.org/ (2007)
18. Höök, M., Aleklett, K.: A review on coal to liquid fuels and its coal consumption. Int. J. Energy Res. **34**, 848–864 (2010). http://dx.doi.org/10.1002/er.1596, http://www.tsl.uu.se/uhdsg/Publications/CTL_Article.pdf
19. Krause, C.: Exxon to abandon a big investment in Qatar. New York Times, 21 Feb 2007. http://www.nytimes.com/2007/02/21/business/worldbusiness/21exxon.html?_r=1&ref=business (2007)
20. Cohen, D.: Questions about the world's biggest natural gas fields. The Oil Drum, 9 Jun 2006. http://www.theoildrum.com/story/2006/6/8/155013/7696 (2006)
21. Aleklett, K., Höök, M., Jakobsson, K., Lardelli, M., Snowden, S., Söderbergh, B.: The peak of the oil age—analyzing the world oil production reference scenario in World Energy Outlook 2008. Energy Policy **38**(3), 1398–1414 (2010). accepted 9 Nov 2009, available online 1 Dec 2009, http://www.tsl.uu.se/uhdsg/Publications/PeakOilAge.pdf
22. Kessler, R.A.: Simmons Oil Monthly—Qatar. 24 Apr 2006, Simmons and Company International, look at: http://www.theoildrum.com/story/2006/6/8/155013/7696 (2006)
23. Shell: Pearl GTL—an overview. http://www.shell.com/home/content/aboutshell/our_strategy/major_projects_2/pearl/overview/ (2010)
24. Bakken formation oil and gas drilling activity mirrors development in the Barnett, U.S. Energy Information Administration, http://www.eia.gov/todayinenergy/detail.cfm?id=3750# (2011)
25. Tight oil, Gulf of Mexico deepwater drive projected increases in U.S. crude oil production, U.S. Energy Information Administration, http://www.eia.gov/todayinenergy/detail.cfm?id=4910 (2012)

Chapter 11

Peak of the Oil Age

In late October 2003 *The Economist* published an article titled, "The End of the Oil Age" [1]. In the article's introduction reference was made to a statement by Sheikh Yamani who served as Saudi Arabia's minister of oil and mineral resources from 1962 to 1986, "The Stone Age did not end for lack of stone, and the Oil Age will end long before the world runs out of oil." The article continued,

> This intriguing prediction is often heard in energy circles these days. If greens were the only people to be expressing such thoughts, the notion might be dismissed as Utopian. However, the quotation is from Sheikh Zaki Yamani, a Saudi Arabian who served as his country's Oil Minister three decades ago. His words are rich in irony.

There were certainly many people who thought that the entire article in *The Economist* was rich in irony or even utopian!

History is made up of many different "ages." For northern Europe, the Viking Age was significant. No precise beginning or end has been defined for the Viking Age, but it is usually described as extending from the start of the ninth century to the end of the twelfth century. This historical age lasted for 300 years. In contrast, a precise beginning can be defined for the Oil Age. The first oilfields in Romania, the Caspian Sea area, and the United States were found in the 1850s and by 1860 petroleum was traded commercially. When *The Economist* discussed the end of the Oil Age it did not mean an end due to shortage. Rather, it discussed a future where, according to current economic models, fuel cells, ethanol, and other forms of alternative energy would become cheaper than oil and so replace it. Peak Oil was not then an accepted reason for a future decline in oil use. When we now discuss "The Peak of the Oil Age," it is limitations in our ability to produce oil that

K. Aleklett *Peeking at Peak Oil*, DOI 10.1007/978-1-4614-3424-5_11,

are decisive. Peak Oil marks the moment when we have reached the high point of the Oil Age and the beginning of its end.

Of the world's oil economists, the one who should have the best knowledge about Peak Oil and future oil production is Fatih Birol, chief economist for the International Energy Agency (IEA). In 2008 he commented on future oil production in an article in the *Independent* newspaper without mentioning "Peak Oil" [2]. However, his statement supports the view that we face an imminent production maximum:

> We are on the brink of a new energy order. Over the next few decades, our reserves of oil will start to run out and it is imperative that governments in both producing and consuming nations prepare now for that time. We should not cling to crude down to the last drop—we should leave oil before it leaves us. That means new approaches must be found soon.

The factor that limits the length of the Oil Age is, of course, access to oil. Obviously, we must find oil before we can use it and we cannot use more oil than we have found. Now, in 2011, we can look back in history and see that the largest volumes of crude oil were found in the 1960s—56 billion barrels (giga barrels, Gb) per annum—and that we now have 800–900 Gb of reserves remaining (see Figs. 3.1 and 11.1). We can also estimate that the oil industry will find an additional 100–200 Gb in the future so that, in total, we will have approximately 1,100 Gb of crude oil left to consume before the Oil Age finally ends. In Chap. 10 we discussed alternative forms of oil production that can compensate for the decline in conventional crude oil production. These alternative forms of production are important for the future, but the decline in the production of conventional crude oil will set the limit for the Oil Age. As discussed in Chap. 6, oil production will not continue into the future at a high rate and then, one year, drop down to nothing. Instead it will decline more or less steadily. If the rate of decline is 2% per annum then we will have consumed all the remaining crude oil (1,100 Gb) by the end of this century. The smoothly declining trend of future consumption shown in Fig. 11.1 is a simplified picture. The reality is that we are far more likely to see rates of consumption that fluctuate above and below this declining trend. In terms of conventional crude oil, it is now accepted even by the IEA and the US Energy Information Administration (EIA) that the world passed its maximal rate of production between 2006 [3] and 2008 [4].

In the middle of the 1990s, I began to plan a course on energy at Uppsala University. While looking for a suitable textbook for the course I read the book, *Energy, Physics and Environment* by E.L. McFarland et al. [5]. In Chap. 4, "The Hubbert model of resource consumption," I became acquainted with Hubbert's peak. At the end of the chapter, a computer program was provided for estimating peak rates of resource production so I loaded it into my computer. I could see that changing the total volume of available oil

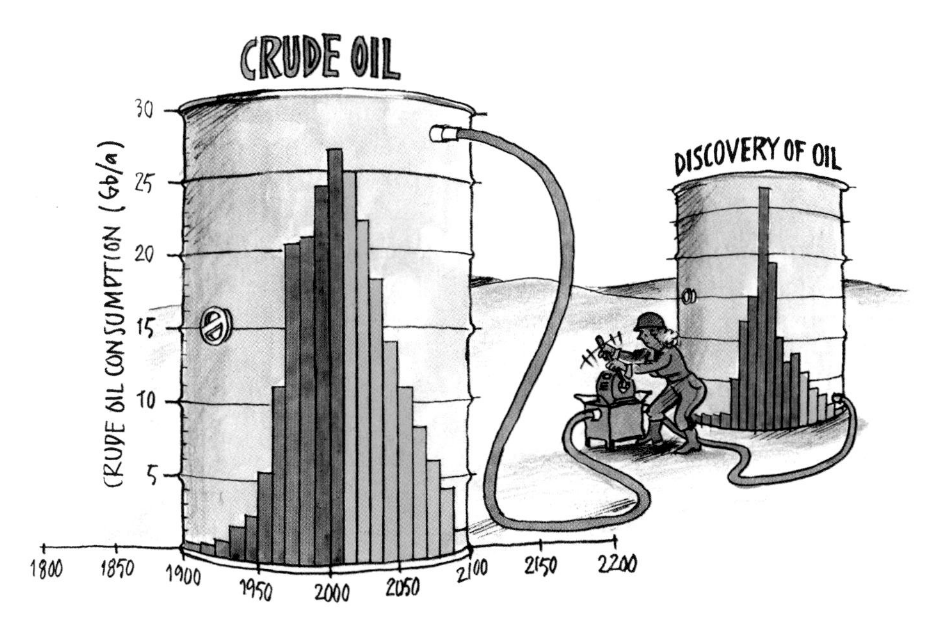

Fig. 11.1 The total volume of crude oil consumed up to the year 2010 was 1,100 Gb [3]. If we add the remaining known reserves of crude oil to the amount that we expect the oil industry to find in the future then we also obtain a volume of 1,100 Gb. In the figure the future decline in crude oil consumption is shown to be rather smooth but, in reality, it will fluctuate. We assume that crude oil consumption will be very limited by the end of this century

changed the date of the peak. However, to my great surprise, I could see that the change in the peak date was only marginal compared to changes in the volume of available oil. To shift the year of peak oil production to 2020 required a massive increase in oil reserves.

As the years passed, my lecture on energy resources became more detailed, and I soon became aware of the work of Dr. Colin Campbell. I contacted him in 2000 when I needed some of his data for a new figure for my lecture. As they say, the rest is history.

At the beginning of the 1990s I was already completely convinced that energy should be at the top of the political agenda but our leading politicians had a different opinion. In Sweden, one debate of that time was whether Russian natural gas should provide part of Sweden's future energy supply. All our political parties were of the view that nuclear power should be shut down, and some thought that a suitable replacement for the lost energy would be natural gas from Russia. I had a different view, and in August 1996 Sweden's leading broadsheet newspaper, *Dagens Nyheter*, published an opinion piece by me warning about the risks of dependence on Russian natural gas [6]. Now my research group has published a peer-reviewed article on Russian natural gas [7], and it shows that the position

I took 15 years ago was completely justified. The majority of Swedish politicians have now also changed their opinion about nuclear power.

Immediately after I had organized the world's first conference on Peak Oil in Uppsala in 2002 [8] Colin Campbell and I were invited to write an article for the scientific journal, *Minerals and Energy*. Michael Lynch, a long-time critic of the Peak Oil idea was also invited to write an article. In this way the journal intended to provide "balance" in the debate on Peak Oil.

Writing a scientific article on Peak Oil that would be subjected to peer-review was a new challenge for me. My previous 30 years of research had involved analyzing the inner characteristics of atomic nuclei. During 20 of those years I had worked with the Nobel Prize-winner Glenn T. Seaborg. He frequently said that, as a researcher, one should become engaged in issues that are important for society, even if they lie outside one's own discipline. His views have influenced me greatly. Dr. Seaborg himself acted as an advisor to ten US presidents. The last investigation he participated in was into the US school system. The report from this inquiry was titled, *A Nation at Risk* [9]. For me to change disciplines from nuclear physics to energy resource analysis was a challenge I accepted in the spirit of Glenn T. Seaborg. Fortunately, the experience that Colin Campbell had accumulated from a lifetime of work in the oil industry compensated for my inexperience. In 2003, the journal *Minerals and Energy* published our article, "The Peak and Decline of World Oil and Gas Production" [10]. This was the first peer-reviewed article about "Peak Oil."

The Peak and Decline of World Oil and Gas Production

The fact that *Minerals and Energy* invited submissions from two diametrically opposed camps in the Peak Oil debate shows how polarized opinion is on future oil production. One camp uses a natural scientific approach to the issue. It observes the factors that determine how much oil exists underground and it insists that the laws of physics must be followed when oil is produced. We can call this the natural science approach. The other, nonnatural science camp follows what we can call the flat-Earth approach. According to the adherents of that camp, the Earth's resources are infinite and production of oil is determined by economic, political, and technological factors.

In the article that Colin Campbell and I co-authored, we examined geological limits, types of rock, how oil becomes trapped in different geological structures, the size of discoveries and when they are reported, recovery factors, future scenarios from the US Geological Survey (USGS), and more. At the end we presented our analysis of future oil production. The 2003

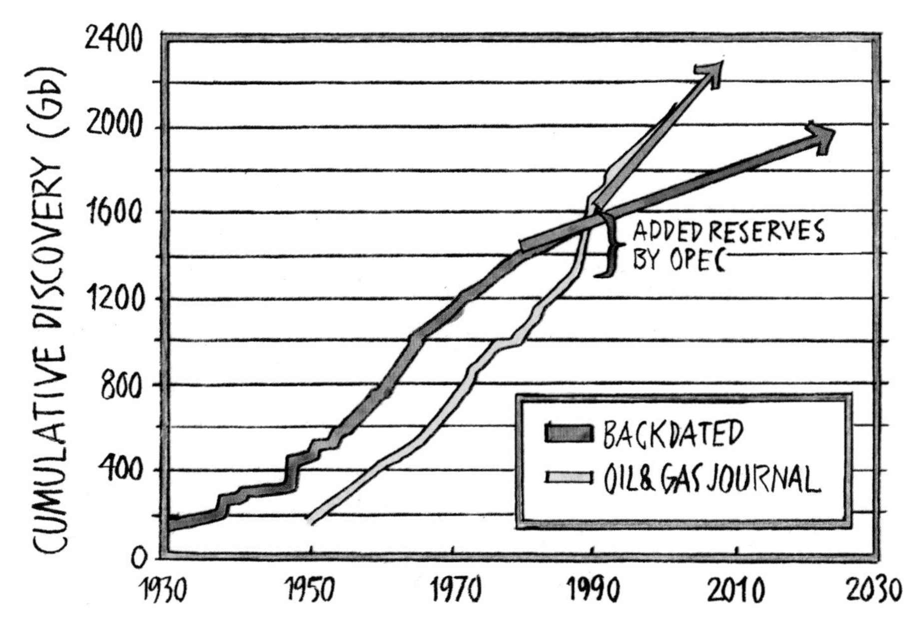

Fig. 11.2 The *Oil & Gas Journal* reports reserve revisions as new discoveries in the year that the revision occurs. This means that an upward revision that occurs in 1988 in a field that was discovered in 1938 is registered as a new discovery in 1988. If, instead, one denotes the upward revision in 1988 as a correction to the original discovery estimate of 1938 then one obtains a revised, "backdated," discovery curve that allows a realistic extrapolation of future discoveries [10]

article in *Minerals and Energy* [10] marks the beginning of the research profile that the Uppsala Global Energy Systems group has developed during the past 8 years at Uppsala University.

If one wishes to extrapolate current trends to predict the future then an accurate and realistic understanding of those trends is essential. At the end of every year the *Oil & Gas Journal* reports on the world's oil reserves [11] and this information is then used by BP for its annual publication, the *BP Statistical Review of World Energy* [12]. In the *Minerals and Energy* article, we discussed the difference between the *Oil & Gas Journal*'s reports on reserves and how we believe these data should be reported. We asserted that upward revisions to an oilfield's reserves should be attributed to the year in which the oilfield was originally discovered: that is, they are a correction to the original estimate of the size of the reserves in an oilfield. This approach is called "backdating" and is the most important factor when estimating future trends. In contrast, the *Oil & Gas Journal* treats upward revisions in oilfield reserve estimates as new discoveries occurring in the year the revision is made.

In Fig. 11.2 the consequences of these two approaches for extrapolation of future oil discovery trends are illustrated. It is obvious that extrapolations made using these two descriptions of "reality" give completely different predictions. By extrapolating the discovery data from the *Oil & Gas Journal* one arrives in a world in 2025 that agrees well with the oil reserve predictions regarded as valid by the USGS [13] (see Fig. 6.4). In contrast, backdating of reserves gives an extrapolation that shows that total crude oil discoveries will be less than 2,000 billion barrels by 2050. Also note that the *Oil & Gas Journal* reserve estimates often remain unchanged year on year, which is implausible as new discoveries of oil never exactly match production in any one year.

Today, the IEA, Cambridge Energy Research Associates (CERA), and others have accepted the backdating approach as the more useful one, yet the *Oil & Gas Journal* and BP cling to their unrealistic reporting methodology. Unfortunately, only BP's reports are generally accessible by the public. This has serious consequences because economists use BP's information when discussing our future.

At the end of the 1980s, the reserves of the OPEC nations increased dramatically. During the 1970s the most influential members of the OPEC cartel had expropriated the assets of the international oil companies and transferred these to national oil companies. They also took possession of the technical data from the previous owners. Then, in 1985, Kuwait suddenly declared its reserves to be 40% larger than previously reported [10]. The motivation behind this increase lies in how OPEC determines what volumes of oil its member nations are allowed to produce. Production volumes are determined partly on the basis of the size of a member nation's oil reserves. When Kuwait increased its declared reserves it was able to increase its rate of oil production at the expense of the other OPEC nations' production.

Three years later, OPEC member Venezuela doubled the size of its reserves by including the heavy oil from the Orinoco Belt (see the section "Production of Heavy Oil from Venezuela's Orinoco Belt", Chap. 10), which had not been previously counted [10]. In response, the other OPEC nations increased the size of their reserves by the following amounts: Abu Dhabi by 197%, Dubai by 186%, Iran by 90%, Iraq by 112%, and, finally, Saudi Arabia by 51%. There is an area lying between Kuwait and Saudi Arabia that is called the "neutral zone" and is administered jointly by both nations. For that area no reserve increase was declared [10].

In 1984 the Middle Eastern members of OPEC reported reserves of 357 Gb but by 1990 the reported reserves had increased to 644 Gb which is a total increase of 80%. During that period no discoveries of new giant fields were reported despite the fact that many such discoveries would be required to give such an increase in reserves. By 2009 Middle Eastern members of OPEC had increased their reported reserves to 716 Gb. Meanwhile, during the last 20 years these nations have produced 143 Gb. The question now is how large OPEC's remaining reserves really are.

If we accept that half of the upward revisions in OPEC's Middle Eastern reserves are real then these reserves would have been 500 Gb in 1990. If we subtract the 143 Gb produced since then and we also assume that recovery factors have increased by an average of 10% (giving a reserve increase of 50 Gb) then Middle Eastern OPEC reserves today would be around 400 Gb. This is 300 Gb less than these nations are currently reporting. In 2009 they produced a total of 7.84 Gb which would be 1% of their reported reserves in that year and 2% if their reserves are actually 400 Gb. Compared with the North Sea where the owners currently produce 6% of the remaining recoverable resources per year, a figure of 1% is unrealistically low. If, instead, the Middle Eastern OPEC nations are producing 2% of their remaining recoverable resources per year then this still would mean that the Middle East is that area of the world with the largest remaining crude oil reserves.

In Fig. 11.2 we can see that the *Oil & Gas Journal* and BP regard the upward revisions of OPEC reserves that occurred during the 1980s as new discoveries that occurred in that decade. (See "Added Reserves by OPEC" in the figure.) In total, OPEC's crude oil reserves were increased by 300 Gb. If we accept that half of this upward revision was real then, according to the backdating approach, this oil was not discovered during the 1980s but was actually discovered when the fields were originally found. In Fig. 6.3 we show how such upward revisions should be added to previously reported reserves. In reality, backdating is the process of correcting previously underestimated reserves.

Another important section in our 2003 *Minerals and Energy* article looked at "creaming curves." If one draws a graph where each year's oil discoveries are added to the total oil ever found up to the start of that year (i.e., a graph of cumulative discoveries over time) then one obtains a curve that can indicate the total amount of oil expected to be discovered in an area. However, if exploration for oil is disrupted for one or more years then this can cause a disruption in the creaming curve that can make the curve difficult to use for mathematical extrapolation of the discovery trend.

An oil well that is drilled when searching for oil is called a "wildcat." When drawing a creaming curve, one can look at cumulative discoveries relative to the total number of wildcats drilled rather than relative to each year that passes. In Fig. 11.3 we show how a creaming curve looks for Norway based on the number of wildcats drilled. The cumulative discovery trend looks similar for many other nations. These creaming curves can be described mathematically as "hyperbolic" and we can extrapolate them to determine the total amount of oil that we expect to discover in a region. The creaming curve will eventually level off as it approaches the maximum

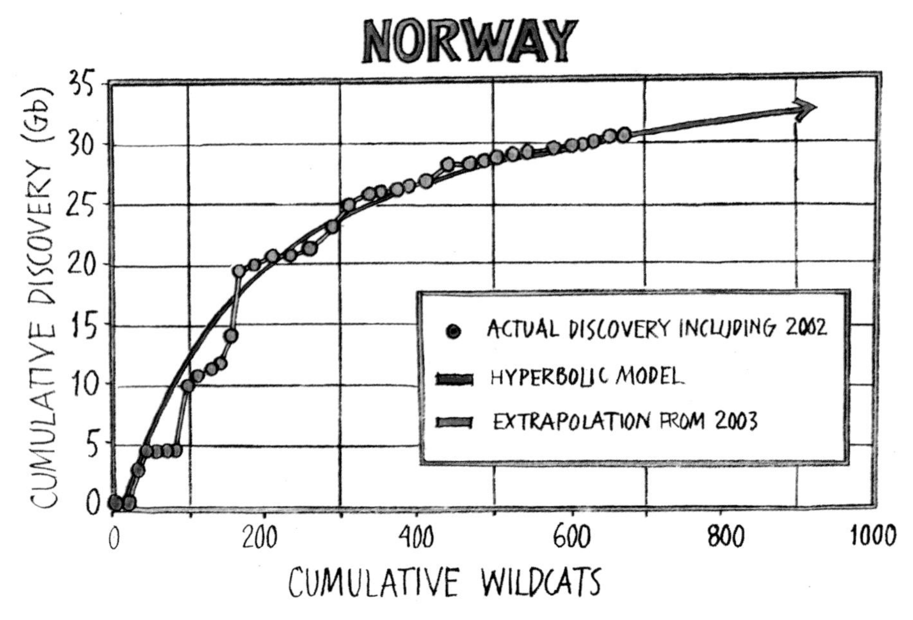

Fig. 11.3 A *creaming curve* for Norway showing actual total/cumulative discoveries including 2002 versus the number of wildcats drilled. A hyperbolic curve has been fitted to the data points [10]

amount of oil that will be discovered, that is, the URR. (Mathematically, we say that the curve approaches the maximum value "asymptotically.") The curve shown for Norway in Fig. 11.3 is based on 670 wildcats and only includes data before 2002. Nevertheless, by fitting a hyperbolic curve to the data points on the graph we can see that Norway's total discoveries are unlikely to exceed 35 Gb of crude oil even if an additional 330 wildcats were to be drilled (so that the total number of wildcats equaled 1,000). The 35 Gb is only slightly more oil than the world currently consumes in 1 year. By 2007 a total of 720 wildcats had been drilled in Norway and the cumulative discoveries then amounted to 31.9 Gb [14]. Our extrapolation of the data from the 670 wildcats drilled before 2002 estimated that 31.7 Gb of oil would have been found after 720 wildcats had been drilled, so our hyperbolic curve model is quite accurate.

Researchers working on Peak Oil are often criticized for not taking into account future discoveries of oil. Fitting hyperbolic curves to discovery data as in Fig. 11.3 is one way that we can predict such future discoveries and determine the URR for a nation, a region, or the world.

To be produced, oil must first be discovered. Therefore, when thinking about future oil production, a very important question is how much crude oil exists in the entire world that can be produced: how much has already been discovered and how much will be discovered in the future, the URR.

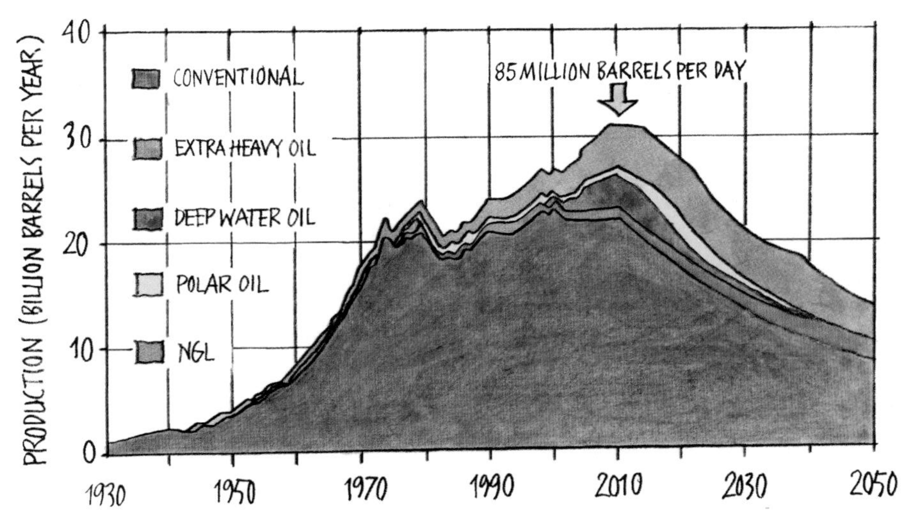

Fig. 11.4 The Oil & Natural Gas Liquids 2003 Base Case Scenario. Historical production of various categories of oil and of natural gas liquids (NGL, liquids from natural gas production) is presented up to 2002 followed by estimates of future production from 2003 to 2050 according to our model [10]

For our 2003 paper in *Minerals and Energy* we predicted the world's URR by analyzing creaming curves for the world's individual nations and then combining this information. This indicated that, worldwide, enough crude oil would be discovered to allow a total of 1,900 Gb of oil to be produced by 2075 [10]. If we examine the various estimates of world URR published since 1965 (Fig. 4 in the article by Aleklett and Campbell [10]) we can see that they fluctuate around a value of 2,000 Gb. When viewing a histogram displaying this collection of mostly similar URR estimates [10], the URR estimate for crude oil of 3,400 Gb made by the USGS in 2000 sticks out like the Empire State Building towering above a crowd of smaller skyscrapers. This is yet one more reason to be critical of the URR estimates published by the USGS.

In the section "The Elephants' Production Rate: Depletion of Remaining Recoverable Resources (DRRR)", Chap. 9, we discussed DRRR (depletion of remaining recoverable resources) which is based on the idea that only a particular fraction of the oil reserves in an oilfield, or a region, can be produced every year. If we apply this idea to the URR estimates for the world's nations we can predict their future production rate. If we then sum together all these future production histories we can estimate the future production history of the entire world. This is the (now widely known) Oil & Natural Gas Liquids 2003 Base Case Scenario figure that we showed in our *Minerals and Energy* article [10] and that is redrawn in Fig. 11.4. New information that became available in the following 8 years has not significantly altered our original 2003 prediction [14].

Our method for analyzing future oil production requires that we make separate estimates for different categories of "oil" such as extra heavy oil, natural gas liquids, crude oil from polar regions, and crude oil from deep water (discussed in Chap. 12). In the analysis we made in the *Minerals and Energy* article using data up to 2002 we concluded that the maximal rate of world oil production—Peak Oil—would be 31 Gb per annum (85 million barrels per day, Mb/d) and would occur in 2010.

When the IEA reports oil production they include "processing gains" which describes the increase in volume that occurs when heavy oil is upgraded to lighter products [15]. The IEA reported oil production for 2010 including processing gains to be 32.1 Gb. Without processing gains it was 31.2 Gb (85.6 million barrels per day) [16]. Our forecast of 31 Gb per annum in 2010 did not include processing gains and so was right on target. It now remains to be seen if oil production will remain at this level, increase, or decline.

International Transport Forum

On June 22, 2007, I received an email regarding a "Round Table Meeting in Paris on 15–16 November at the OECD/International Transport Forum." The email was sent by Stephen Perkins, head of the OECD/ECMT Joint Transport Research Centre in Paris. (ECMT is the European Conference of Ministers of Transport.) He wrote:

> Dear Professor Aleklett,
>
> We are organising a Round Table meeting of experts in November to examine the short and long term outlooks for oil prices and oil supply and the implications for transport policy. Please see the outline attached. I would like to have one session to focus on the Peak Oil—economic resource debate. I read with great interest your paper with Colin Campbell in *Minerals and Energy* and I wonder if you would be able to present the arguments you made there in our round table and prepare a short paper (20–30 pages) to brief participants in detail beforehand.

The article that Colin Campbell and I wrote for *Minerals and Energy* was the first peer-reviewed article ever to discuss Peak Oil and it felt very satisfying that the article had now motivated my invitation to Paris. Accompanying the invitation was a detailed description of the *Round Table Meeting* that was to be held on the theme of "Oil Dependence: Is Transport Running Out of Affordable Fuel?" I read that they had also invited three other people to write papers that would be the basis for discussion at the *Round Table Meeting*. The additional three were Lawrence Eagles, head of the oil industry and markets division of the IEA, Dr. David Greene, Oak Ridge National Laboratory Center for Transportation Analysis, United States, and Professor Kenneth Small, University of California Irvine, United States. At that time Lawrence Eagles was responsible for producing the IEA's annual *Medium-Term Oil and Gas Markets* report that one can obtain if one is willing to pay the IEA €250. David Greene had produced several reports on transport for the US Department of Energy. Kenneth Small is a professor of economics and an expert in transport economics. From his list of achievements one could read that he had acted as an advisor to both the World Bank and the European Commission. I was very pleased to have the opportunity to represent Uppsala University and its academic research for this eminent group. After further discussion with Stephen Perkins we agreed that my written paper would address "Peak Oil and the Evolving Strategies of Oil Importing and Exporting Countries" [17]. We also agreed that I would write an additional paper with the title "Reserve Driven Forecasts for Oil, Gas and Coal and Limits in Carbon Dioxide Emissions" [18].

The Round Table Meeting in Paris

The *Round Table Meeting* on oil prices and supply and the implications for transport policy was one of several such meetings preceding the *First International Transport Forum* in Leipzig in May 2008. It had been arranged that, in association with the Transport Forum, the transport ministers of the OECD nations would meet to discuss current issues. When my papers for the *Round Table Meeting* were ready I sent copies to Sweden's Minister of Transport. Somehow the news of my impending visit to Paris to attend the meeting was given to Sweden's ambassador in Paris, Mats Ringborg. He contacted me to say that he wished to invite me, the authors of the other papers, and the leadership of the International Transport Forum to dinner on the evening before the *Round Table Meeting*. I felt very honored to receive his invitation.

The meeting was held in the IEA's quarters in Paris. The round table was, in reality, oval in shape and could accommodate 40 people seated around it (Fig. 11.5). Experts on oil had been invited to attend from many different corners of the world. One of these was my opponent Michael Lynch but ASPO's secretary at that time, Roger Bentley, had also been invited.

Fig. 11.5 Presentation of the report *Peak Oil and the Evolving Strategies of Oil Importing and Exporting Countries* at the *Round Table Meeting* in Paris on November 15, 2007 [17] (Second from the left at the far end of the table is Michael Lynch and second from the left in the foreground is Roger Bentley. I am standing and the fourth seat to my right is occupied by Paul Portney, the chair for the meeting)

The chair for the meeting was Professor Paul Portney from the University of Arizona in the United States. After his introductory remarks, it was time for the papers that had been prepared for the meeting to be presented by their authors. The papers were:

- Price instability, the determinants of oil prices in the short term, Lawrence Eagles, head of the oil industry and markets division of the IEA.
- Peak-oil and the evolving strategies of oil importing and exporting countries, Professor Kjell Aleklett, Uppsala University, Sweden.
- The determinants of oil prices and supply in the long term, Dr. David Greene, Oak Ridge National Laboratory, Center for Transportation Analysis, United States.
- Long-run trends in transport demand, fuel price elasticities, and implications of the oil outlook for transport policy, Professor Kenneth Small, University of California, Irvine, United States.

The meeting lasted for 2 days so I cannot describe it in detail here. In any case, our papers are still available online to read as is the summary written by Stephen Perkins, the head of the Joint Transport Research Center of the OECD and the International Transport Forum [19]. The next significant event for increasing awareness of Peak Oil was the International Transport Forum in Leipzig that followed the *Round Table Meeting*.

International Transport Forum in Leipzig

During the past half-century Europe's ministers of transport have met regularly to discuss and make decisions regarding an integrated transport system. However, increased globalization and international traffic made it desirable to broaden this discussion to include other nations. To satisfy this need, the OECD's 28 member nations decided to create a new organization within the OECD called the International Transport Forum (ITF). Since its formation they have also invited several non-OECD nations to participate and today 52 nations are members of the ITF. In practical terms the ITF functions as a think-tank for transport policy. Every year they organize a forum in Leipzig. The *Round Table Meeting* in Paris was held in preparation for the very first *International Transport Forum*. The theme for that forum was transport and energy. It was the energy part of the theme that led the OECD to contact the Global Energy Systems group at Uppsala University. I was very pleased that our entire research group was invited to the event in Leipzig. We were all quite excited during the trip to Leipzig as we contemplated how our work on Peak Oil might be received.

The first day in Leipzig consisted of four workshops and an open forum at which my research group had been invited to speak. We had 90 min to present our research under the theme of "The Future of Energy Supplies." My PhD students Bengt, Kristofer, and Mikael were very pleased to be able to present their research. However, the highpoint of the Transport Forum was the discussion between political decision makers and key people in the transport sector that occurred on the second day.

Dr. Rajendra Pachauri is the chairman of the United Nations Intergovernmental Panel on Climate Change (UN IPCC). He was the first to speak at the discussion on the second day and he delivered the expected message on global warming. He presented many well-known statistics. During the twentieth century the sea level rose by 17 cm, snow-cover on land decreased markedly, and it became warmer. He also emphasized that the expected changes in climate will affect us in various ways. After his presentation I had an opportunity to exchange a few words with him and I gave him copies of the papers I had written for the *Round Table Meeting* in Paris (and the ITF). One of the two papers described how there are problems with the carbon dioxide emissions scenarios used by the UN IPCC in their predictions of future climate change but this was a topic that Pachauri did not want to discuss with me (the emissions scenarios will be discussed in Chap. 17).

At the discussion on the second day of the forum in Leipzig we also heard a presentation by Yvo de Boer who was the chief negotiator for shaping a revised Kyoto Protocol. He emphasized the importance of regulation.

As examples he mentioned the regulations regarding emissions by industry and from automobiles. It was only when regulations were put in place that behavior regarding emissions changed.

Pekka Himanen is a professor from Helsingfors in Finland and one of the world's young, new thinkers in economics. However, he did not discuss economics as much as dignity. The world's inhabitants should regard themselves more as members of one family because, of course, we would always want to help a member of our own family.

The discussion included a panel debate between politicians and representatives of industry. There is much that could be said about the panel discussion but I want to highlight the most sensational statement of the proceedings that came from Norway's then minister of transport Liv Signe Navarsete. She told us that we may need to adapt to a future of economic decline rather than growth. She recognized, indirectly, that Peak Oil is imminent. The panel discussed the price of oil that, in May 2007, was still rising and they discussed the possibility that the price might rise to US$300 per barrel. The CEO of Airbus, Thomas Enders, said without wavering that "If the price per barrel rises to US$200 it would mean the collapse of the entire aviation industry. No aviation company can cope with that price." Coincidentally, my research group was, at that time, studying the effects of Peak Oil on aviation. Enders' statement had a very significant impact on my view of future oil prices.

The high point of the second day of the *Transport Forum* was Chancellor Angela Merkel's speech. Her message was that, when we look to the future, we must begin a new chapter. She noted that, although all the world's people should have a right to the same opportunities in life, there is a global shortage of energy. She stated that action on global issues should be directed by the United Nations. She did not mention Peak Oil but her statements agree well with the concluding words in my paper titled, "Reserve Driven Forecasts for Oil, Gas and Coal and Limits in Carbon Dioxide Emissions" [18]: "Climate change is current with more change to come, and furthermore, climate change is an enormous problem facing the planet. However, the world's greatest problem is that too many people must share too little energy."

The third day of the *Transport Forum* consisted of summaries, conclusions, and thanks to all the participants. Personally, I can state that the email I received from the OECD's Stephen Perkins and the invitation to the *International Transport Forum* had a very great influence on me. It meant recognition that the research we were (and are) conducting at Uppsala University is significant and influential, and that we had become actors at an international level. On the invitation to the forum it was written that Peak Oil would be discussed but when the panel debate between ministers and industry representatives had ended there had been no discussion of the issue. A vice-president for Shell who represented the oil industry at the discussion, stated that oil production could not satisfy demand. However, such a statement is circuitous and gives far too weak a warning about the impact of Peak Oil. Also, the fact that a CEO in the aviation industry could state that oil at $130 per barrel is too expensive and that $200 per barrel will kill the aviation industry should be seen as a cry for help rather than as recognition that Peak Oil is occurring. The representatives of the automobile industry should have directly addressed the Peak Oil issue but, instead, it was camouflaged in discussions of CO_2 emissions, energy-efficient products, future hybrid and electric vehicles, and biofuels. The closest that any politician came to acknowledging Peak Oil was the statement by Norway's minister for transport on negative economic growth, an idea that was determinedly dismissed on the last day by the World Bank's representative.

The authority that most definitely should have raised the issue of Peak Oil at the forum was the IEA but it had not progressed to that point at that time. At the conclusion of the forum, the former CEO of the IEA, Claude Mandil, summarized the panel discussion and the entire meeting. He declared that the future must be global and integrated and that everything will be expensive. He even stated that a high oil price could have a positive impact on the future but he said nothing about resource shortages. Instead he described how high oil prices were caused by a combination of other factors. I have mentioned previously that the IEA's current prediction for oil production in 2030 must be decreased. However, Mandil only mentioned global solutions to oil demand, infrastructure solutions such as greater efficiency and that the IEA needed better data so that their future prognoses could be more certain. It is remotely possible that Mandil's failure to mention Peak Oil could be explained by poor quality data in the IEA's databases at that time.

Climate change was discussed at the forum and everyone wanted to see emissions of CO_2 reduced. However, the tone of the discussion was mainly that the OECD was doing a fine job of reducing emissions and that it was others that must now make the greatest reductions. Pekka Himanen drew attention to this arrogant attitude by comparing it to that of a man who is photographed coming out of the bedroom of his neighbor's wife but who nevertheless denies having been in there.

I left Leipzig with the feeling that we lack leadership on the future's most serious issues: energy and climate. There was no one who would stand up and say, "I have a dream." Instead, only nightmares were discussed in Leipzig. In the future, Peak Oil will be a politician's best friend because Peak Oil will reduce CO_2 emissions from oil even if the world's nations cannot agree on how to reduce fossil fuel consumption.

World Energy Outlook 2008

In September 2004, at the *10th Annual Energy Conference of The Emirates Center for Strategic Studies and Research* in Abu Dhabi, I met the chief economist for the IEA, Fatih Birol. We discussed the IEA's *World Energy Outlook 2004* report [20] that was to be released 2 months later. Fatih Birol remarked that I and ASPO would probably be quite pleased. My interpretation of his remark was that Peak Oil would be mentioned in the report. When *WEO 2004* was released we heard from the *New York Times* and other sources that "The report predicts that world oil demand will grow about 50%, to 121 million barrels a day, by 2030," [21] in other words, no sign of Peak Oil.

If one reads Chap. 3 of *WEO 2004* in detail then it becomes apparent that the IEA begins to discuss crude oil reserves and resources in the same way that Colin Campbell and I did in our article, "The Peak and Decline of World Oil and Gas Production" [10]. They discuss backdating, creaming curves, and Peak Oil. What was lacking was a cipher or decoding device that could, in plain language, translate what was described in *WEO 2004*. Therefore, I wrote *The Uppsala Code* and published it on ASPO's website [22].

The IEA was established by the OECD. Because it was the OECD that gave me the task of writing a report on Peak Oil for the *Round Table Meeting* in Paris held in late 2007, I had great hopes for a breakthrough for the Peak Oil issue in the *World Energy Outlook* report for 2008 [23]. It is true that the *WEO 2004* prediction for 2030 of oil production at 121.3 million barrels per day (Mb/d) had been reduced in *WEO 2006* to 116.3 Mb/d and was reduced again in *WEO 2008* to 106.4 Mb/d. These were steps closer to the truth but *WEO 2008* still contained no hint of Peak Oil.

The *WEO 2008* report was a turning point for the IEA. For the first time, they did not follow their usual practice of estimating future oil production based on projected economic growth and the oil it would require. Instead they made a "bottom-up" analysis of production by examining currently producing oilfields, oilfields that were known but not yet in production, and estimates of future oilfield discoveries. In *WEO 2008* the IEA had, quite correctly, estimated that total crude oil production from fields in production in 2008 would decline by 6% per year. However, when they estimated future crude oil production from known but yet to be exploited oilfields they exaggerated the rate of production increase. The same is true for the undiscovered fields that they predicted would be found. I saw that there was need for a detailed analysis of the estimates that had been made in *WEO 2008*. We were able to make this analysis by extracting data from the graphs the IEA had published in *WEO 2008*. Our analysis of *WEO 2008* is titled "The Peak of the Oil Age." It was accepted for publication (and became available online) in the journal *Energy Policy* only days before the release of *WEO 2009*. The article was officially published in printed form in 2010 [24]. It clearly showed that the IEA's estimates of future oil production in *WEO 2008* were unrealistic. A decisive part of our analysis in "The Peak of the Oil Age" was calculation of rates of depletion of remaining recoverable resources (DRRR: see the section "The Elephants' Production Rate—Depletion of Remaining Recoverable Resources (DRRR)", Chap. 9).

In "The Peak of the Oil Age" analysis we used the exact same number of fields and the same volumes of oil reserves as in *WEO 2008*. However, when examining liquids produced during natural gas production (natural gas liquids, NGL) the IEA had neglected to compensate for the lower energy content of NGL compared to crude oil (see the section "NGL, Natural Gas Liquids", Chap. 10). Therefore, we recalculated these in "barrel of oil equivalents." The result was a reduction in NGL production from 19.8 to 14.9 Mb/d in 2030. We also used realistic rates of production of crude oil from fields (based on our research into DRRR) and realistic rates of production from Canada's oil sands. When our recalculations were complete we observed that a "realistic" rate of oil production in 2030 (possible but still very optimistic) was actually 75.8 Mb/d rather than the 106.4 Mb/d predicted by the IEA. For those who are interested, "The Peak of the Oil Age" article is available from the journal *Energy Policy* or is freely available from my research group's website [24]. Here I show only the final figure from this article (see Fig. 11.6).

In Fig. 11.6 one can see that the Oil Age's maximum ever rate of production can occur at any time between now and 2014. The peak of the Oil Age is truly in sight!

The "Whistleblower"

On November 9, 2009 the journal *Energy Policy* accepted our "Peak of the Oil Age" article for publication. On that same day the front-page headline of the *Guardian* newspaper in the United Kingdom was "Key Oil Figures Were Distorted by US Pressure, Says Whistleblower." The subtitle was, "Watchdog's Estimates of Reserves Inflated Says Top Official." The "watchdog" they were referring to was the IEA [25]. When I was in Paris 2 years earlier for the *Round Table Meeting* (see above) I was contacted by a high-ranking official in the IEA who told me much the same thing. The official's comments were partly what inspired me to write the "Peak of the Oil Age" article.

The future oil production prognosis that the IEA presented in *WEO 2009* was identical to what they published in 2008 so our "Peak of the Oil Age"

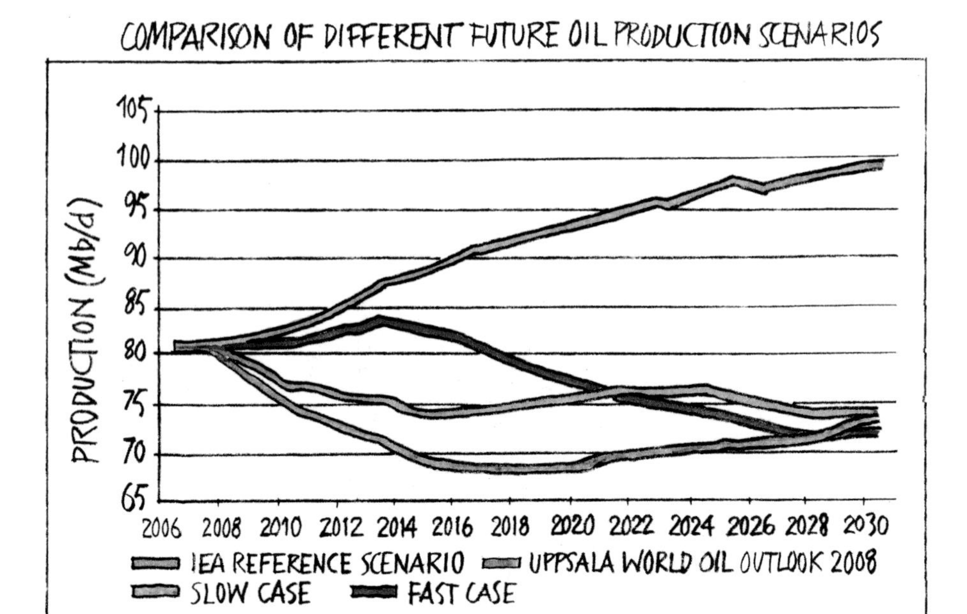

Fig. 11.6 Comparison of different future oil production scenarios. The Uppsala world oil outlook is based on normal speeds of oilfield development and rates of oil production and the slow and fast cases feature alternative development speeds for the fields yet to be developed with all other factors being equal [24] (In the figure, the volume of NGL has been converted to barrel of oil equivalents)

article analyzing *WEO 2008* was still directly relevant. Terry Macalister, who wrote the *Guardian*'s story on "the whistleblower," contacted me and I was able to explain that our article supported the whistleblower's claims. Terry then wrote a second article for the *Guardian* under the headline, "Oil: Future World Shortages Are Being Drastically Underplayed, Say Experts." "Swedish academics slate IEA's report as 'political document' for countries with vested interest in low prices—Oil production 'likely to be 75 m barrels a day rather than 105 m'" [26]. Terry's second article showed our graph predicting future oil production side by side with the prognosis from the IEA. (Fig. 11.6 is a composite of the two figures.)

The news of our "Peak of the Oil Age" article spread rapidly around the world. Canada's most widely read newspaper, the *Globe and Mail*, published a well-written article with the title, "Is the World Awash in Oil?" [27] I would like to quote the following text from that article to summarize "The Peak of the Oil Age:"

Mr. Aleklett and his co-authors use essentially the same data as the IEA but interpret it in a different way. They and the IEA are in agreement on most issues. They all agree that the oil fields now in production are quickly running out of puff (that is, their "depletion rates" are high). They all agree on the estimated oil volume in the fields yet to be developed and to be discovered. Where they differ is on productivity of the new fields—the ones that, according to the IEA, will more than fill the gap as the old fields amble off to reserve heaven.

History, Mr. Aleklett says, shows that the new fields, generally smaller, are less productive than old ones—note the virtual freefall in production rates from the North Sea fields, which reached peak output in 2000. Another reason is development pace, or lack thereof. The yet-to-be-developed reserves in the WEO report cover 1,874 fields of various sizes that would have to come into production in the next 20 years.

"That is something like eight fields per month coming on stream," Mr. Aleklett's report notes. "Even if the oil exists, it is questionable whether the necessary investment needed to produce such a rapid pace of development can be achieved in a timely fashion.

His conclusion is shocking: Production in 2030 will be about 76 million barrels a day. That's about one-third less than the WEO's figure and some ten million less than current production. Peak Oil, he says, is already here.

The International Energy Agency Does It Again

In this section we examine a large amount of numerical data on oil production. If you would like avoid this analysis of the IEA's future oil production scenarios then you can skip forward to the next section (Executive Summary of the Analysis of Crude Oil Production in *World Energy Outlook 2010*) where the results of our analysis are summarized.

Our "Peak of the Oil Age" analysis of the oil production scenarios in *WEO 2008* was published almost simultaneously with the release of *WEO 2009* by the IEA. As expected the IEA's 2009 WEO report presented the same prognosis as in 2008. However, when *WEO 2010* was released I read it with great anticipation. Had they responded to our criticism of *WEO 2008*? Had they corrected the errors we found?

Surprisingly, the answer was "no." The old oil production prognosis remained under the new title of the "Current Policies Scenario" although it had been adjusted down by a few percent. In their "New Policies Scenario" the IEA's predicted oil production in 2030 had been reduced by 12 Mb/d but there was no sign of a peak in oil production up to 2035.

WEO 2004 predicted oil production in 2030 of 121.3 Mb/d. During the 6 years since then the IEA has reduced its 2030 prognosis by 25 Mb/d which is more than the production from two Saudi Arabias. However, to approach

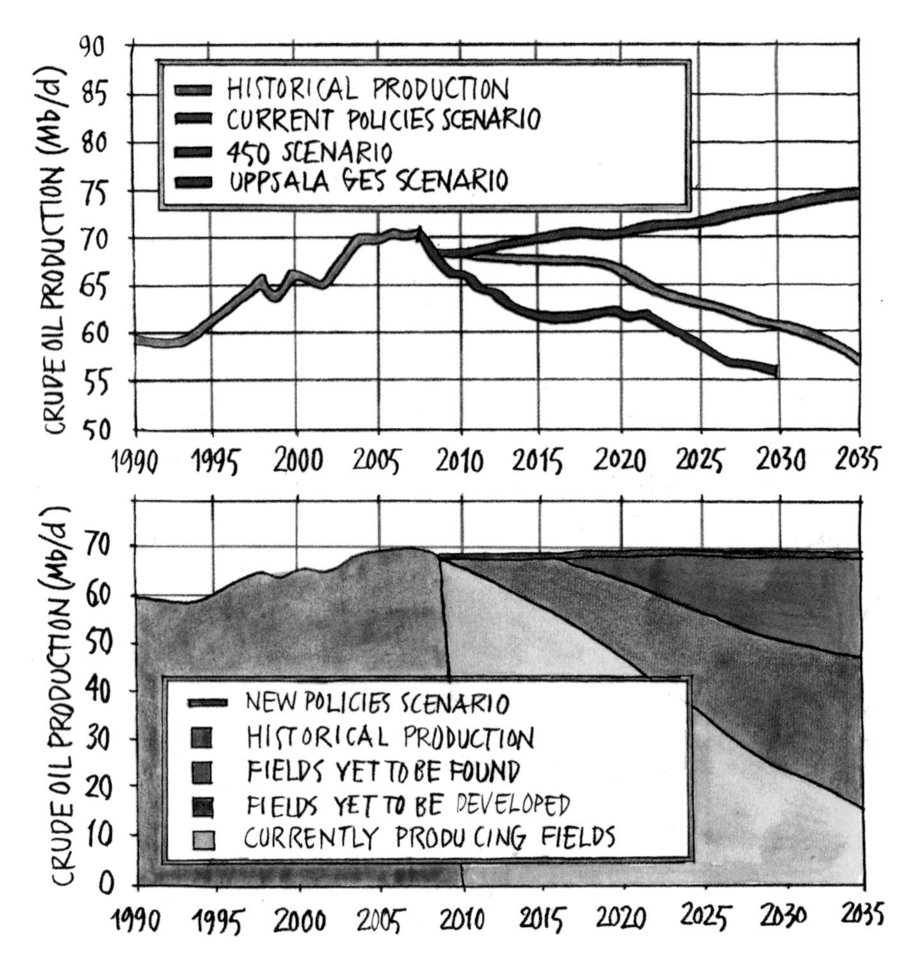

Fig. 11.7 The *upper part* of the figure shows the three scenarios discussed in *WEO 2010*. The "Current Policies Scenario" is nearly identical to that presented in *WEO 2008* whereas the "New Policies Scenario" and the "450 Scenario" are new. The Uppsala Global Energy Systems Scenario, "UGES," shows the crude oil part of the oil production scenario we presented in our analysis of *WEO 2008* [24]. The *lower part* of the figure shows the components that comprise the crude oil production of the "New Policies Scenario" in *WEO 2010* [4]

reality they still needed to remove one or more additional Saudi Arabias. The climate change negotiations in Copenhagen had inspired the IEA to present a new scenario named "450 ppm." The crude oil production predicted under the various scenarios is shown in Fig. 11.7. (The "450 ppm" scenario is named "450 Scenario" in that figure.) Future crude oil production as predicted in 2009 by the Global Energy Systems research group at Uppsala University is also presented in Fig. 11.7.

In the *WEO 2010* report there is a figure (Fig. 3.19 in the report) that illustrates from where the IEA believes the oil for its "New Policies Scenario" will come. The sources of crude oil are divided into currently producing oilfields, oilfields yet to be developed, and oilfields not yet found. In Fig. 11.7 we show information derived from that part of the figure concerning crude oil production. In *WEO 2010* crude oil production in the near future from currently producing fields is a little greater than was predicted in *WEO 2008* but soon the expected decline begins. *WEO 2008*'s prediction of production from oilfields yet to be developed was unreasonably high at 29 Mb/d in 2023. In *WEO 2010* the prediction for 2023 had decreased by 10 Mb/d. So 29 − 10 = 19 Mb/d of production in 2023 from oilfields yet to be developed. This is still unrealistic but *WEO 2010* now also predicted around 30 Mb/d from these oilfields in 2035. This is pure fantasy.

In *WEO 2010* the IEA predicted that oilfields yet to be developed will produce around 170 Gb of oil by 2035. According to our research, a production level of 30 Mb/d in 2035 could only be achieved if large oil reserves still remained in that year (because only a certain fraction of reserves can be produced in 1 year). Our research shows that these reserves would need to total 170 Gb in 2035 to permit this production. If those remaining reserves are added to the 170 Gb to be produced up to 2035 from oilfields yet to be developed then the total current reserves in oilfields yet to be developed would need to be 340 Gb. In *WEO 2008* the IEA stated that there were 257 Gb in oilfields yet to be developed so we can see that a production rate of 30 Mb/d in 2035 is completely unrealistic.

For production from oilfields not yet found, the IEA uses the same unrealistic numbers in *WEO 2010* as it did in *WEO 2008*. Obviously the IEA has not understood the lesson from our "Peak of the Oil Age" article. Because repetition can assist the learning process I repeat the lesson below.

In the autumn of 2010 I contacted Leif Magne Meling from Statoil to see if he had an updated version of his analysis of "recovery factors" (see the section "Recovery Factor: The Amount of the OOIP That Can Be Produced", Chap. 6). He told me that the version he had shown at the meeting in Stockholm was still current but he then sent me information from an additional presentation, a statistical analysis of how many oilfields were discovered each year and the average size of each field [25]. I have presented this information graphically in Fig. 11.8.

The number of fields in Leif's analysis was approximately 13,800. During the first 20 years of the twentieth century an average of ten oilfields was found per year. By the 1930s and until the end of the Second World War the number of fields discovered increased to 25 per year. After the Second World War the oil industry expanded rapidly. The 1960s was the golden age of oil discovery when 200 oilfields were found per year. As I described in

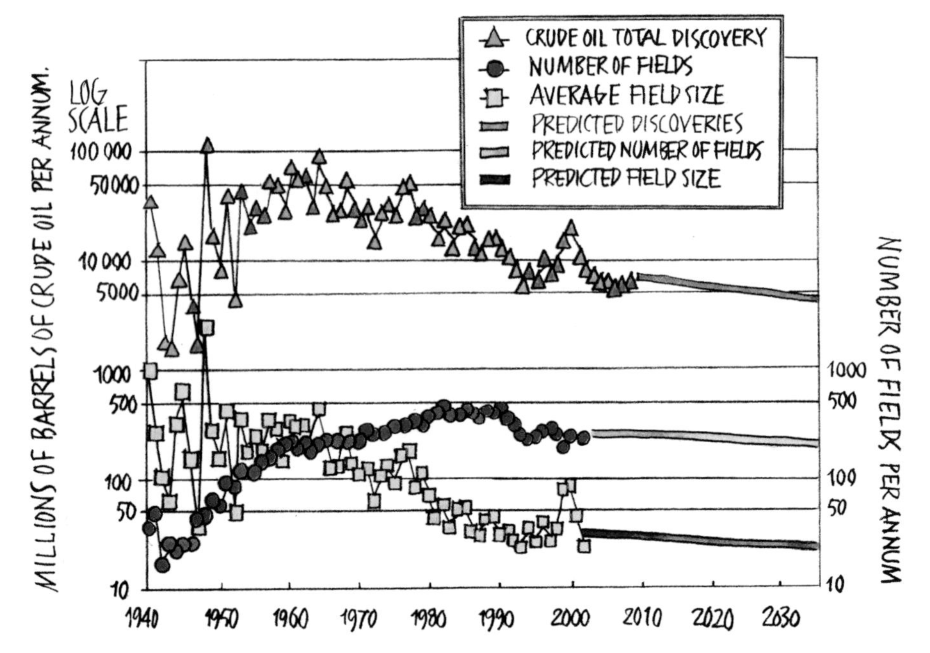

Fig. 11.8 Yearly discoveries of crude oil have been divided up into the numbers of fields found per year and the average size of the discovered fields per year [28]. Both these quantities show declining trends so future volumes of oil to be discovered must also decrease. The numbers of fields found per year and the average size of the discovered fields per year will continue to vary from year to year as has been seen in the past. However, these fluctuations will be around declining trends

Chap. 3, an average of 56 Gb per year of reserves was discovered during the 1960s. This gives an average discovered field size of 280 million barrels. During the 1980s new technology was developed that assisted oil discovery and the number of fields discovered per year grew to 360. This was, in fact, an average of one field discovered per day for 10 years. During the 1990s the number of fields discovered per year decreased to 240 per year and today there is no doubt that we are on a declining trend. The data in Leif's analysis ended at 2002.

If we now wish to estimate the number of new oilfields that will be discovered between 2010 and 2040 we can look at Leif's statistics and start with 240 per year in 2000. In the 1990s the rate of oilfield discovery was decreasing by approximately 1% per year. If we begin with 240 new oilfields discovered in 2000 then by 2040 the rate of oilfield discovery will have declined to 160 oilfields per year, and the total number discovered between 2010 and 2040 will be 5,800 fields.

The average size of discovered oilfields is also now clearly trending downwards. The exception was the years around 2000 and one can ask what caused that temporary increase. The Soviet Union collapsed at the beginning of the 1990s and a number of new nations were then formed. The Soviets had known that oil existed in these nations but they had no suitable export route for the oil at that time. Building pipelines through Afghanistan had been one idea for facilitating export but Osama Bin Laden had stopped that with the support of the United States. All the new, post-Soviet nations that had the potential to export oil opened themselves up for exploration and a number of large fields were subsequently found. At the same time, new deep water oil production technologies made exploration in deeper water worthwhile. These two factors meant that the average size of discovered fields increased for a number of years before declining back to the preceding trend (see "average field size" in Fig. 11.8).

New seismic exploration technology is making it possible to find smaller fields. If we assume that the average size of a discovered field in 2010 was 30 million barrels and that this is trending downwards by 1% per year, then the average size of a discovered field in 2040 will be around 20 million barrels. Under these conditions the oil industry will find 140 Gb of oil reserves during the 30 years between 2010 and 2040. In the 20 years to 2030 we can assume that they will discover 100 Gb of reserves and this is the volume that it is possible for them to put into production by 2035 (because 5 years—and usually longer—is often required to put a newly discovered oilfield into production; see below). When we estimate the numbers of fields that will be put into production up to 2040 there is one additional factor to consider. It is that isolated small fields are not profitable to exploit because of the production infrastructure required. In contrast, small fields adjacent to larger fields can be profitable. The number given should therefore be treated as an upper limit.

The IEA estimates that production in 2035 from oilfields yet to be found will be 21.8 Mb/d (8.0 Gb per year). With a maximum rate of DRRR of 10% this means that reserves of 80 Gb are required in 2035 to produce oil at this rate. The IEA estimates that, by 2035, the yet to be found oilfields will have produced 92 Gb so this means that $92+80=172$ Gb of reserves must be discovered between 2010 and 2030. However, the trend indicates that only 100 Gb will be discovered. In *WEO 2008* the IEA estimated that the volume of oil reserves in fields yet to be found would be 110 Gb by 2030. Therefore, it is obvious that the IEA is presenting unrealistic production estimates from these fields for the period up to 2035.

In the article, "The Peak of the Oil Age," we used a "maximum depletion model" to estimate production from future discoveries [24]. The model assumes that it takes some years to reach maximal production from a discovered oilfield. By studying fields that have recently been brought into production we can see that oil producers commonly plan for a maximum annual production equivalent to 10% of a field's estimated URR. The inevitable decline in production normally begins before 50% of a field's URR is produced. For the model we used a figure of 45% of URR produced for the start of decline. We set the rate of production decline after this point at 10% per year.

Fields that are discovered in proximity to already producing fields can be put into production quite rapidly whereas fields in new areas require more than 5 years for production to begin. In the maximum depletion model we made the assumption that, on average, a field would be put into production only 4 years after discovery. Using the assumptions in our maximum depletion model and the assumptions we described above regarding discovered volumes we see that production from fields yet to be found follows the IEA's prognosis until 2025 when it reaches 13 Mb/d. However, then the production levels off and subsequently declines so that we see a maximal production of around 11 Mb/d in 2035. In contrast, in *WEO 2010* the IEA sees twice that rate, 22 Mb/d, being produced in 2035 from fields yet to be found. This is a difference in production rate equivalent to Saudi Arabia's current production.

Executive Summary of the Analysis of Crude Oil Production in *World Energy Outlook 2010*

To summarize the detailed analysis of future oil production that we presented in the section "The International Energy Agency Does It Again" above we can say that, in *WEO 2010*, the IEA has once again presented production estimates that are unrealistic. Part of the baggage they have carried over from *WEO 2008* is a scenario called the "Current Policies Scenario." For 2035, this scenario predicts total world oil production of 107 Mb/d. However, they quickly abandon this scenario and introduce a new one, the "New Policies Scenario" that predicts only 99 Mb/d of oil production in 2035. Nevertheless, Peak Oil is still not part of the picture presented by the IEA. They have also introduced a scenario that represents what is needed for a future maximum atmospheric CO_2 concentration of 450 ppm. With this "450 ppm" scenario oil production in 2035 is down to 81 Mb/d which means that the peak of oil production will have been passed (because current production is ~87 Mb/d).

The issue that will be decisive for our future is expected crude oil production. The review of the IEA's predictions that we presented in the section "The International Energy Agency Does It Again" shows that, once again, the IEA has exaggerated future oil production. The IEA's chief economist Fatih Birol has said on a number of occasions that in 2030 we will need to have brought on line new oil production equivalent to four times Saudi Arabia's capacity. In *WEO 2010* the IEA believes this new oil production can occur but our analysis shows that two of those new Saudi Arabias are exaggerated expectations. If the IEA made use of the information that we have published in peer-reviewed scientific articles then they would no longer be able to sweep Peak Oil under the mat.

Reflections Regarding "The Peak of the Oil Age"

In recent years while lecturing on Peak Oil there have been many times when I have observed one or more people in the audience experience what we call their "peak moment." That is the moment when they finally, and usually suddenly, become convinced that Peak Oil is reality. (Often a rather shocked expression spreads over their face and they look very uneasy.) In this chapter I have described my Peak Oil journey which started with including oil production in my university lectures on energy and has taken me as far as leading a research group totally focused on Peak Oil and the limits of the other fossil fuels. When I began my journey I was already an established research physicist working in another area so it was only natural for me to seek to publish our Peak Oil research in peer-reviewed scientific journals. It is important to note that it was one of these scientific publications that stimulated the interest of the OECD in Peak Oil. To date we have published over 20 peer-reviewed articles on Peak Oil issues.

As for when Peak Oil will occur there are many influential factors at play. However, from a longer-term perspective, it is the rate at which oil can flow through the pores in rock underground that sets the upper limit for the rate of production of crude oil. The prognosis that Colin Campbell and I published in 2003 gave 2010 as the year in which oil production would peak. In the PhD thesis that Fredrik Robelius defended in 2007 four future scenarios were presented [29]. The scenario labeled as "worst case" describes a production plateau of around 85 Mb/d from 2005 until 2012 whereas that labeled "best case" has maximal oil production at 93 Mb/d occurring around 2018. In our "Peak of the Oil Age" article we analyzed the data presented by the IEA and concluded that Peak Oil would occur sometime between 2008 and 2014.

Whether we are currently at Peak Oil or will be in a few years has little significance for how we need to plan our future. In the *Hirsch Report* prepared for the US Department of Energy [30] they calculated that, to avoid very serious negative consequences from declining oil production, one must begin adaptation 20 years before decline begins. Therefore, one conclusion from our research results is that we should have begun preparing for, and adapting to, oil decline more than a decade ago.

Finally, I must return to the comment made by Sheikh Yamani that, "The Stone Age did not end for lack of stone, and the Oil Age will end long before the world runs out of oil." There is debate about why the Age of the Vikings came to an end but my impression of the arguments is that eventually the Vikings' ships and weapons became outmoded compared to the newer better weapons and fortresses of their intended victims. The same argument applies to the Stone Age. Stone became meaningless as an instrument for exercising power when people began to use various metals. The Viking Age lasted for 300 years and it looks as though the Oil Age will also span 300 years, a relatively short event in the course of human history.

The question Sheikh Yamani needs to answer is which energy resource can become so much more useful than oil that we are no longer motivated to produce oil from oilfields. If we compare the energy content per unit volume of oil to that of renewable sources of energy then nothing can replace oil. Uranium and thorium have higher energy contents per unit volume than oil but it is unrealistic to think that nuclear energy could replace oil's role in transport and agriculture. The technological progress that we have experienced from the beginning of the nineteenth century until today was founded on our development of methods to exploit cheap, abundant energy resources. The dramatic technological progress that we have seen since the Second World War has been based principally on oil. The Oil Age will end because oil is running out, not because oil is replaced with some other energy source. If Sheikh Yamani meant by his comment that the Oil Age will end while producible oil remains then he was wrong. However, if he meant that the Oil Age will end while most of the world's oil remains unproduced because we could not extract it then he is correct. But that does not mean that we have something superior to oil waiting for us just around the corner.

References

1. The end of the oil age. Economist, 23 Oct 2003. http://www.economist.com/node/2155717 (2003)
2. Birol, F.: Outside view: "we can't cling to crude: we should leave oil before it leaves us". The Independent, 2 Mar 2008, http://www.independent.co.uk/news/business/comment/outside-view-we-cant-cling-to-crude-we-should-leave-oil-before-it-leaves-us-790178.html (2008)

3. World crude oil production 1960–2009. Energy Information Administration, 2011. http://www.eia.doe.gov/aer/txt/ptb1105.html (2011)
4. WEO: World Energy Outlook 2010. International Energy Agency, November 2010. http://www.worldenergyoutlook.org/2010.asp (2010)
5. McFarland, E.L., Hunt, J.L., Campbell, J.L.: Energy, Physics and Environment. Wuerz Publishing Ltd., Winnipeg (1994). ISBN 0-920063-62-4
6. Aleklett, K.: Ryssland kan strypa vår energi, DN debatt, Dagens Nyheter (in Swedish), 5 Aug 1996
7. Söderbergh, B., Jakobsson K., Aleklett, K.: European energy security: an analysis of future Russian natural gas production and exports. Energy Policy **38**, 7827. http://www.tsl.uu.se/uhdsg/publications/Russian_Gas_Article.pdf (2010)
8. International Workshop on Oil Depletion, Uppsala, 23–25 May 2002. http://www.peakoil.net/aspo-conferences/iwood-2002-uppsala (2002)
9. A Nation at Risk: The Imperative for Educational Reform, April 1983. http://www2.ed.gov/pubs/NatAtRisk/index.html (1983)
10. Aleklett, K., Campbell, C.: The peak and decline of world oil and gas production, Minerals and Energy—Raw Materials Report, **18**, 5–20. http://www.ingentaconnect.com/content/routledg/smin/2003/00000018/00000001/art00004, http://www.peakoil.net/files/OilpeakMineralsEnergy.pdf (2003)
11. Oil & Gas Journal: http://www.ogj.com/index.html (2012)
12. BP Statistical Review of World Energy, BP, June 2011. http://bp.com/statisticalreview (2011)
13. U.S. Geological Survey: World Petroleum Assessment 2000—Description and Results. U.S. Department of the Interior, U.S. Geological Survey. http://pubs.usgs.gov/dds/dds-060/ (2000)
14. Campbell, C.J., Heapes, S.: An Atlas of Oil and Gas Depletion. Jeremy Mills Publishing Limited, Huddersfield (2008). ISBN 1-906600-42-6
15. Oil refining involves the upgrading of heavy oil into lighter products, which reduces their density and gives rise to an increase in volume for a given amount of energy content. Processing gains as a share of overall supply increase slightly in all three scenarios as a result of more upgrading of oil feedstocks in response to the shift in demand towards lighter products such as diesel and gasoline
16. International Energy Agency: Oil Market Report, 15 Mar 2011. http://omrpublic.iea.org/currentissues/full.pdf (2011)
17. Aleklett, K.: Peak oil and the evolving strategies of oil importing and exporting countries. Discussion Paper No. 2007–17, December 2007. Joint Transport Research Center, Paris, France. http://www.internationaltransportforum.org/jtrc/DiscussionPapers/DiscussionPaper17.pdf (2007)
18. Aleklett, K.: Reserve driven forecasts for oil, gas and coal and limits in carbon dioxide emissions, Discussion Paper No. 2007–18, December 2007. Joint Transport Research Center, Paris, France. http://www.internationaltransportforum.org/jtrc/DiscussionPapers/DiscussionPaper18.pdf (2007)
19. Oil dependence: is transport running out of affordable fuel?. OECD/ITF Joint Transport Research Centre Discussion Papers 2008/5. http://www.internationaltransportforum.org/jtrc/roundtables.html (2008)

20. WEO: World Energy Outlook 2004. International Energy Agency, Oct 2004. http://www.iea.org/textbase/nppdf/free/2004/weo2004.pdf (2004)
21. Mouawad, J.: Oil demands can be met, but at a high price, energy agency says. The New York Times, 28 Oct 2004. http://www.nytimes.com/2004/10/28/business/28oil.html?ex=1100004658&ei=1&en=230f4d77af7d6ab2 (2004)
22. Aleklett, K.: International Energy agency accepts peak oil—an analysis of Chapter 3 of the World Energy Outlook 2004. http://www.peakoil.net/uhdsg/weo2004/TheUppsalaCode.html (2004)
23. WEO: World Energy Outlook 2008. International Energy Agency, November 2008. http://www.iea.org/textbase/nppdf/free/2008/weo2008.pdf (2008)
24. Aleklett, K., Höök, M., Jakobsson, K., Lardelli, M., Snowden, S., Söderbergh, B.: The peak of the oil age—analyzing the world oil production reference scenario in World Energy Outlook 2008. Energy Policy **38**(3), 1398–1414 (accepted 9 Nov 2009, Available online 1 Dec 2009, http://www.tsl.uu.se/uhdsg/Publications/PeakOilAge.pdf) (2010)
25. Macalister, T.: Key oil figures were distorted by US pressure, says whistleblower. The Guardian, 9 Nov 2009. http://www.guardian.co.uk/environment/2009/nov/09/peak-oil-international-energy-agency (2009)
26. Macalister, T.: Oil: future world shortages are being drastically underplayed, say experts. The Guardian, 12 Nov 2009. http://www.guardian.co.uk/business/2009/nov/12/oil-shortage-uppsala-aleklett (2009)
27. Reguly, E.: Is the world awash in oil? Globe and Mail, 12 Nov 2009. http://www.theglobeandmail.com/globe-investor/awash-in-oil/article1360337/ (2009)
28. Meling, L.M.: Data available at: How and for how long it is possible to secure a sustainable growth of oil supply (PowerPoint presentation), 27 Feb 2004. http://www.bfe.admin.ch/php/modules/publikationen/stream.php?extlang=en&name=en_830170164.pdf (2004)
29. Robelius, F.: Giant oil fields – the highway to oil: giant oil fields and their importance for future oil production, Uppsala Dissertations from the Faculty of Science and Technology, ISSN 1104–2516; 69. http://uu.diva-portal.org/smash/record.jsf?pid=diva2:169774 (2007)
30. Hirsch, H.L., Bezdek, R., Wendling, R.: Peaking of World Oil Production: Impacts, Mitigation, & Risk Management, February 2005. Report to US Department of Energy. See also: http://www.netl.doe.gov/publications/others/pdf/Oil_Peaking_NETL.pdf (2005)

Chapter 12

Oil from Deep Water: The Tail End of Extraction

It is the first week of September 2006 and the telephone rings. An enthusiastic reporter from Sweden's national television network (SVT) tells me that CNN, Fox, the *New York Times*, and virtually the rest of the world's media are reporting that a new gigantic oilfield has been found in the Gulf of Mexico. I am invited to visit Stockholm the next morning to participate in a discussion about the fantastic discovery. Of course, I will be asked whether the world can now stop worrying about Peak Oil.

The background to the media's excitement was that Chevron had reported a discovery of oil in the sedimentary layers from the later Tertiary period in a deep geological formation that they named "Jack" [1]. This find was made when they drilled exploratory well number 2, Jack-2. According to the newspaper and TV reports the discovery was in the range of 3–15 billion barrels of oil (Gb). From the information available it could not be excluded that the discovery might even be larger. That would make it the U.S.'s largest oilfield! I promised the reporter to get back to him but first I needed to find out a little more information about the new giant field.

To make their discovery, the oil company had anchored a floating drilling platform in an area where the water depth was 2,100 m (2.1 km). They had then drilled to a total depth of 8,598 m (8.6 km) below sea level. As a comparison, one can mention that Sweden's highest mountain is Kebnekaise, the peak of which is 2,100 m above sea level. The world's tallest mountain, Mount Everest, reaches 8,848 m above sea level (see Fig. 12.1). The first thought this knowledge brings to mind is that the international oil industry must really be desperate to be looking in such extreme environments for new oil discoveries. If the opportunity had existed to find oil on land

K. Aleklett *Peeking at Peak Oil*, DOI 10.1007/978-1-4614-3424-5_12,

Fig. 12.1 Jack is a geological formation dating from the later Tertiary period that lies deep under the sea floor in deep water in the Gulf of Mexico. To discover this giant field of, reportedly, 10 Gb of oil Chevron had anchored a drilling platform in 2,100 m of water with the help of "tension legs." At a depth of 8,598 m below sea level the well produced an oil flow of 6,000 barrels of oil per day. A water depth of 2,100 m is similar to the height of Sweden's highest mountain and the depth of the well below sea level can be compared with the height above sea level of the world's highest mountain, Mount Everest

instead, then they naturally would have been looking there. Occasionally it is said by some that the international oil companies (IOCs) have been forced out into deep water because they have been denied access to particular areas. The reality is that they have already drilled virtually everywhere on land where favorable geological formations are known to exist. Another thought aroused by Chevron's tale of discovery of the Jack oilfield was whether the field would ever really contribute ten billion barrels (Gb) of oil to world oil production, equivalent to about 4 months of global consumption.

If Jack was really a giant field of 10 Gb it would only shift the date of Peak Oil forward by a few months.

I rang back to SVT and told them that, of course, I was interested in coming in to discuss the new discovery and that it might shift the date of Peak Oil by 4 months. Soon they rang back to tell me that the idea of a breakfast TV interview at SVT had been canceled. So now, 5 years later, what has happened to Jack?

Deep, Deeper, Deepest: Oil Production in Deep Water

During the Cretaceous period 90 million years ago, sedimentary layers containing kerogen were formed in an area northeast of Africa and in the new geological basin that was formed as Africa and South America began to move slowly away from each other (see Fig. 4.3). Millions of years later Africa began to collide with Asia causing the land surface in the northeast to rise. The sedimentary layers thus came closer to the surface in the area that later became the Middle East. Meanwhile, the sediments in the basin west of Africa were divided into two as the separation between Africa and South America grew and grew. One half of the ancient basin now makes up the continental shelf off South America's east coast and the other half became the continental shelf off the west coast of Africa. In both the Middle East and the remnants of the ancient basin now found off the South American and African coasts, favorable geological structures exist in which to find oil. To find oil in the Middle East one does not need to explore in deep water but the remnants of the ancient basin that was located west of Africa now lie in deep water off the South American and African coasts (see Fig. 4.4).

As knowledge grew about the geological preconditions necessary for the discovery of oil it was realized that oil might be found in some coastal areas. The development of technology to allow oil production from areas under water created two classes of oil production, land-based or "onshore" production and sea-based or "offshore" production. The first step towards offshore production was taken as early as 1896 when jetties were extended out from the coast at Santa Barbara, California. From these they could drill into an offshore area of the Summerland Oil Field in a water depth of 11 m. During the 1930s they began to build platforms for drilling and oil production in the intertidal zone along the coasts of Louisiana and Texas. However, the first big step into offshore production was not taken until 1947 when, in the Gulf of Mexico, the first platform was built that was so far offshore it could not be seen from land [2].

By the time the IOCs were forced to leave the Middle East, oil had been discovered under the North Sea, and development of offshore production technology became an IOC priority. However, water depths in the North Sea are not extreme and the oil industry has steadily extended its capabilities to explore in deeper and deeper water. Their new frontlines for exploration and development have become the Gulf of Mexico, off the east coast of Brazil and off Africa's west coast. Discoveries in ever-deeper water require new technical solutions so technological development has been driven forward very rapidly. Production in water deeper than 500 m (1,640 ft) is usually regarded as deep water production but sometimes this term is also used for production where the water depth exceeds 1,000 ft. Production in depths greater than around 1,500 m is regarded as "ultra-deep." Now the capabilities of the oil industry are approaching drilling in water depths of 3,000 m. We now look at some production platforms used at water depths of 1,000, 1,500, 2,000, 2,500, and 3,000 m.

The largest deep water field in the Gulf of Mexico is "Mars-Ursa" lying below 1,034 m (3,392 ft) of water. The field was discovered in 1989 by Shell and came into production 7 years later in 1996. The proven reserves (1P) are given as 1.21 Gb. It was production from this field that, for the first time, was performed in water depths of greater than 1,000 m. A simplified description of the platform is given in Fig. 12.2.

The "ultra-deep" production threshold was first reached by ExxonMobil's "Hoover-Diana" platform when it began production in 2000. The water depth is 1,463 m (4,800 ft) and the reserves of the oilfield being produced by Hoover-Diana are estimated to exceed 300 Mb of oil and natural gas (oil-equivalent barrels). This means that the oilfield is not classed as a giant, but the production platform is gigantic! A simplified description of the platform is given in Fig. 12.2.

The second largest oilfield in the Gulf of Mexico is "Thunder Horse" with proven reserves of 0.64 Gb. This oilfield, and the third largest, "Atlantis," are both controlled by BP [3]. Atlantis lies under a water depth of over 2,000 m. A simplified description of the Atlantis production platform is given in Fig. 12.2, and the following text has been published describing the field's production infrastructure:

> Discovered in 1998, the Atlantis Field development is designed to utilize one of the deepest moored semi submersible platforms in the world. The water depth at the semi PQ (Production Quarters) location is 7,074 ft (2,156 m). It is designed to process 200,000 barrels of oil per day (b/d) and 180 million standard cubic feet per day (mscf/day). First oil was achieved in 2007. Oil and gas are transported to existing shelf infrastructure via the BP-operated Caesar (oil) and Cleopatra (gas) pipeline systems. BP operates the development (56% interest), with co-owner BHP Billiton owning the balance. The field is being developed via a semi-submersible PQ facility, supporting a network of wet-tree subsea wells.

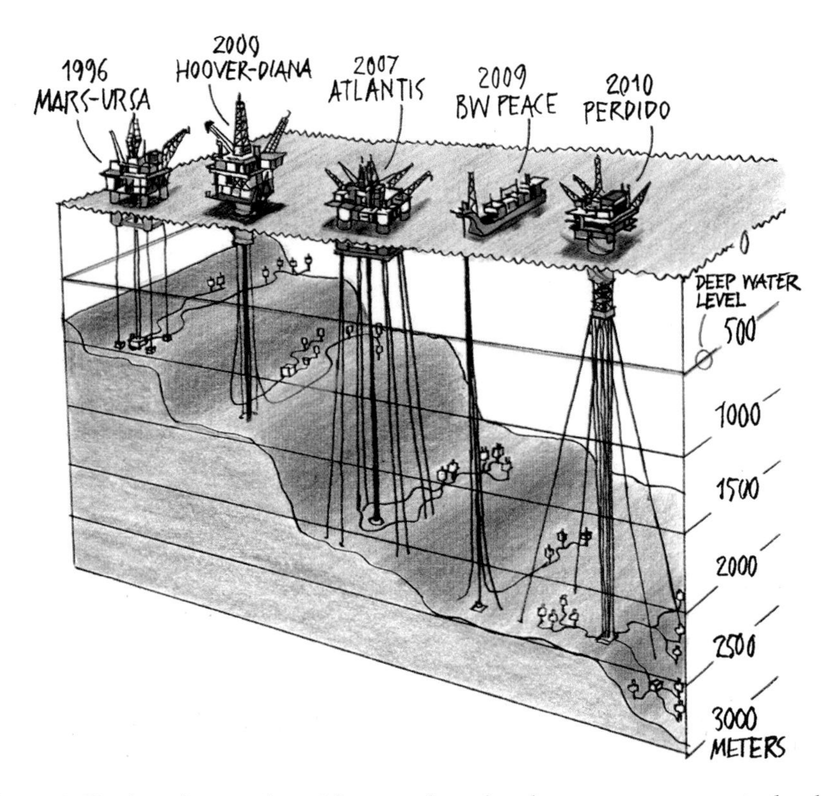

Fig. 12.2 During the previous 20 years there has been an enormous technology development rate for deep water oil discovery and production. In the illustration, depth below sea level is marked in 500 m intervals. In fact, 500 m is the threshold at which "deep water" production is said to begin. Some examples of platforms are shown including when they were put into operation. Production in 1,000 m of water was achieved in 1996 and by the new millennium in 2000 oil companies were producing oil in 1,500 m of water. Seven years later it was shown to be possible to produce oil at depths greater than 2,000 m and in 2009 an FSPO reached down to 2,500 m. In 2010 a permanent platform for production from this depth was installed with some production units lying as deep as 2,900 m under water

The next step into deeper water takes us to the east coast of Brazil and the oilfield named Tupi (which was renamed Lula in December 2010). The field is a so-called "sub-salt" field and more information is given regarding this in the section below, "Deep water production in Brazil". Petrobras has reported that the field contains 6.5 Gb of oil which makes Lula the world's largest deep water field. Pilot oil production of the field at 100,000 barrels per day was performed by hiring a FPSO-unit (floating production, storage, and offloading). In 2009 the FPSO unit "BW Peace" was coupled to the field's "Christmas tree" (see Fig. 7.7) at a depth of 2,500 m below sea level and a new record depth for deep water production was achieved [4]. A simplified description of the FSPO is given in Fig. 12.2.

After the 2,500 m production depth was reached, it took only 1 year before the IOCs approached operating at a depth of 3,000 m with the production platform "Perdido." This time it was Shell's turn to take the main credit for the achievement [5]. The platform, which is nearly as large as the Eiffel Tower, was built in Finland and transported to the Gulf of Mexico for fitting with equipment before it was chained to the sea floor 2,400 m below. However, the production wells exist at a depth of 2,900 m. A simplified description of the platform is given in Fig. 12.2, and the following information is provided by Shell [6]:

> The facility is capable of handling 130,000 barrels of oil equivalent per day. To get the oil and gas to market required installing 77 miles of oil export pipelines and 107 miles of gas export pipelines in a remote part of the Gulf of Mexico over very rugged sea floor terrain to connect to the existing offshore pipeline infrastructure. The Perdido Development has already set a world water depth record in drilling and completing a subsea well 9,356 feet (1.77 miles) below the water's surface. The project intends to drill an even deeper well at 9,627 feet.

We have now taken a fantastic technological journey into the oceans' depths. It is this technology that many people consider will, in the next 25 years, place Peak Oil on history's bookshelf where it can gather dust.

Production on the Outer Edge

The three deep water locations mentioned earlier, the Gulf of Mexico and the areas off the east coast of Brazil and off the west coast of Africa, are the main regions of the world where this type of oil production occurs. Technological innovation has driven the development of oil production in these regions but the conditions for oil production vary markedly among them. In the Gulf of Mexico it is market forces that drive oil production whereas in Brazil production is controlled by Brazil's national oil company Petrobras. Off the west coast of Africa it is mainly the international oil companies that hold sway but China's national oil company also wants access to these resources. More and more nations along Africa's west coast are finding resources that might be suitable for oil production and the area is still underdeveloped.

Geologically there are also differences between the Gulf of Mexico and the other two regions. However, it is most interesting to study the similarities between the geological layers off Brazil's east coast and off Africa's

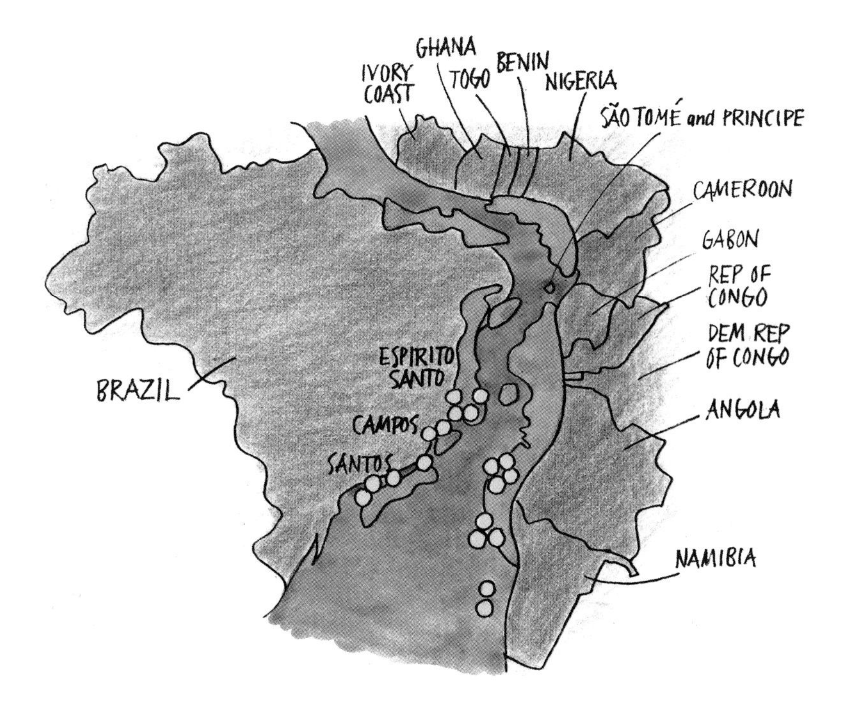

Fig. 12.3 The land areas that are now South America and Africa began to separate and formed a basin 135 million years ago. This created the conditions for formation of marine sedimentary layers containing kerogen. As time passed the basin was broken up and part of it now lies off Brazil's east coast and the other part lies off Africa's west coast. In the sedimentary layers along Brazil's east coast there is a salt layer beginning at about 3,000 m beneath the seabed. The salt layer is about 1,000 m thick and, under this, they have found oil. A similar layer of salt has been found off the west coast of Angola and a geological comparison between the two areas shows comparable sedimentary layers. (The location of the geological tests that showed this are marked as *yellow spots* on the map.) The question is whether oil exists underneath the salt layer off the coast of Angola [7]

west coast. In Fig. 4.3 we saw that South America and Africa separated from each other between the Jurassic and Cretaceous periods and the sedimentary layers lying between them containing kerogen were formed during the Cretaceous period. A description of how these sedimentary layers became divided geographically is given in Fig. 12.3. The fact that they have found oil under a thick salt layer off the coast of Santos in Brazil and that a similar layer of salt exists off the coast of Angola is of interest to some geologists who assert that oil must also exist under the Angolan salt layer [7].

Deep Water Production from the Gulf of Mexico

In the analysis that Global Energy Systems made of oil production from the Gulf of Mexico we chose a water depth of 500 m as the upper limit for deep water production. (The U.S. Energy Information Administration Agency, EIA, uses 200 m as the upper limit [8].) The Minerals Management Service (MMS) of the U.S. Department of the Interior provides publicly available data on discoveries, production, and so on for the Gulf of Mexico. From these statistics, we have assembled a database of deep water production. In total there are 172 registered deep water fields of which some only produce natural gas. The numbers of fields with registered resources are 91 and, of these, 69 were in production in 2009. The reserves reported by the MMS are proven reserves (1P). Because 0.5 Gb is the lower limit for a giant field this means that, among the producing fields in our database, there are four giant fields. Three of the giant fields and 16 lesser fields ("dwarf fields") can be classed as ultra-deep. The proven reserves for the 91 fields amount to 6.85 Gb. For fields in production the 1P reserves amount to a total of 5.89 Gb [9].

In 1989 the first deep water field, named "Jolliet," came into production. It has 1P reserves of 0.037 Gb. However, large-scale deep water production first began in 1996 when the giant field Mars-Ursa began production with 1P reserves of 1.21 Gb. From the available data we can put together a history of oil production in the Gulf of Mexico (Fig. 12.4). We can see that deep water production between the depths of 500 and 1,500 m has already passed its peak. The fact that Gulf of Mexico deep water oil production is currently sitting on a plateau of production is due to the application of new technology making it possible to produce oil from ultra-deep fields. Production from the giant fields in this area is especially significant for maintaining this production plateau.

The production infrastructure of individual fields in the Gulf of Mexico is dimensioned for different rates of maximal flow. Mars-Ursa (1P reserves of 1.21 Gb) maximally produced around 270,000 barrels per day (b/d) whereas Thunder Horse (1P reserves of 0.64 Gb) produces 250,000 b/d. Atlantis (1P reserves of 0.56 Gb) produces 200,000 b/d. Mars-Ursa showed maximal production between 2001 and 2004 but production has since fallen to 150,000 b/d. The other fields can be expected to experience similarly short periods of maximal production.

Mars-Ursa reached maximal production in 2004 at 272,614 b/d. The total production at the end of the peak year was 51% of proven reserves. By the end of 2009 77% of its 1P reserves had been produced. It is now time to make a new estimate of "proven and probable" reserves, 2P. Based on our

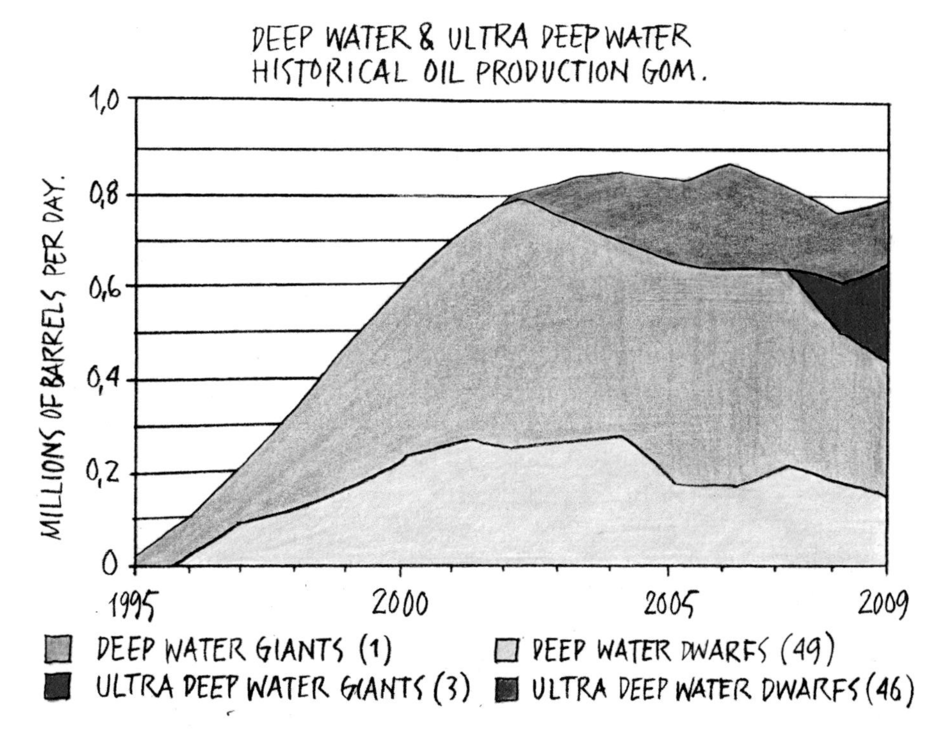

Fig. 12.4 Oil production from the 69 deep water fields in the Gulf of Mexico. The largest deep water field in this area is Mars-Ursa that saw its production plateau between 2001 and 2004. All the deep water fields in the Gulf of Mexico have a combined maximal production of just under 800,000 b/d. Production from ultra-deep oilfields has compensated for the declining production from other fields up until 2009 but the question is when ultra-deep production will reach its maximum and deep water production from the Gulf of Mexico (GOM) as a whole will decline

research proven and probable reserves amount to 1.67 Gb which is an increase of 38% over the original estimate [9].

The Gulf of Mexico, with proven reserves of 5.89 Gb, currently produces 800,000 b/d. The proven reserves that are not yet in production amount to 0.98 Gb and an important question is what rate of oil production these reserves can provide in the future. Chevron and partners reported that half of these reserves that are not yet in production (i.e., 0.5 Gb) will be in production by 2014 with a maximal rate of 170,000 b/d when Jack-St. Malo comes into production. Production from the remaining 0.5 Gb of these reserves by other oil companies can be estimated to show a similar production history. We conclude that total oil production from deep water in the Gulf of Mexico will never exceed 1 million barrels per day (Mb/d). Presumably it will show a production plateau of around 800,000 b/d up until 2020.

Deep Water Production in Brazil

In Brazil the national oil company Petrobras has the main responsibility for oil exploration and production. The fact that they have found giant oilfields in the Santos Basin under a 1 km thick layer of salt, the "pre-salt" areas, has led the Brazilian government to introduce special regulations for these areas. Four proposals for new legislation have been presented to the National Congress of Brazil and the first two of these became law in July 2010. The first piece of legislation created a new authority, Petrosal, that will administer production from the pre-salt areas. The second piece of legislation will strengthen the government's power over Petrobras by giving Petrobras production rights equivalent to 5 billion barrels of oil in exchange for greater government ownership of the company. The third and fourth pieces of legislation were passed by Brazil's National Congress in December 2010. They will create a fund to administer the profits from the pre-salt areas and a new system for distributing rights to explore for oil and produce it, the Production Sharing Agreement (PSA). Now they are introducing a law that Petrobras will have full responsibility for all production from pre-salt areas and will own at least 30% of these discoveries. This legislation guarantees that the Brazilian nation will receive income from future oil production from the pre-salt oilfields.

When discovery of the first giant oilfield in the pre-salt areas was reported there were many who doubted the size of the discovery. However, in December 2010 Petrobras submitted a "declaration of commerciality" to ANP, Brazil's national petroleum association and the fields that had been known as Tupi and Iracema were renamed Lula and Cernambi, respectively. The total amount of recoverable oil and gas in oil equivalents was given as 6.5 Gb for Lula and 1.8 Gb for Cernambi. In turn, ANP has now reported that Brazil's proven reserves (1P) increased in 2010 by 10.65% to 14.25 Gb and proven and probable reserves (2P) have now increased by 34.57% to 28.47 Gb. This means that Brazil's 2P-reserves are now similar in size to 1 year of global oil consumption at 30 Gb [10].

During the years 1984, 1985, and 1987 three giant oilfields were discovered in Brazil's deep water area. When production from these began slowly in 1987, Brazil's total 2P-reserves were 5.35 Gb [11]. By 2000 production from deep water had reached 772,000 b/d and Brazil's 2P-reserves had exceeded 10 Gb. From 2000 to 2009 production increased by 1 Mb/d to reach 1,772,000 b/d. Exploration had met with success and before they added the pre-salt oil, 2P reserves had already reached 20 Gb. It is these reserves discovered after 1999 that were exploited for deep water production in 2009 and this constituted approximately 60% of Brazil's oil production.

From the Petrobras website we can find that Brazil's oil production in January 2011 was 2,069,342 b/d and that they have found a new oilfield in the pre-salt area off the coast of Santos [12]. We are told that the exploration well that discovered the new oilfield was drilled in a water depth of 2,134 m and the oil was found 4,900 m below the sea bed. The quality of the oil is good (26° API) and BP is a 25% participant in the consortium that will produce the oil.

It is still too early to make any statements regarding future production of all the oil that exists in the pre-salt geological formation. However, when these expensive projects need international finance the OOIP (original oil in place) will become public knowledge many years before production begins. At the moment they have started pilot production of 100,000 b/d and they plan to connect an additional FPSO to the production infrastructure. According to Petrobras' business plan to 2014, they will invest US$33 billion with the aim of reaching a production level of 4,000,000 b/d by 2020 (1.5 Gb per annum). One quarter of this production is to come from the pre-salt areas [13].

To understand Brazil's future production we can compare it with Norway inasmuch as their original resources were similar in size [14]. In Norway, it took 10 years to double production to a maximum of 3 Mb/d and this was followed by rapid decline. The reserves that Brazil has today can be expected to follow a similar developmental path. Brazil's oil production from deep water may be 3 Mb/d in 2020 but its production level in 2030 critically depends on the size of the oil discoveries made during the next 10 years.

Deep Water Production off the West Coast of Africa

In the past 20 years the deep water area off the west coast of Africa has been for the international oil companies what the Klondike was for gold prospectors at the turn of century around 1900. Exploration permits and bribes both provide access to production rights. In the Klondike, security was provided by Sheriff M. M. "Si" Tanner [15] but for West Africa NATO has taken on the role of sheriff [16].

The National Energy Policy Report issued by the Office of Vice President Richard Cheney on May 16, 2001 stated, "West Africa is expected to be one of the fastest-growing sources of oil and gas for the American market. African oil tends to be of high quality and low in sulfur … giving it a growing market share for refining centers on the East Coast of the U.S."

The following year, the Washington, DC-based African Oil Policy Initiative Group conducted a symposium titled, *African Oil: A Priority for U.S. National Security and African Development*, with the participation of American legislators, policy advisers, the private sector, and representatives of the State Department and Defense Department. At the symposium Congressman William Jefferson said, "African oil should be treated as a priority for U.S. national security post 9-11. I think that … post 9-11 it's occurred to all of us that our traditional sources of oil are not as secure as we thought they were."

Officially the governments that invaded Iraq in 2003 have asserted that there were motivations other than oil for their actions. However, in the case of West Africa and the Gulf of Guinea it has been declared clearly that NATO's bases there have been built to guarantee secure oil production for those international oil companies that have their head offices in NATO nations.

According to the *BP Statistical Review of World Energy* [17] the following West African nations produced oil in 2009: Angola, Cameroon, Chad, Congo, Equatorial Guinea, Gabon, and Nigeria. Their combined oil production was 4.85 Mb/d. Deep water production exists primarily off the coasts of Angola and Nigeria. In 2009 production from these areas was 1.14 and 1.15 Mb/d, respectively. Of the total original reported 2P-reserves of 13.17 Gb in deep water off Angola's coast, 10.33 Gb remain to be produced. Nigeria had equivalent discoveries of 15.37 Gb and they now have 10.53 Gb left to produce [11]. By observing the existing discovery trends Colin Campbell has estimated that future discoveries in both nations will amount to 4–5 Gb each. When these new finds are included he predicts that, in 2013, Angola and Nigeria will reach plateau production of 1.64 Mb/d and 1.40 Mb/d, respectively, which can be compared with the plateau production of 0.8 Mb/d in the Gulf of Mexico (see Fig. 12.2). Current reserves and future discoveries can support plateau production for about 10 years but by 2030 oil production in Angola and Nigeria will definitely be past their peaks.

Angola has a pre-salt area similar to that off the coast of Santos in Brazil (see Fig. 12.3). If they find oil there then the possibility exists for a second phase in Angola's deep water oil production. In that case Angola's plateau of oil production might be extended until 2030 or later.

The Tale of Jack

On the front page of the *Wall Street Journal* (*WSJ*) of September 5, 2006 the discovery of the giant oilfield Jack in the Gulf of Mexico was reported. One day later there was no self-respecting media outlet that did not have broad coverage on the fantastic discovery including comments from significant personalities in the oil industry. But before we look at the media coverage we should first look at what Chevron itself had to say about the discovery [1].

> Chevron Corporation (NYSE: CVX) announced today that it successfully completed a record setting production test on the Jack #2 well at Walker Ridge Block 758 in the U.S. Gulf of Mexico. The Jack well was completed and tested in 7,000 feet of water, and more than 20,000 feet under the sea floor, breaking Chevron's 2004 Tahiti well test record as the deepest successful well test in the Gulf of Mexico. The Jack #2 well was drilled to a total depth of 28,175 feet.
>
> The test was conducted during the second quarter of 2006 and was designed to evaluate a portion of the total pay interval. During the test, the well sustained a flow rate of more than 6,000 barrels of crude oil per day with the test representing approximately 40 percent of the total net pay measured in the Jack #2 well. Chevron and its co-owners plan to drill an additional appraisal well in 2007.

Chevron also stated that the oil-bearing geological formation named Jack was discovered in 2004 but they did not say anything about the total size of the discovery.

Russell Gold from Austin, Texas, is an energy reporter for the *Wall Street Journal*. He published an article on Jack on the same day that Chevron released its announcement. It is clear that Gold assumes that Chevron's press release and his article will be published on the same day by the *WSJ* so he must have been fed the news from Chevron directly [26].

> Chevron Corp. and partners Devon Energy Corp. and Statoil ASA are expected to announce today the first successful oil production from the region, a 300-mile-wide swath of the Gulf that lies below miles of water and deep within a bed of ancient rocks geologists call the lower tertiary.

Gold reported the news of the successful test well and also reported information about the size of the discovery:

> Chevron and Devon officials estimate that the recent discoveries in the Gulf of Mexico's lower-tertiary formations hold more than three billion barrels' and perhaps as much as 15 billion barrels' worth of oil and gas reserves. If the industry succeeds in finding 15 billion barrels of oil, it would boost the nation's current reserves of 29.3 billion barrels by 50%.

It was this news that spread like wildfire around the globe and that also influenced decisions about future American energy policy. The following day the news about the discovery being of 3–15 Gb in size was reported by the *New York Times* [18], the *Washington Post* [19], *Houston Chronicle* [20], CNN [21], and practically all other media outlets. In the *Houston Chronicle* we could read:

> If projections hold up, the Jack field could eventually provide 11 percent of all the oil produced in the U.S. between 2012 and 2014, Cambridge Energy Research Associates (CERA) said Tuesday.

The *Washington Post* has the headline "U.S. Oil Reserves Get a Big Boost" and in the article they quote Daniel Yergin, the chairman of CERA, "This looks to be the biggest discovery in the United States in a generation, really since the discovery of Prudhoe Bay 38 years ago."

In the *New York Times* the headline sang "Big Oil Find Is Reported Deep in Gulf" and in the accompanying article we were told:

> Chevron, Devon Energy and Statoil ASA, the Norwegian oil giant, reported that they had found 3 billion to 15 billion barrels in several fields 175 miles offshore, 30,000 feet below the Gulf's surface, among formations of rock and salt hundreds of feet thick.

The test well that was reported to give a flow rate of 6,000 b/d had now become a discovery of 3–15 Gb of oil. They had also consulted CERA's chairperson Daniel Yergin, whom many regard as the highest authority on oil and who is the author of the prize-winning book on oil, *The Prize*. Yergin said to the *New York Times*, "This is frontier stuff. Success at these depths in the Gulf of Mexico would facilitate ultra-deepwater exploration elsewhere in the world because it will have proven the technology and capabilities."

We can now see that CERA had become as important in the coverage of the Jack discovery as the original press release from Chevron. CERA is a consultancy firm frequently used by oil companies and its opinions and advice usually command a high price. For the Jack story CERA's Energy Strategy Group published its opinions free of charge. On September 6, the same day as the *Houston Chronicle, Washington Post, New York Times*, and the rest of the world was reporting on Jack, CERA announced [22]:

> The successful drilling test at the Jack discovery off the coast of Louisiana is a significant development for the future of U.S. oil and gas supply. Up to 800,000 b/d of light, sweet crude oil and 1 billion cubic feet per day of natural gas could begin flowing from this reservoir in 2012–2014.
>
> - It is possible that as much as 2–3 billion barrels have been discovered in this play, and that additional oil resources will be identified in nearby geological structures.
> - The find confirms the impact of technology on the ultradeepwater frontier, including advances in seismic exploration and the evolution of new drilling rigs.
> - This represents the largest potential gas supply addition in the Gulf of Mexico, which has been characterized by declining production and a paucity of gas-directed drilling.

Let us conclude this cavalcade of news on Jack with the story that CNN spread around the world and that was the reason that *SVT* rang me to discuss Peak Oil at the start of this chapter [21]:

> Successful test by Chevron partners in deep Gulf waters could rival Alaska in potential supply; U.S. reserves may swell 50 percent.

Furthermore, they discussed how world markets had reacted to the news of the gigantic oil discovery:

> The news sent oil prices lower, with U.S. light crude for October delivery sinking 69 cents to $68.50 on the New York Mercantile Exchange.

The share prices of the companies involved in the discovery also did well:

> Shares of the three partners in the test well known as "Jack-2" rose sharply in trading Tuesday. Chevron (up $1.51 to $66.34), which owns a 50 percent stake, jumped 3 percent, while Devon Energy (up $7.99 to $72.14) soared nearly 12 percent, and Norwegian oil company Statoil's U.S. shares added about 2 percent (up $0.66 to $28.17). Devon and Statoil each own 25 percent stakes.

The oil industry's leading journal is the *Oil & Gas Journal*. On September 11, it summarized the hysteria around Jack in an article titled, "The Jack-2 Perspective" [23]. The article begins by stating that the "general media" had shown great interest in Jack but that, in the future when production from Jack is underway, the coverage from September 2006 will have been forgotten. Nevertheless, they believe that the Jack discovery will have great significance for the oil industry:

> Estimates of recoverable volumes for the deepwater Lower Tertiary play reach as high as 15 billion bbl of crude and other liquids. It's too soon for such numbers to be anything but speculation. If supply foretold by the Jack-2 results remains uncertain, however, the well clearly has engaged the imagination of the news media and therefore the public. It thus offers the industry a chance to help move energy politics beyond its fixation on the price of gasoline and its orientation to futile goals.

Despite its position as the oil industry's leading journal, the *Oil & Gas Journal* is, nevertheless, realistic about the future:

> By the time production from discoveries related to Jack-2 might be hitting stride 5 years from now, rising US demand and depletion of fields now on production will have increased the US need for foreign oil by perhaps 1.5 million b/d. That number represents extension of the past 5 years' average net effect on the crude oil balance of demand gains and production declines. It's what the new area would have to produce just to get import dependency back to its level of 2005, still far shy of self-sufficiency.

Just over 1 month later on October 12, Jack was mentioned at presidential level when U.S. President George W. Bush spoke at a conference on renewable energy. The *Oil & Gas Journal* reported [24]:

> "As you can tell, I'm excited about new technologies. But I think we've got to be realistic about the timing. And in order to become less dependent on foreign sources of oil, we've got to explore for oil and gas in our own hemisphere in environmentally friendly ways. And one of the interesting technological developments is the capacity to find oil in unique places," he said.
>
> Referring to the recent Jack-2 discovery made by Chevron Corp. and its partners in the deepwater Gulf of Mexico, Bush said it was accomplished with new technologies, "which enable us to go to new places, and they enable us to be wise stewards of the environment."

There is no doubt that President Bush had complete faith in new technologies:

> I understand there's a big debate about whether or not you can explore for oil and gas and protect the environment. I believe you can. And I understand that as we transition to the ethanol era . . . or the hydrogen area, we must also find oil and gas in our own hemisphere if the objective is to become less dependent on foreign oil.

President Bush advanced the opinion that it was time for the House of Representatives and the Senate to submit proposals for the opening up of new areas to exploration:

> And I believe Congress needs to get the bill to my desk as quick as possible. So when you finish the elections, get back and let me sign this bill so the American people know that we're serious about getting off foreign oil.

A common thread running through the reportage is that Jack will make the United States less dependent on imported oil. CERA and other voices predicted that oilfields might be found in this geological formation larger than the U.S.'s largest in Texas of just over 5 Gb. In October 2010, 4 years after President Bush hoped that the United States would become less dependent on imported oil, Chevron and its partners reported what became of Jack [25]. In a press release from October 23 we can read that they had decided to invest in production of oil and gas from Jack and from a nearby field named "St. Malo." The project will require an investment of $7.5 billion and it will consist of three underwater production centers for production from Jack and St. Malo that will then be coupled together at a "hub." The combined reserves of the fields are estimated to be 0.5 Gb and the production rate is to be 170,000 barrels of oil and 42.5 million cubic feet of natural gas per day.

Jack and St. Malo will provide less than 1% of the oil that the United States uses every day. In the section "Deep Water Production from the Gulf of

Mexico" we showed that new production was needed to maintain production from deep water in the Gulf of Mexico at a level of 800,000 b/d. Jack and St. Malo will contribute to this but the U.S.'s need to import oil will remain. The fact that inhabitants of the United States use, on average, twice as much oil as the people of Europe indicates that decisions other than opening up new areas for exploration are what is needed in the United States. Every year that passes without the United States addressing the issue of its rate of oil consumption is a year lost for preparation for the future.

Have We Reached the End of the Road?

When we summarize deep water production up until 2020 we see that the Gulf of Mexico can contribute 0.8 Mb/d, Brazil can contribute 3 Mb/d, Angola and Nigeria can contribute 3 Mb/d, and the rest of the world can give 1.6 Mb/d [11]. That means that, in 2020, we can expect total oil production from deep water to be 8.4 Mb/d which is an increase of 2.4 Mb/d over the level in 2009. According to the IEA's *World Energy Outlook* report for 2010, total conventional crude oil production was 68.1 Mb/d in 2009 and a decline of 6% per year means that production from fields that were in production in 2009 will have fallen to 34.5 Mb/d by 2020. To maintain crude oil at a constant level until 2020 means that new projects giving production equivalent to 33.6 Mb/d must be brought on line. However, not even 10% of the necessary increase can be provided by production from deep water.

Deep water is the last outpost of global oil production. The production journey that began in the United States and Russia in the 1850s has now reached the end of the road. In *WEO 2010* the IEA asserts that Peak Oil for conventional crude oil occurred in 2006 and that production has declined since then. Production at the limits of technological feasibility such as in deep water entails risks as demonstrated by the Deepwater Horizon catastrophe. There, BP and its contractors had problems with the "Christmas tree" that resulted in leakage of approximately 5 million barrels of oil (see Fig. 12.5).

We do not yet know all the environmental consequences caused by the oil spilled after the loss of the Deepwater Horizon platform in the Gulf of Mexico. However, for the international oil companies the catastrophe meant only a short pause before they took on new projects. It is difficult to get the leadership of BP to make statements regarding the Deepwater Horizon incident but on March 7, 2011 BP's chairman Carl-Henric Svanberg was one of the main speakers at a conference in Malmö, Sweden on the handling of oil leaks. The conference was organized by the World

Fig. 12.5 The oil platform Deepwater Horizon burns. Despite a massive firefighting effort the platform could not be saved and it sank to the bottom of the Gulf of Mexico

Maritime University and the International Maritime Organization. In his speech Carl-Henric Svanberg noted that the oil industry must continue with deep water drilling, including in the Arctic, to satisfy the world's future energy needs.

We can now state that deep water oil production can only make a marginal contribution to global oil production and that it will only serve to make the future decline of global oil production a little less steep. For the IOCs deep water might be their last hope for securing their value on the stock market.

References

1. Chevron Announces Record Setting Well Test at Jack. Chevron, 5 Sept 2006. http://www.chevron.com/chevron/pressreleases/article/09052006_chevronannouncesrecordsettingwelltestatjack.news (2006)
2. Robert Lamb.: How offshore drilling works, howstuffworks; http://science.howstuffworks.com/environmental/energy/offshore-drilling.htm (2012)
3. BP Atlantis Field Fact Sheet: http://www.bp.com/liveassets/bp_internet/globalbp/STAGING/global_assets/downloads/A/abp_us_atlantis_fact_sheet_081209.pdf (2009)
4. Petrobras tenders for Tupi FPSO: http://www.upstreamonline.com/live/article154718.ece (2008)
5. Plumbing the depth. The Economist. http://www.economist.com/node/15582301 (2010)
6. Shell Successfully Installs Perdido Topsides: http://www.shell.us/home/content/usa/aboutshell/media_center/news_and_press_releases/2009/perdido_042209.html (2009)
7. Geological similarities with Brazil's pre-salt attract investments to Africa. Offshore, July 1, 2010. http://www.offshore-mag.com/index/article-display/9989570614/articles/offshore/volume-70/Issue_7/latin-america/Geological_similarities_with_Brazils_pre_salt_attract_investments_to_Africa.html (2010)
8. EIA: US Energy Information Administration (2011)
9. Lundén,U.: Deepwater oil production – land of confusion. Diploma thesis, Uppsala Global Energy Systems. http://www.physics.uu.se/ges/en/publications/diploma-theses (2011)
10. Brazil's oil reserves believed to be close to 30 billion barrels. BrazzilMag. http://www.brazzilmag.com/component/content/article/95-february-2011/12515-brazils-oil-reserves-believed-to-be-close-to-30-billion-barrels.html (2011)
11. Campbell, C.: Deepwater data base. Private communications (2011)
12. Our oil production grew 5.4% in January. Petrobras, 11 Feb 2011. http://www.petrobras.com.br/en/news/our-oil-production-grew-5-4-in-january/ (2011)
13. US Energy Information Administration: Pre-salt oil. http://www.eia.gov/countries/cab.cfm?fips=BR&scr=email (2012)
14. Höök, M., Aleklett, K.: A decline rate study of Norwegian oil production. Energy Policy **36**, 4262–4271. http://www.physics.uu.se/ges/en/publications/a-decline-rate-study-of-norwegian-oil-production (2008)
15. Klondike Gold Rush, 6 Aug 2009. http://www.nps.gov/history/history/online_books/klgo/hpd1/app1.htm (2009)
16. Rozoff, R.: Militarization of energy policy: U.S. Africa command and the Gulf of Guinea. Global Research, 9 Jan 2011. http://www.globalresearch.ca/index.php?context=va&aid=22699 (2011)
17. BP Statistical Review of World Energy, June 2011. http://bp.com/statisticalreview (2011)
18. Oil find is reported deep in Gulf. New York Times, 6 Sept 2006. http://www.nytimes.com/2006/09/06/business/worldbusiness/06oil.html (2006)

19. Mufson, S.: U.S. oil reserves get a big boost. Washington Post, 6 Sept 2006. http://www.washingtonpost.com/wp-dyn/content/article/2006/09/05/AR2006090500275.html (2006)
20. Hensel, B. Jr.: New oil field deep in the Gulf a potential giant. Houston Chronicle, 6 Sept 2006. http://www.chron.com/disp/story.mpl/front/4165848.html (2006)
21. CNN: Major U.S. oil source is tapped. CNN 6 Sept 2006. http://money.cnn.com/2006/09/05/news/companies/chevron_gulf/ (2006)
22. Cambridge Energy Research Associates: Significant new deepwater oil play confirmed: biggest in 38 years? http://www.cera.com/aspx/cda/public1/news/articles/newsArticleDetails.aspx?CID=8322
23. The Jack-2 perspective: Oil Gas J., 11 Sept 2006. http://www.ogj.com/index/article-display/271344/articles/oil-gas-journal/volume-104/issue-34/regular-features/editorial/the-jack-2-perspective.html (2006)
24. Bush: More US E&P vital in alternative fuel transition, Oil Gas J., 23 Oct 2006. http://www.ogj.com/index/article-display/275284/articles/oil-gas-journal/volume-104/issue-40/general-interest/bush-more-us-eampp-vital-in-alternative-fuel-transition.html (2006)
25. Chevron Sanctions Jack/St. Malo Project in the Gulf of Mexico. Chevron, 23 Oct 2010. http://www.chevron.com/chevron/pressreleases/article/10212010_chevronsanctionsjackstmaloprojectinthegulfofmexico.news (2010)
26. Gold, R.: In Gulf of Mexico, industry closes in on new oil source, 5 Sep 2006; Page A1. http://www.cob.unt.edu/firel/Kensinge/Fina5170/Selected%20Cases/JackField20060905WSJ.pdf (2006)

Chapter 13

Peeking at Saudi Arabia: "Twilight in the Desert"

Keynote speakers at the world's first-ever Peak Oil conference held in Uppsala, Sweden (May 22–23, 2002) were Ali Samsam Bakhtiari from Iran and Matt Simmons from the United States. The Association for the Study of Peak Oil and Gas, ASPO, was interested to hear Dr. Bakhtiari speak because, in 2001, he had published a critical review of OPEC's future production capacity which concluded that, "Relying on OPEC to provide the oil required for increasing world demand over the next two decades doesn't seem justified…" [1]. In 2002 Matt Simmons led Simmons and Company International, a world leading investment bank serving the energy industry. At that time he was particularly interested in conventional natural gas production in the United States. Later (and partly inspired by his attendance at the conference) he became increasingly interested in future global oil production.

When Dr. Bakhtiari came to Uppsala he told us that he was on the list of people under surveillance in Iran and that he assumed his visit to Uppsala would move him closer to the top of that list. Nevertheless he was happy to attend because, as he put it, "I like to be part of history." I am wiser now but at that time I did not know that by promoting awareness of Peak Oil we were playing with fire. That Ali Samsam Bakhtiari could be risking his safety by attending the world's first Peak Oil conference (and that he would do so willingly) was difficult for me to conceive.

Matt Simmons published his famous book on Saudi Arabian oil production, *Twilight in the Desert*, in 2005 [2]. I had a fair amount of contact with him while he was writing his book and I know that the amount of work that went into finding references and ferreting out information from hundreds

K. Aleklett *Peeking at Peak Oil*, DOI 10.1007/978-1-4614-3424-5_13,

Fig. 13.1 Drilling for oil in the desert as the sun sets

of articles followed by assembling these pieces of the puzzle into a coherent whole made his book worthy of acceptance as a PhD thesis. Instead, it became an international bestseller that was translated into several languages and has influenced many people's views on Peak Oil.

It weighs heavily upon me and many others that these two champions of Peak Oil, Ali Samsam Bakhtiari and Matt Simmons, are no longer with us. (We lost them suddenly and unexpectedly to heart attacks in 2007 and 2010, respectively). Matt's book described doubts about Saudi Arabia's future production capacity and Ali stated the same, "How much can the kingdom really deliver? Could it really supply 14 million barrels per day in 2010 and 22 million barrels per day by 2020, as explicitly predicted by both the International Energy Agency (IEA) and the US Energy Information Administration" [1]? (Fig. 13.1).

The Tällberg Forum and the Debate About Saudi Arabia's Oil

One especially impressive memory I have from my trip to Abu Dhabi in 2004 is of the beauty of desert sunsets. However, breathtakingly beautiful sunsets are also a feature of midsummer evenings in the Dalarna region of

Sweden. In June 2006 I was in Tällberg, Dalarna, a bedazzlingly green cultural landscape dotted with traditional red Swedish cabins where the sun drifts slowly down over a body of water named Siljan. However, it was not the breathtaking sunsets over Siljan that drew me to Tällberg. Rather, I was invited to the Tällberg forum to participate in a discussion titled "The New Landscape of Human Security." This debate was opened with a speech by Prince Turki Al-Faisal who, in 2006, was Saudi Arabia's ambassador in Washington [3]. As I see it, Peak Oil will mean an enormous change in our future cultural landscape and I was very keen to discuss another kind of sunset with Prince Turki Al-Faisal: the sunset described by Matt Simmons in his book, *Twilight in the Desert* [2].

I can best describe the Tällberg forum as a forum where the world's non-governmental organizations (NGOs) and political and business leaders are gathered under one roof for an open discussion of important future global issues [4]. In reality the roof is that of a large tent. In 2006 the theme for the forum was "How on Earth Can We Live Together?" I asked Prince Turki Al-Faisal's secretary if it were possible to have a personal meeting with the prince and a time was arranged for the morning before the afternoon on which "The New Landscape of Human Security" would be discussed. When it was time for our meeting I found that it had been expanded into a lunch with the prince's wife as additional company. During the lunch we had an interesting discussion about Peak Oil and other relevant issues. As expected, the prince did not see Peak Oil as imminent, and especially not for Saudi Arabia.

The CSIS Seminar, *Future of Global Oil Supply: Saudi Arabia*

The oil production of a nation is, naturally, enormously important for that nation's economic welfare. However, if that nation produces so much oil that it is an exporter then that oil production also becomes important for the rest of the world. Saudi Arabia is the world's largest oil exporting nation and in 2010 they controlled approximately 16% of the oil on the world export market. For that reason an understanding of Saudi Arabia's oil reserves and production and their energy policy is crucial for all our futures.

Matt Simmon's interest in Saudi Arabia inspired the Center for Strategic and International Studies (CSIS) in Washington to organize a symposium with the theme, *Future of Global Oil Supply: Saudi Arabia*. The symposium was held in February 2004. In addition to Matt (who was then the chairman of Simmons and Company International), CSIS invited Mahmoud Abdul-Baqi, vice president, exploration, Saudi Aramco (Saudi Arabia's national oil company) and Nansen Saleri, manager, reservoir management, Saudi Aramco.

These two Saudi Aramco officials made a joint presentation. Copies of the presented material and an audio recording of the event can be found on the CSIS website [5]. The interesting aspect of the presentation from Saudi Aramco was that it revealed new information on Saudi Arabia's oil production. When I learned of this I wondered whether Saudi Aramco might reveal more, so I wrote a letter to the company's leadership asking whether it might be possible to obtain more detailed information on oil production and reserves. The answer I received was a copy of the February 24 presentation with a note stating that this represented Saudi Aramco's public data [6].

In 2004 there were 85 oilfields in Saudi Arabia made up of 320 discrete reservoirs. An example of an oilfield made up of several reservoirs is shown in Fig. 7.2 where nine of the reservoirs of the Bab field in Abu Dhabi are illustrated. The 2004 information from Saudi Aramco stated that Saudi Arabia's oil production capacity was 10 million barrels per day (Mb/d) which amounts to 3.65 billion barrels of oil (Gb) per year. According to the publicly available data every year the oil produced by Saudi Arabia is replaced by new reserves. In 2004 Saudi Aramco had nine seismic research teams (see Fig. 5.4) and 48 rigs for drilling new production wells and exploratory wells. Saudi Arabia declares that it has found 700 Gb of OOIP (oil originally in place) and that, in 2004, 260 Gb of this was reserves. A very interesting piece of information is that, in 2004, only 131 Gb of the reserves were developed and contributing to production. At a production rate of 3.65 Gb per year (including condensate and natural gas liquids) this means that they produced approximately 2.8% of their productive reserves in 2004. In the presentation at CSIS on February 24, 2004 Saudi Aramco revealed that it strives for low production rates from individual fields because this is one of the requirements for high recovery factors. The CSIS information has given us a snapshot of oil production in Saudi Arabia from one moment in time and a question is whether we can use this information to estimate that nation's future oil production.

In Chap. 11 we discussed the scientific article, "The Peak and Decline of World Oil and Gas Production" [7], and how OPEC nations in the Middle East dramatically increased their oil reserve estimates between 1985 and 1989 because these estimates determined how much oil they could produce under OPEC's quota system. The last among the nations to revise its oil reserve estimates upward was Saudi Arabia. Matt Simmons' critical analysis of Saudi Arabian oil production—based on over 200 publications from The Society of Petroleum Engineers—also raises doubts about Saudi Arabian oil reserves [2]. Also, the IEA's *World Energy Outlook 2005* report (*WEO 2005*) contains a detailed analysis of Saudi Arabia that can be interpreted as raising questions about that nation's oil reserves [8]. Below, we use publicly available information including that revealed in the 2004 CSIS presentation to investigate whether we can make any predictions about future Saudi Arabian oil production.

Oil Production and Oil Reserves in Saudi Arabia

The analysis we make in this section Oil Production and Oil Reserves in Saudi Arabia" and the next section "The Ten Giant 'Elephants'" is quite detailed and if you wish to skip over these details you can proceed directly to the section "Summary of Oil Production and Oil Reserves in Saudi Arabia" where the information is summarized.

The copy of the 2004 CSIS presentation that Saudi Aramco sent me contains some additional pages not found in the published version of the presentation. One of these pages has the title "Discovered Oil Resources" and states that the oil Saudi Arabia originally possessed underground, the OOIP, was 700 Gb. I have presented the information from this page in Fig. 13.2. By January 1, 2004 Saudi Arabia had produced 99 Gb, its remaining proven reserves were 260 Gb, and possible additional reserves were

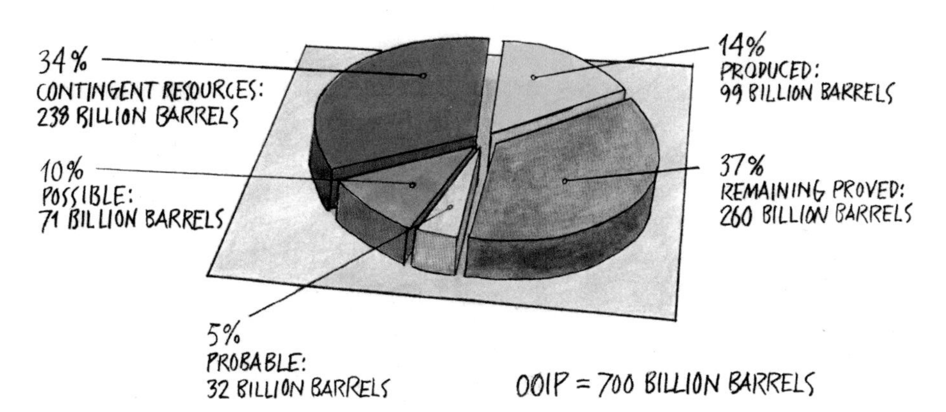

Fig. 13.2 Saudi Aramco stated in 2004 that the volume of oil found in Saudi Arabia (OOIP) was 700 Gb of which 99 Gb had been produced, 260 Gb existed as reserves, 32 Gb were probable reserves, and 71 Gb were possible reserves. The volume of oil that they regarded as not commercially viable to produce ("Contingent Resources") was 238 Gb. According to the regulations set by the Society of Petroleum Engineers (SPE) OOIP must also include oil that is "unrecoverable." However, no such oil is described. Possibly Saudi Aramco included this volume in Contingent Resources

32 Gb. This means that the remaining 2P reserves were 292 Gb (260 + 32 = 292 Gb) and that also means that the official recovery factor for Saudi Arabia in 2004 was 56% (99 + 260 + 32 = 391/700 = 0.56 = 56%). When Saudi Aramco discusses that it is possible to find an additional 200 Gb as OOIP they are relying on the analysis released by the US Geological Survey (USGS) in 2000 [9] and we have previously (in Chap. 6) seen how this is dubious. According to the USGS, by 2025 Saudi Arabia can expect to discover an additional 87 Gb as 2P reserves in previously unknown oilfields but, to date, they are falling far short of that goal.

The kerogen-containing sedimentary layers in Saudi Arabia originate mainly from the Jurassic and Cretaceous periods (see Fig. 4.3). Geographically they are found in the eastern area of the nation and under Saudi Arabia's territorial waters in the Persian Gulf. The layers are thicker the more eastward one moves.

Close to the Persian Gulf is Ghawar, the world's largest field of conventional crude oil. Its astonishing size is equivalent to a motorway 26 km wide stretching from Paris to Brussels (Ghawar is illustrated in Fig. 9.3). The heart of oil production from Ghawar is a limited area in the northernmost part containing one third of Ghawar's oil, Ain Dar–Shedgum. Saudi Aramco's CSIS presentation in Washington gave detailed information on production from this area.

Geologically, the Ain Dar–Shedgum area consists of two anticlines and the OOIP was given as 68.1 Gb. By 2004 they had produced 26.9 Gb from this area representing a recovery factor of 40%. This means that, in that part of Ghawar in 2004, the recovery factor was at a level that Meling estimated would be the average for the world in the future (Fig. 6.2). Remaining proven reserves were given as 13.9 Gb with 3.4 Gb given as future possible reserves. If these reserves eventuate then the recovery factor in this part of the oilfield will be 65%. It is high recovery factors such as these that Saudi Aramco is counting on in future and that many are skeptical about.

In their *World Energy Outlook 2005* report the IEA published a special analysis of nations in the Middle East and North Africa (MENA). The IEA's analysis of Saudi Arabia makes up a significant part of the report [8]. Saudi Arabia has approximately 90 known oilfields and around 30 of these are in production. Ten of these 30 oilfields possess proven reserves of 10 Gb or greater. In the world as a whole only 30 oilfields of this size exist. That one third of these oilfields lie in Saudi Arabia demonstrates how remarkable that nation's oil wealth is. Although 90 oilfields are known to exist in Saudi Arabia, more than 80% of the nation's oil exists in only the ten largest fields. Saudi Arabia's future production will be largely dependent on these ten oilfields. In Table 13.1 we describe these ten fields and their characteristics as reported by the IEA and Uppsala Global Energy Systems group (UGES) [10].

Table 13.1 Data for Saudi Arabia's ten largest oilfields[a]

Field	Discover/in production year	URR IEA Gb)	URR UGES (Gb)	2P 2004 (Gb)	Prod. 2004 (Mb/d)	C.P. 2004 (Gb)	Prod. 2030 (Mb/d)
Ghawar	1948/1951	147.0	66–150	86.3	5.772	60.7	3.0
Safaniyah	1951/1957	55.0	21–55	39.6	1.728	15.4	1.7
Manifa	1957/1964	23.1	11–23	22.8	0.050	0.3	0.9
Shaybah	1968/1998	21.5	7–22	20.7	0.492	0.8	0.8
Zuluf	1965/1973	20.0	11–20	18.2	0.407	1.8	0.4
Abqaiq	1941/1946	18.5	13–19	5.5	0.434	13.0	0.3
Berri	1964/1967	18.4	10–25	15.3	0.212	3.1	0.2
Khurais	1957/1963	17.0	13–19	16.8	0.000	0.2	1.2
Marjan	/1973	10.0		9.3	0.223	0.7	0.2
Qatif	/1946	10.0		9.2	0.100	0.8	0.2
Ten fields		340.5	152–333	243.7	9.418	96.8	8.9
Saudi		388.7		292.1	10.35	109.3	

[a]When a field was discovered and when production began, the upper and lower limits for URR obtained by the Global Energy Systems research group [10] and given in World Energy Outlook 2005, and 2P reserves at the end of 2004 [8]. Also shown is production for 2004 and cumulative production (C.P.) up to and including 2004 as well as estimated production in 2030 [8]

The Ten Giant "Elephants"

The Abqaiq Oilfield

To make an estimate of future Saudi Arabian oil production we must make particular assumptions for most of the oilfields except Abqaiq. This oilfield was found in 1941, began producing oil in 1946, and reached maximal production in 1973. There is sufficient information available to make a more detailed analysis of Abqaiq and with this as a starting point we can then expand our analysis to the other oilfields.

Abqaiq lies east of the northern section of Ghawar and is a long, thin perfect anticline with the highest point of the anticline (the shallowest point of the field) at the southern end of the oilfield (see Fig. 13.3). Approximately 77% of the oil exists in a layer of sedimentary rock given the name of Arab-D. Under this reservoir exists a layer of rock that is impermeable to oil and under that exists another sedimentary layer containing oil, a reservoir named Hanifa. This deeper layer contains 23% of the oil and lies at the southern end of the oilfield.

In the article "Abqaiq and Eat It Too" Joules Burn has collected various data available for Abqaiq [11]. In Fig. 13.3 we summarize and illustrate this publicly available information about Abqaiq. At the top of the figure you can see a satellite capturing images. When the image was captured upon which this illustration is based there were four rigs present drilling wells for oil production and water injection. Wells drilled for water injection have been marked as blue dots and because these are typically drilled at the periphery of this oilfield they indicate the size of the field. The red well indicates where natural gas was used for pressurization earlier in the oilfield's production history (see below). If we compare the earlier red gas pressurization well with the very large number of later blue water pressurization wells we can begin to appreciate both the enormous size of the field and the fact that water pressurization to maintain production has required commensurately enormous levels of investment. In the two reservoirs Arab-D and Hanifa illustrated in Fig. 13.3 the concentration of oil is indicated as it was reported in 2003. We can see that there is no oil in the center of the Arab-D reservoir. This is because the field was pressurized using natural gas during the period 1954–1978. Technically, pressurizing an oilfield with natural gas is easier than pressurization with water inasmuch as fewer wells are needed. Natural gas is used to pressurize an oilfield from above the oil layer rather than below the oil layer as occurs for water pressurization. For a domed structure such as Abqaiq, oil is concentrated at the central upper area and this means that gas pressurization produces oil from the most concentrated part of the oilfield at high rates. The international oil companies that were responsible for production from Abqaiq before Saudi Aramco took over used natural

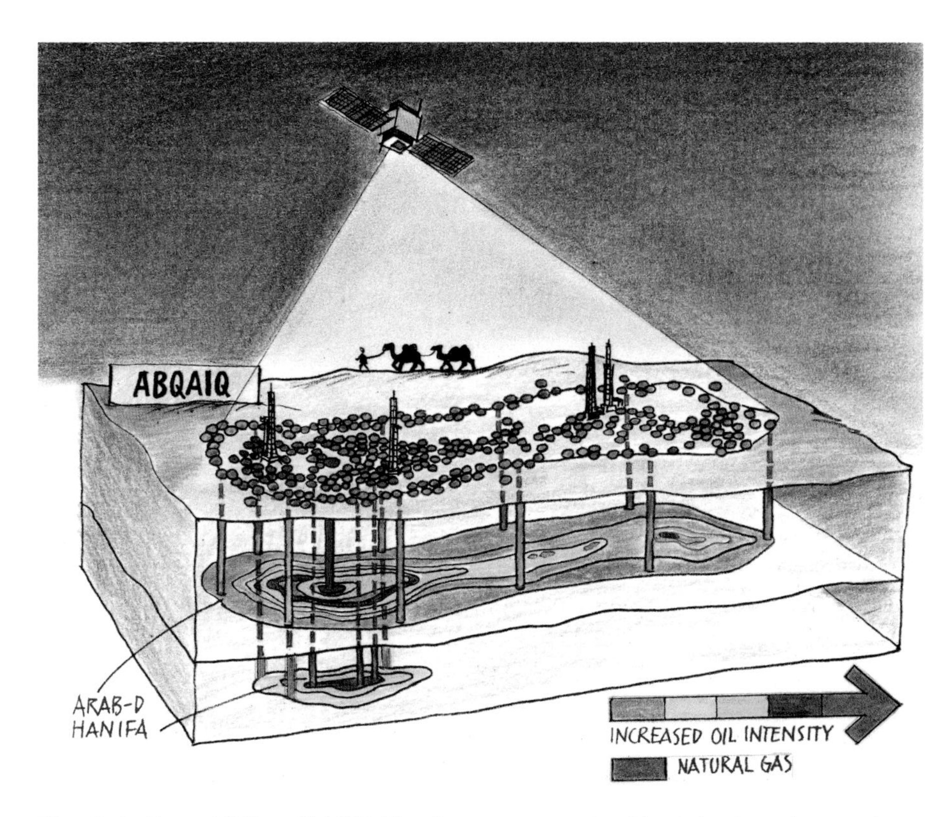

Fig. 13.3 From 1954 until 1978 Abqaiq was pressurized by injection of natural gas at the top of the anticline. The result was that they "skimmed the cream from the milk" and by 1973 production had increased dramatically to 1.1 Mb/d. Production then fell by 8% per year until Saudi Arabia began to reduce its oil production for political reasons. During that period of political tension wells were drilled at the oilfield's periphery to allow pressurization by injection of water. When political tensions eased and Saudi Arabian oil production increased again Abqaiq was ready for a new phase of oil production. In the illustration, wells for oil production are colored *green* and wells for water injection are colored *blue*. Note that there are also *blue* wells within the oilfield for pressurization of the Hanifa formation. The distribution of oil in the field in 1993 is indicated for the Arab-D and Hanifa oil-bearing sedimentary rock layers. One can see the height of the water in the field (*blue color*) and also that the injected natural gas (*red color*) has forced out the oil that originally existed in the top of the dome. When the satellite image was taken upon which this illustration is based [11] four drilling rigs were present and these are indicated (The oil distribution profile for Hanifa is not public knowledge, therefore the coloration of that part of the illustration is only an artist's impression)

gas injection to maximize oil production with the least investment. During the period of gas pressurization the field achieved its highest production rate (in 1973) at 1.1 Mb/d. The subsequent phase of oilfield development used water injection but this required a far greater number of wells.

Detailed production data for the two reservoirs of Abqaiq were presented at a conference in Houston in 2007 and from that information we have created a database that can be used for detailed calculations [12].

After production began in 1946 Abqaiq reached a first minor peak in 1951 at 0.5 Mb/d. It then dropped below 0.3 Mb/d in the mid-1950s as it completed the first phase of oil production (see Fig. 7.4). Pressurization of the field was begun in 1956 and production then shot up to reach a maximum of 1.1 Mb/d in 1973. When Saudi Arabia reduced oil production for political reasons in the 1980s, Abqaiq's production rate had already fallen by 8% per year for 6 years. Production was then kept at a lower level determined by the prevailing political reality until the 1990s. During this period of low production Saudi Aramco began to expand the system for pressurization of Abqaiq with water by drilling wells along the outer edge of the field. Then, as political tensions eased in the 1990s, Saudi Arabia's oil production regained momentum and by 2003 it had increased to 8.1 Mb/d. Until 1997 Abqaiq had contributed approximately 8% of annual Saudi Arabian oil production but after that its contribution declined every year. Thus, Abqaiq had reached the point of irreversible decline and this means that for the first time ever we can calculate the characteristic production parameters of this representative Saudi Arabian oilfield.

When we analyze the detailed production history for the two reservoirs of the Abqaiq oilfield we see that there have been problems with production from Hanifa whereas production from Arab-D has progressed well [12]. For Arab-D the rate of decline after maximal production in the post-1980s phase is 4.6% and it is this rate that we use as the depletion of remaining recoverable resources parameter, DRRR (see the section"The Elephants' Production Rate: Depletion of Remaining Recoverable Resources (DRRR)", Chap. 9). If we use 14.3 Gb as the URR for Abqaiq [12] we find that the calculated production parameter values are too high. This means that there exists more recoverable oil in the oilfield than Saudi Aramco estimates. An additional 700 million barrels probably exists. If we continue to assume that the OOIP for Arab-D is 22.8 Gb [11] then this would mean a recovery factor of 65%. We can state that the numbers given above for Abquiq by Saudi Aramco are not overestimates.

At the presentation at CSIS in Washington, Saudi Aramco made it very clear that their intention is to maximize the oil recovery from their fields by producing oil at low rates. This practice is clearly demonstrated by the history of oil production from Abqaiq and we can now see that it is a successful policy. We have previously discussed how oil reservoirs usually consist of sandstone or limestone. In Saudi Arabia the reservoirs in general are limestone and these generally give higher recovery rates if they are not stressed during production. If Abqaiq had been a sandstone reservoir we presumably would not have seen such high recovery rates.

The Shaybah Oilfield

The most recent supergiant field to be put into production is Shaybah. At the CSIS symposium in Washington Saudi Aramco revealed information showing that production from Shaybah appears to resemble that of Abqaiq. Originally they chose to produce oil from Shaybah at 0.5 Mb/d but production will eventually be increased to 0.8 Mb/d. From the data on Shaybah found in Table 13.1 and using the assumption that Shaybah's production parameters are similar to those of Abqaiq's we can now investigate when we can expect production from Shaybah to begin to decline. Because oil production from the field has used modern technology from the start we can expect that production will proceed more efficiently than occurred for Abqaiq. Therefore, we assume that a maximum of 4.6% of Shaybah's recoverable oil will be produced per year (i.e., a DRRR of 4.6%). If the oilfield had been situated under deep water we would have assumed a higher DRRR number but for this analysis we accept the production profile that is typical practice for Saudi Aramco. Our analysis indicates that Shaybah can maintain production at this low rate of DRRR until it begins to decline in 2060. Employment at the Shaybah oilfield must be regarded as one of the world's most secure jobs!

The Ain Dar-Shedgum Section of Ghawar

Before we attempt to analyze Ghawar in its entirety we first look at the Ain Dar-Shedgum section of the oilfield. Production from this section is 2 Mb/d. Based on 2P estimates of recoverable oil, 26.9 Gb has been produced so far and 17.5 Gb remains to be produced. If we assume that the maximal rate of DRRR will be 4.6% then this section of Ghawar should enter its decline phase in 2019 and production in 2035 will be less than 1 Mb/d.

The Khurais and Manifa Oilfields

So far we have discussed two and a half fields (Abqaiq, Shaybah, and the northern part of Ghawar). Next on our list are Khurais and Manifa. To develop these fields Saudi Aramco has sought international finance because the investments required are so enormous. In our article, "The Peak Of The Oil Age," we discuss development of these oilfields [13]. In Table 13.1 the 2P reserves for Khurais are given as 16.8 Gb. However, when the media discuss the project for the development of this field (that also includes development of the smaller field Abu Jifan), a volume of 27 Gb is named. If we assume that 27 Gb is the OOIP and we accept the IEA's URR value for Khurais of 17 Gb then we obtain a recovery factor of 63% which is greater than Saudi Arabia's official average value of 56% for all fields, but of the same order as for Abqaiq. According to Saudi Aramco's description of this project, they are aiming for a production volume of 1.2 Mb/d. Using this volume and the fact that 0.2 Gb has already been produced from the field, we can estimate the field's future DRRR and we see that this reaches the crucial 5% level in 2030 after which the oil production volume should decline.

Manifa is an offshore oilfield off the northeast coast of Saudi Arabia. This field has the heaviest oil of all the Saudi Arabian giant fields and the sulphur content is also very high. It has been difficult to find buyers for this low-quality oil. This is one of the reasons why Saudi Aramco is now building its own new refinery that, according to the plans, will have the capacity to process 0.9 Mb/d when completed. Originally, the refinery should have begun processing oil in 2011 but the 2008 financial crisis and reduced oil demand have led to delays in completion. We presume that production from the refinery will now begin in 2012 and that it will take 3 years until it reaches maximal production of 0.9 Mb/d, equal to the possible future production from Manifa. Information varies on the size of the Manifa oilfield and it is also difficult to judge how much of the OOIP is producible. To assume a recovery factor of 50% from this field containing heavy oil without any supporting data would be unrealistic. In *WEO 2005* Manifa's URR is given as 22.8 Gb but Saudi Aramco cites a volume of 10 Gb. Our analysis of the available literature reveals a lower limit of 11 Gb and that is the number we use in our analysis. Using these assumptions we predict that Manifa will show plateau production until 2023. The big question regarding Manifa's future production, that could reach a rate higher than 0.9 Mb/d, is whether the oil industry will build sufficiently many refineries that can handle Manifa's heavy oil. In *WEO 2005* the IEA suggests that this will not occur.

The Safaniyah Oilfield

Safaniya is the world's largest offshore oilfield and is one of the cornerstones of Saudi Arabia's oil production. If Safaniya's URR is 55 Gb (the upper limit in our analysis) then this would be the world's third largest oilfield. In *WEO 2005* this oilfield is stated as having a peak of production at around 2 Mb/d in 2000 and that production has since decreased to 1.7 Mb/d which is also given as the oilfield's maximum sustainable production level. Considering the oilfield's size, it is technically possible to maintain a production level of 1.7 Mb/d until 2035 but the continuous need for drilling to maintain production is enormous on this gigantic offshore field. Like most other oil companies, Saudi Aramco accepts that oilfields that have been in production for extended periods have a natural decline rate of 6% per annum if additional investments are not made. The IEA does not believe that Saudi Aramco will be able to perform sufficient drilling to maintain production and that future production will be closer to 1.33 Mb/d.

The Zuluf, Berri, Marjan, and Qatif Oilfields

Investment projects of the scale described for Khurais and Manifa above are called "megaprojects." In 2004, Zuluf (0.407 Mb/d), Berri (0.212 Mb/d), Marjan (0.223 Mb/d), and Qatif (0.100 Mb/d) had a combined production of 0.94 Mb/d. In the future, all four fields may become the subject of large technological investment, megaprojects, such that their combined production may increase by several hundred thousand barrels per day. However, it is a fact that these fields already require massive investment simply to maintain their current production levels. We assume that Zuluf, Berri, and Marjan will retain the level of production they had in 2004 as their maximum sustainable production but that production from Qatif will increase somewhat such that the future combined production level from the four fields will be 1.0 Mb/d. If the recovery factor is the only limiting factor for these oilfields then there will be no problem in maintaining this production level and there might also be possibilities for increasing production. The determining factor will be how much Saudi Aramco is willing to spend.

The Ghawar Oilfield

We have now discussed nine of Saudi Arabia's ten supergiant oilfields. The only remaining oilfield to discuss is Ghawar. When analyzing Ghawar it is apparent that production from its northern section, Ain Dar–Shedgum, will decrease by 2035 and Saudi Aramco admitted as much in their presentation at CSIS in Washington. Ghawar recently underwent a revamp and Saudi Aramco intends to continue with this expansion of activity. To pressurize an area equivalent to a 26 km wide motorway between Paris and Brussels is an enormous undertaking and the only water mass of sufficient size available is saltwater in the Persian Gulf. We cannot be certain what will happen when saltwater is injected into the oilfield but we know that when salt forms a sedimentary layer it is impervious to oil. The problem for the future of production from Ghawar might be that Saudi Aramco must constantly move its injection wells farther and farther into the field. If we analyze the sustainability of production from Ghawar assuming a maximum annual DRRR of 4.6% from that part of Ghawar currently producing 3.3 Mb/d, then we see that this oilfield can maintain plateau production until 2022 and production will then decline to 1.8 Mb/d by 2035. Certainly, there is no doubt that Ghawar will begin to decline before 2035.

Total Saudi Arabian Production

Our detailed analysis described above has been based on the assumption that Saudi Arabia has a maximum rate of DRRR of 4.6% which is the DRRR rate that we noted for the Arab-D reservoir of Abqaiq. If we combine the future production rates that we determined for the individual supergiant fields, then we predict that Saudi Arabia's ten supergiant fields will have a total production of 8.9 Mb/d in 2035 which is a decrease of 0.5 Mb/d compared with production in 2004. For 2012 we calculate that the production capacity for these ten fields is 10.9 Mb/d.

Saudi Aramco says that its production capacity in 2012 will be 12.0 Mb/d. That means that the remaining 20 smaller fields currently in production will produce a total of 1.1 Mb/d compared with 1.0 Mb/d in 2004. This projected increase in production is quite realistic. If we examine Saudi Aramco's projected production under a scenario where they hold production at the maximum sustainable level then we see no problem with this occurring when the assistance of the 20 smaller fields and the exploitation of yet undeveloped fields is considered. Production from these sources can also be expanded in order to compensate for the decline in production from the supergiant fields.

Summary of Oil Production and Oil Reserves in Saudi Arabia

Abqaiq was the first Saudi oilfield and was put into production in 1946. Its production has now entered the "natural decline" phase. The maximal annual DRRR for the Arab-D reservoir of Abqaiq is 4.6%. The available information on Abqaiq's production allows us to assess its reserve size and we conclude that the stated reserves should be increased by 0.7 Gb. This means that the final recovery factor for the Arab-D reservoir of Abqaiq may be 65% which is higher than the average recovery factor for the combined oilfields of Saudi Arabia.

At the presentation at CSIS in Washington the representatives of Saudi Aramco stated that the company planned production from oilfields in order to give maximal sustainable capacity (MSC). For Abqaiq the maximal rate of DRRR that gives MSC is 4.6%. In our analysis we have used this DRRR rate for Saudi Arabia's other fields. If Saudi Aramco wished to produce oil at a faster rate than 12 Mb/d then the DRRR parameter would need to be increased. In the rapidly exploited North Sea we saw a record high rate of DRRR of 6.0%. If Saudi Arabia depleted its remaining recoverable reserves at that rate then they could reach a production level of 15 Mb/d. A DRRR rate higher than 6.0% has never been seen for any oil-producing region, which implies that 15 Mb/d is the maximum possible daily rate of oil production from Saudi Arabia.

Saudi Arabia has the world's largest reserves of crude oil and they state that their proven reserves are 260 Gb. They may also have an additional 32 Gb that could become proven reserves. Using these values we can calculate that the recovery factor for the nation as a whole will be 56%. One important factor when discussing Saudi Arabia's recovery factors is that the reservoirs are of limestone and that these limestone reservoirs normally have higher recovery factors (than, e.g., sandstone reservoirs). Judging by the production performance of Abqaiq, a national recovery factor of 56% is not an unrealistic assumption.

In 2004 Saudi Arabia produced 3.9 Gb of crude oil and had 131 Gb as developed reserves at the start of that year. This means that they produced 3.0% of their developed reserves. If they did not develop any new reserves this means that, at that annual rate of production, they would reach the critical DRRR of 4.6% within 10 years and production would then begin to decline by at least 4.6% per annum thereafter. This means that they must now invest in developing those reserves within existing fields that are currently undeveloped. Inasmuch as 83% of Saudi Arabia's proven reserves exist in its ten largest fields it is predominantly these fields that require massive investment for development.

The Uppsala Global Energy Systems research group has made a special study of Saudi Arabia's ten supergiant fields that, together, possess over 80% of that nation's proven and possible reserves (2P; see Table 13.1). In 2004, production from these fields was 9.4 million barrels per day (Mb/d) and our MSC analysis shows that production from them can reach 10.9 Mb/d in 2012 and then fall to 8.9 Mb/d by 2030. The remaining 20 oilfields that were in production in 2004 were producing 1.0 Mb/d at that time and this must rise to 1.1 Mb/d in 2012 if Saudi Arabia is to attain its officially stated production capacity of 12.0 Mb/d. To maintain production at 12 Mb/d until 2030 the volume from oilfields other than the ten super-giants must increase to 3.1 Mb/d. Our MSC analysis shows that production at 12.0 Mb/d is possible until 2029. At the presentation at CSIS in Washington, Saudi Aramco identified 2033 as the point in time when 12.0 Mb/d of MSC production could no longer be sustained. That our analysis arrives at a date of 2029 for this timepoint is, in fact, in remarkably close agreement with Saudi Aramco's date considering the paucity of public data on which our analysis is based and our unique methodology (based on our discoveries regarding DRRR behavior for giant fields; see Chap. 9).

Our analysis of projected Saudi Arabian production in 2030 can be compared with that made by the IEA in *WEO 2005*. The IEA projected Saudi Arabian oil production at 18.2 Mb/d of which 3.4 Mb/d is to come from upward revision of reserves in previously discovered oilfields and from completely new fields to be found by 2030. The volume of new reserves needed by 2030 is around 50 Gb. We regard this as unrealistic.

Without intervention (additional investment) Saudi Arabia's production from fields that were producing in 2004 would decline by 6% per year [8]. This is the same rate of decline that we found in our analysis of global oil production [14]. Without additional investment, production that was 10.35 Mb/d in 2004 would be 2.0 Mb/d in 2030. The projects in Khurais, Manifa, and Shaybah compensate for a decline of 2.9 Mb/d but more large projects are needed. Thousands of wells must be drilled for those fields that are currently in production. All this work will require costly investment. When we state from our analyses that it is possible to produce 12.0 Mb/d in 2030 this is based on the assumption that financing will be found for these projects.

WikiLeaks and Oil Reserves of Saudi Arabia

On 8 February 2011 the *Guardian* newspaper published an article on Saudi Arabia's oil reserves, "WikiLeaks cables: Saudi Arabia cannot pump enough oil to keep a lid on prices—US diplomat convinced by Saudi expert that reserves of world's biggest oil exporter have been overstated by nearly 40%" [15]. The article discussed a confidential diplomatic cable sent in November 2007 from the US ambassador in Riyadh to the US Department of Energy and the CIA. The subject of the cable was given the title, "Former Aramco Insider Speculates Saudis Will Miss 12.5 Mb/d in 2009" [16]. The insider was Dr. Sadad al-Husseini who had formerly held the position of executive vice president for exploration and production at Saudi Aramco. This cable was one of the thousands of confidential US diplomatic cables published by Wikileaks on its website in late 2010 and early 2011 with selected cables published by leading newspapers such as the *Guardian*.

Those with any knowledge of Saudi Arabia's oil reserves realized immediately that someone had misunderstood the cable's contents. The *Guardian* wrote, "The cables, released by WikiLeaks, urge Washington to take seriously a warning from a senior Saudi government oil executive that the kingdom's crude oil reserves may have been overstated by as much as 300 billion barrels—nearly 40%." As we discussed earlier, Saudi Arabia's official oil reserves are 260 Gb. That the *Guardian* could assert that oil reserves are "overstated by 40%" after subtracting 300 Gb from official reserves of 260 Gb shows that John Vidal, their environment editor, has very limited knowledge of Saudi Arabia's reserves.

Three days after that article appeared in the *Guardian* I received an email from Dr. Sadad al-Husseini in which he expressed his frustration over what had been reported:

> Greetings from Dhahran and hope this e-mail finds you well.
>
> The recently released WikiLeaks note on comments attributed to me at year end 2007 had so many misrepresentations that I thought I would issue a press release to correct them. I attach below the text I sent to Thomson Reuters for their release and hope you'll find it interesting in terms of how discussions can be re-structured to fit specific agendas or media purposes. Meanwhile best personal wishes and hope this kind of news is the least of our worries.
> Sadad

The fact that WikiLeaks published the cable from Riyadh to Washington and that we have access to Dr. Sadad al-Husseini's press release [17] means that we can make the following interesting comparison between the information that was released to the public and what was reported confidentially to, among others, the CIA.

First, we can conclude from the cable that Dr. Sadad al-Husseini met "CG and Econoff" where CG is the Consul General John Kincannon. The identity/role of Econoff is not explained in the cable. According to al-Husseini this was not a planned meeting but only a chance encounter during which, among other things, Saudi Arabia's oil reserves were mentioned. This means that notes were not taken of the meeting and the observations in the cable were written after the event.

An interesting part of the cable is the following report:

> In a December 1 presentation at an Aramco Drilling Symposium, Abdallah al-Saif, current Aramco Senior Vice President for Exploration and Production,

> reported that Aramco has 716 billion barrels (bbls) of total reserves, of which 51% are recoverable. He then offered the promising forecast—based on historical trends—that in 20 years, Aramco will have over 900 billion barrels of total reserves, and future technology will allow for 70% recovery.

Compared with the information given at the CSIS in Washington in 2004 it appears that, by 2007, Saudi Aramco had not changed its view on its oil reserves and resources (700 Gb as discovered OIIP and 200 Gb yet to be found), but CG mistakenly states, "900 billion of total reserves" when these would be, in fact, OOIP resources. Dr. Sadad al-Husseini agreed with the generally accepted estimate of 360 Gb for "original proven reserves, oil that has already been produced or which is available for exploitation based on current technology." He also discussed the importance of investment in people, infrastructure, and management to achieve an output of 12.5 Mb/d in 2009. These discussions were important in 2007 but we now know that, in fact, Saudi Aramco did make the necessary investments. From our analysis in the previous sections of this chapter we know that a production level of 12 Mb/d can be achieved and maintained if the necessary investments continue to be made. The duration of a Saudi Arabian plateau of production at 12.5 Mb/d will be determined by the extent of investment.

In 2004 Saudi Aramco stated that a production plateau at 12 Mb/d would end in 2033 whereas our analysis concluded that decline would begin in 2028. In 2028 the cumulative oil produced would be a little more than half of the 360 Gb stated as total possible production. It is also interesting to note that al-Husseini believes that the rate of oil production from the world's currently producing crude oil fields is currently declining each year by 4 Mb/d which is the same result that we reported [14] and that the IEA presented in *WEO 2008* [18].

The conclusion we can draw from examining the information in the diplomatic cable is that it did not state anything that was not previously known but it contains a little confusion regarding reserves and resources. It was this confusion that the *Guardian* interpreted as al-Husseini disagreeing with Saudi Aramco's reserve estimates when, in fact, they are in agreement. The fact that the leaked cable was originally a secret document together with *The Guardian*'s mistaken interpretation resulted in a "news" story that circled the globe and caused many people to form views on Saudi Arabian oil reserves that are not relevant. Such is the power of the media.

What Saudi Arabian politicians say on future production is most often found as fragmented statements in various newspaper articles. Therefore, it can be interesting to read one statement in its entirety that was made on May 24, 2005 in San Francisco by Saudi Arabia's minister of petroleum and mineral resources, H. E. Ali bin Ibrahim Al Naimi [19].

> Saudi Arabia has initiated a number of mega oil projects which will significantly increase its production capacity to both meet demand and maintain spare capacity. These projects represent a combined production capacity of more than 3 MBD, part of which will be utilized to offset natural decline and the rest to expand capacity. By 2009, we expect our maximum sustainable capacity to rise from the current 11.0 to 12.5 Mb/d.
>
> Additional projects have been identified and can be advanced, as needed, to meet any new supply requirements. In fact, the Kingdom has evaluated a production capacity scenario of 15 Mb/d, which can be implemented when dictated by market demand.

It is interesting to note that the production target for 2009 of 12.5 Mb/d existed as early as 2005 and that plans existed for 15 Mb/d of production capacity. However, the belief that Saudi Arabia could "meet any new supply requirements" must be regarded as an exaggeration.

Saudi Arabia has now undertaken its megaprojects and our analysis has shown that it is technically possible for their production to reach 15 Mb/d but that this would have negative consequences for their sustainable long-term production. In the past, future scenarios by the US Energy Information Administration (EIA) and the IEA have been promoted where Saudi Arabia would produce as much as 22 Mb/d but after intensive criticism that projection is no longer made. As the November 2007 cable said of al-Husseini, "He stated that the IEA's expectation that Saudi Arabia and the Middle East will lead the market in reaching global output levels of over 100 Mb/d is unrealistic, and it is incumbent upon political leaders to begin understanding and preparing for this "inconvenient truth"". In their *2010 World Energy Outlook* report the IEA has suggested a plateau of Saudi Production at 14.6 Mb/d [20].

Twilight in the Desert

In the Middle East sunset passes quickly whereas the summer sunset in Swedish Tällberg is very long. However, the sunset for Saudi Arabian oil production will more closely resemble that of a Scandinavian summer. For example, the oilfield Shaybah was put into production at the end of the 1990s under a production plan that should see it produce oil at a steady pace until 2060. If Saudi Aramco wished to double Shaybah's production to 1.6 Mb/d from 2015 then sunset for Shaybah would arrive already in 2025 but this would mean abandonment of the company's MSC-strategy and would cause Shaybah's production history to be comparatively short in relation to the field's size. It is sometimes commented that Saudi Arabia's production could be expanded to 15 Mb/d. Technically, this is possible but production at this level could only be maintained for a limited time.

It can be difficult to comprehend how immensely large an oilfield must be for it to be able to produce oil at a steady rate for 50 years before beginning to decline. The fact that the Arab-D reservoir in the Abqaiq oilfield has now entered its natural decline phase means that, for the first time, we can obtain the data required to make projections of the behavior of the Saudi Arabian supergiant oilfields. These data show that it is not unrealistic to regard 292 Gb as Saudi Arabia's 2P reserves. However, it is difficult to accept 260 Gb as Saudi Arabia's proven reserves when they have only developed a limited amount of these (131 Gb in 2004).

When examining possible future Saudi Arabian production to 2035 knowledge of the exact size of their reserves is less important than the comprehension that, although oil production could be held at 12 Mb/d for this period, a doubling of this rate would be damaging for later Saudi Arabian production. Oil production at that very high level is not consistent with Saudi Aramco's stated production policy. When the IEA and the EIA make prognoses of future oil production they should keep this in mind.

Ali Samsam Bakhtiari was quite justified in questioning projections of future Saudi Arabian production at 14 Mb/d in 2010 and 22 Mb/d in 2020 [1]. Also, the twilight in the Saudi desert described by Matt Simmons can very well become reality when only a limited proportion of the reserves are developed and enormous investments are required to meet planned production targets. Investors in the oil industry have every reason to be skeptical of the IEA's and EIA's predictions regarding future Saudi Arabian oil production.

During my discussion with Prince Turki Al-Faisal in Tällberg I stated that, "To not care about Peak Oil is to put one's head in the sand." The reply I received was, "Do you mean that we are putting our head in Saudi Arabia's desert?" It is important to note that Saudi Arabia does indeed make careful analysis of its oil production reality by (figuratively) putting its head in the sand (to assess its reserves) and that it is not willing to sacrifice its future by overproducing its oilfields. Nature and the laws of physics show that gentle and steady production of oil maximizes production from oilfields. Even at an increased oil production rate of 12 Mb/d Saudi Arabia would still be the last nation on Earth to pass Peak Oil; all other nations will pass their peak of oil production long before it. However, this means that no other nation will be able to export oil to Saudi Arabia when falling production means that Saudi Arabia can no longer supply its own needs. Therefore, it is in Saudi Arabia's long-term interest to produce oil at a rate that maximizes recovery of oil from its fields.

I hope that, one day, I will have the opportunity to visit Abqaiq and Ghawar and maybe even Shaybah way out in the desert so that I can experience again the natural wonder that a sunset in the desert can be.

References

1. Samsam Bakhtiari, A.M.: OPEC's evolving role: OPEC capacity potential needed to meet projected demand not likely to materialize. Oil Gas J., 9 July 2001. http://www.ogj.com/index/article-display/106545/articles/oil-gas-journal/volume-99/issue-28/special-report/opecs-evolving-role-opec-capacity-potential-needed-to-meet-projected-demand-not-likely-to-materialize.html (2001)
2. Simmons, M.: Twilight in the Desert: The Coming Saudi Oil Shock and the World Economy. Wiley, Hoboken (2005)
3. Prince Turki Al-Faisal: Address at Tällberg Forum 2006. Royal Embassy of Saudi Arabia, Washington, DC, 30 June 2006. http://www.saudiembassy.net/archive/2006/speeches/page27.aspx (2006)
4. Tällberg Foundation: About the Forum. http://www.tallbergfoundation.org/TÄLLBERGFORUM/Abouttheforum/tabid/161/Default.aspx (2011)
5. CSIS—Center for Strategic and International Studies: Future of Global Oil Supply: Saudi Arabia (presentations and audio). http://csis.org/event/future-global-oil-supply-saudi-arabia
6. Abdul Baqi, M.M., Saleri, N.G.: Fifty-year crude oil supply scenarios: Saudi Aramco's perspective, Saudi Aramco, 24 Feb 2004, CSIS, Washington, DC. http://csis.org/files/media/csis/events/040224_baqiandsaleri.pdf (2004)
7. Aleklett, K., Campbell, C.: The peak and decline of world oil and gas production. Minerals and Energy—Raw Materials Report, vol. 18, pp. 5–20. http://www.ingentaconnect.com/content/routledg/smin/2003/00000018/00000001/art00004, http://www.peakoil.net/files/OilpeakMineralsEnergy.pdf (2003)
8. WEO: World Energy Outlook 2005. International Energy Agency (IEA), November 2005. http://www.worldenergyoutlook.org/2005.asp (2005)
9. US Geological Survey: World Petroleum Assessment 2000—Description and Results. US Department of the Interior, US Geological Survey. http://pubs.usgs.gov/dds/dds-060/ (2000)
10. Robelius, F.: Giant oil fields—the highway to oil: giant oil fields and their importance for future oil production. Uppsala Dissertations from the Faculty of Science and Technology, ISSN 1104–2516; 69. http://uu.diva-portal.org/smash/record.jsf?pid=diva2:169774 (2007)
11. JoulesBurn: Abqaiq and eat it too (or, more geological analysis of potential Saudi depletion). The Oil Drum, May 2008. http://www.theoildrum.com/node/3923 (2008)
12. Saleri, N.G.: The past and predicted future production from Abqaiq (Nansen G. Saleri is former head of Reservoir Management in Saudi Aramco), CERA Week

in February 2007 (The grapf can be found at, Burn, 2008). http://www2.cera.com/ceraweek2007/home/1,3277,,00.html (2007)
13. Aleklett, K., Höök, M., Jakobsson, K., Lardelli, M., Snowden S., Söderbergh, B.: The peak of the oil age—analyzing the world oil production reference scenario in World Energy Outlook 2008. Energy Policy, **38**(3), 1398–1414 (accepted 9 Nov 2009, Available online 1 Dec 2009, http://www.tsl.uu.se/uhdsg/Publications/PeakOilAge.pdf) (2010)
14. Höök, M., Hirsch, R., Aleklett, K.: Giant oil field decline rates and their influence on world oil production. Energy Policy, **37**(6), 2262–2272. http://www.sciencedirect.com/science/article/pii/S0301421509001281; http://www.tsl.uu.se/uhdsg/Publications/GOF_decline_Article.pdf (2009)
15. Vidal, J.: WikiLeaks cables: Saudi Arabia cannot pump enough oil to keep a lid on prices. The Guardian, 8 Feb 2011. http://www.guardian.co.uk/business/2011/feb/08/saudi-oil-reserves-overstated-wikileaks (2011)
16. Former Aramco Insider speculates Saudis will miss 12.5 Mb/d in 2009. US Ambasy in Riyadh, 10 Dec 2007, Ref: Riyhad 1950. http://www.guardian.co.uk/business/2011/feb/08/oil-saudiarabia?intcmp=239 (2007)
17. Al-Husseini S.: Press release, 11 Feb 2011. http://aleklett.wordpress.com/2011/02/12/press-release-by-dr-sadad-al-husseini/ (2011)
18. WEO: World Energy Outlook 2008. International Energy Agency, November 2008. http://www.iea.org/textbase/nppdf/free/2008/weo2008.p (2008)
19. Al-Naimi, A.: Globalization & the Future of the Oil Market. World Affairs Council for Northern California and the Council on Foreign Relations in San Francisco, California, at Banker's Club of San Francisco, California, 24 May 2005. http://www.susris.com/articles/2008/ioi/080822-naimi-oil.html (2005)
20. WEO: World Energy Outlook 2010. International Energy Agency, November 2010. http://www.worldenergyoutlook.org/2010.asp (2010)

Chapter 14

Russia and the USA: The Oil Pioneers

The volume of oil the world currently uses in 3 weeks, 1.7 billion barrels, is equal to all the oil produced during the nineteenth century. The first reported oil production—of 2,000 barrels per year—came from Romania in 1857. However, many regard the beginning of the Oil Age as occurring on August 27, 1859, when the drilling team of Edwin Drake reached a total depth of 21 m (69.5 ft). Near the end of the day the drill bit slipped when it hit a new formation so the team decided to stop and continue the next day (see Fig. 14.1). On the following day, Drake's driller Billy Smith looked into the hole and was surprised and delighted to see crude oil rising up. Drake was summoned and the oil was brought to the surface with a hand pitcher pump. The oil boom in Pennsylvania had begun. By 1860 the United States had overtaken Rumania in oil production and it remained the world's leading producer until 1898 when Russia took the lead (with the help of the Nobel brothers' company Branobel, see the section "The Discovery of Oil Seeping Out of the Ground", Chap. 5).

After the First World War the United States overtook Russia (then part of the Soviet Union) in oil production and by 1930 the United States was producing 2.3 million barrels per day (Mb/d) whereas the USSR only managed 175,000 barrels per day. In 1971 the United States reported what was to be its highest ever level of oil production (peak production) at 9.3 Mb/d. By that time the USSR had achieved 6.2 Mb/d. Three years later the Soviet Union exceeded the US production level and from 1978 until 1990 the USSR produced more than 10 Mb/d with a peak production of 11.5 Mb/d in 1987. Today, after the dissolution of the Soviet Union,

K. Aleklett *Peeking at Peak Oil*, DOI 10.1007/978-1-4614-3424-5_14,

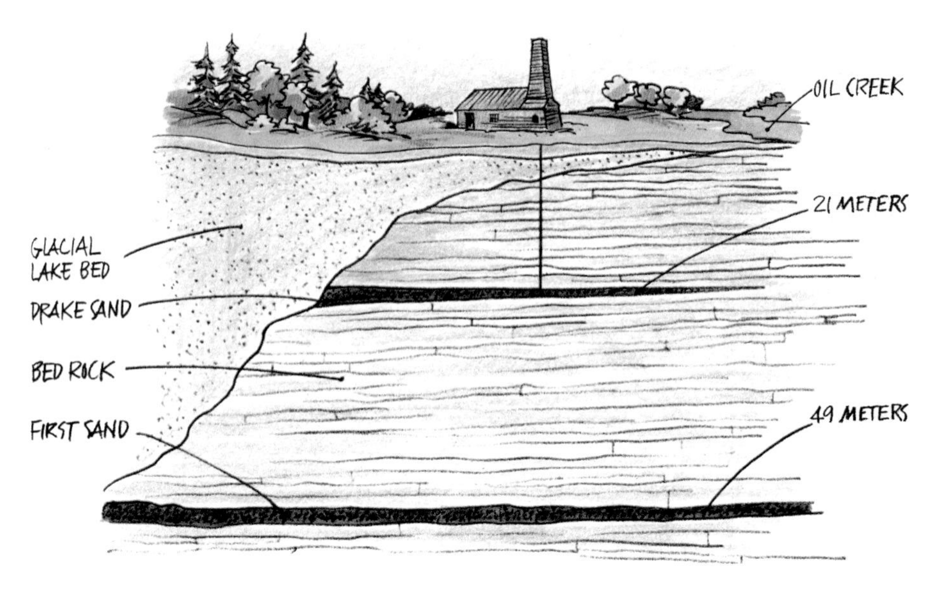

Fig. 14.1 When Drake drilled his well on the bank of Oil Creek he was fortunate because a small pocket of oil-bearing rock existed immediately below at a depth of 21 m (Drake Sand). If he had drilled a little farther up the river he would have needed to reach a depth of 49 m to find oil (First Sand). The largest oil reservoir in that area existed at 122 m

independent Russia produces approximately 12% of the world's oil and the United States contributes 9% of world production. Nevertheless, the United States leads the world in oil consumption by using 22% of the oil produced every day whereas Russia only uses 3%. This means that the United States is the world's largest importer of oil whereas Russia is the world's second largest exporter after Saudi Arabia. During the entire twentieth century oil was central to the development of both the United States and Russia and it will continue to be a crucial factor during the current century.

When the Soviet Union collapsed in 1991 there were many who dismissed Russia as a future great power. When President Bush received President Putin for the first time in 2001 at his ranch in Texas, Putin met an arrogant host who did not consider that Russia might one day reclaim great power status. Oil has changed that situation completely. An important aspect of Peak Oil is what it will mean for the future of the United States and Russia.

Historic Oil Production and Consumption in the United States

Another legendary oilman, Hamilton McClintock, owned property along Oil Creek in Titusville, Pennsylvania. Oil seeped into Oil Creek from a natural oil well on his property. Using buckets to collect the precious flow he was able to fill 20–30 barrels with oil each year. In August 1861 he decided to drill a first well, McClintock No. 1, to see whether he could find more oil and, indeed, he found an oilfield at a depth of 189 m. His rate of oil collection subsequently leapt to a (then) astonishing 5,000 barrels per day (b/d). Now 150 years later the oil flow has fallen to only 12 barrels per month but the McClintock No. 1 well is still producing oil, making it the world's oldest continually producing oil well.

A commonly asked question is "When will oil run out?" Using McClintock No. 1 as an example we can answer, "Oil will not run out during the next 100 years but, like the McClintock No. 1 well, all the world's other oil wells will reach a production maximum and then decline." Up until 1930, exploration revealed ever larger and more numerous oilfields in the United States, but after the Second World War fewer and mainly smaller fields have been found. (The exception to this is the Prudhoe Bay oilfield in Alaska.) Figure 14.2 illustrates this declining discovery trend for the states of the United States lying south of Canada (the "US Lower-48"). The small increase in discovered volumes around the year 2000 is due to the development of new technology that permitted exploration in deeper water. However, compared with what was found on land, the deep water discoveries represent a very limited volume (see Chap. 12). In total, approximately 200 billion barrels (Gb) of crude oil have been discovered in the United States which is about 10% of the oil reserves discovered globally.

In Chap. 2 we discussed the Hubbert model as a method for predicting oil production and the peak of US oil production in 1971. The oil discovery and production curves for the United States are very widely known but our discussion of the history of Russian and US oil production would be incomplete if we did not examine them here. In Fig. 14.2 we show these classic curves and we also show how the number of oil wells being drilled each year changed during the course of the twentieth century.

In 1956 King Hubbert famously predicted the peak year of US oil production by fitting a logistic curve to the United States' previous production history [1]. When he did this discoveries of oil in the United States were already on a downward trend. King Hubbert estimated that the ultimately recoverable resources (URR) for the US Lower-48 states would be between

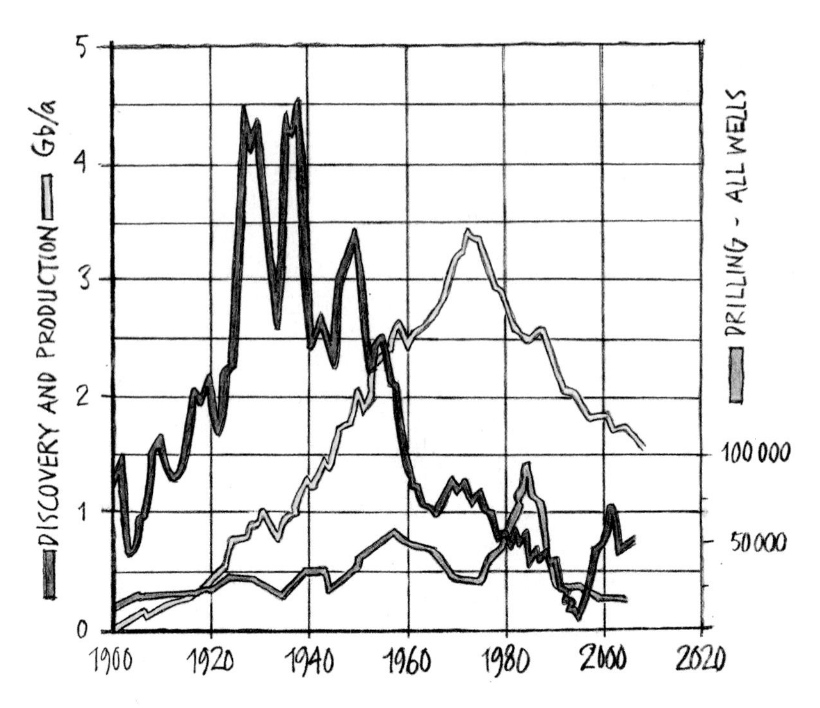

Fig. 14.2 Oil discoveries and production in the states of the United States lying south of Canada (the "US Lower-48"). Note the significant increase in the number of new wells being drilled during the 1980s and that this resulted in only a marginal increase in oil production and did not halt the fall in the annual volume of oil discovered (Data for this illustration were provided by Jean Laherrère [2])

150 and 200 billion barrels (Gb). He used these URR values to fit logistic curves to the history of US oil production and thus predict when the peak of that production would occur. It was his prediction based on 200 Gb of URR that correctly saw US Peak Oil occurring in 1971. Now that the United States is approaching the sunset of its oil production we know that the 200 Gb value is close to the actual URR for the Lower-48.

The rate of oil production from an oilfield is dependent on the number of wells drilled into it. Figure 14.2 shows how many oil wells were drilled each year in the United States from 1900 until 2005. The strong increase in oil production after the Second World War required a significant increase in the annual rate at which new wells were drilled. Around 1960 oil imports began to compete with the United States' domestic oil production leading to a decline in drilling activity. When the price of oil reached record levels in the 1980s the rate of drilling new wells increased again and reached record levels. This resulted in a limited increase in oil production but production levels never reached those attained in the peak year of 1971. The idea has been advanced by some that the peak of US oil production in 1971 was due, in part, to the declining frequency of drilling new wells during the 1960s. There may be an element of truth in this but we can see that the massive expansion of drilling during the 1980s resulted in only a small increase in production. If the rate of drilling new wells had been kept constant during the 1960s it is likely that Peak Oil for the United States would have been delayed by only 1 or 2 years.

The statistics for drilling of new oil wells in the United States also include exploratory wells drilled when searching for new oilfields. As one might expect from current economic models, the frequency of drilling increases when the price of oil rises. However, the fact that discoveries of new oil declined while the frequency of drilling increased during the 1980s is not consistent with current economic theory. Some resources are limited, first regionally and then globally, regardless of market forces.

New technology has made it possible to explore for and exploit oil in deep waters. At the end of the 1990s this opened up a new province for exploration in the Lower-48, the Gulf of Mexico, and the rate of oil discoveries increased temporarily. Compared with previous discoveries, those made in the Gulf of Mexico are relatively small but they have been very important for the share price and total market value of the international oil companies. Today, 25% of US oil production comes from fields under the waters of the Gulf of Mexico (see Chap. 12 which examines deep water production in detail).

North of the Arctic Circle in Alaska lies the Prudhoe Bay oilfield. This is the United States' largest oilfield. It was discovered at the end of the 1960s but first came into production in 1977 when the Alaska pipeline was completed. The pipeline's flow capacity from Pump Station Number 1 at the town of Prudhoe Bay across Alaska to the city of Valdez is 2 Mb/d. Approximately 10 years after production at Prudhoe Bay began it contributed 1.5 Mb/d to the pipeline and associated fields filled the pipeline to maximum capacity. Oil production from Prudhoe Bay began to decline in 1988 and now total Alaskan production has declined so far that the pipeline operates at only half its maximum capacity.

Historic Oil Production and Consumption in Russia

During most of the twentieth century Russia was part of the Soviet Union but even during Soviet times most of the USSR's oil production occurred in Russia. Both Russia and the Soviet Union reached maximum oil production in 1983 at 11.4 and 12.2 Mb/d, respectively. This means that only 7% of Soviet

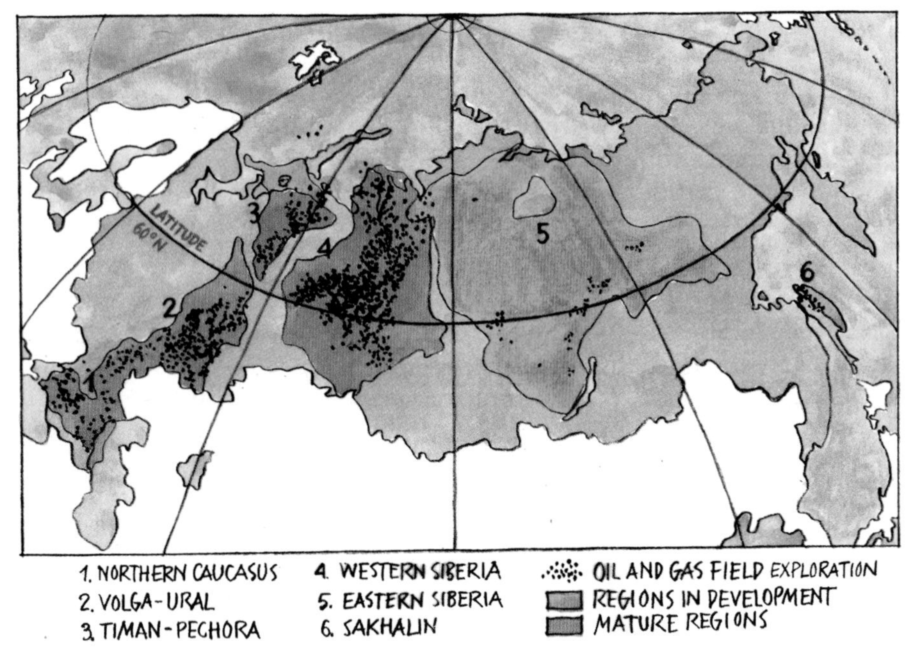

Fig. 14.3 Exploration statistics for Russia. The map shows Russia's six primary regions for oil exploration and the number of exploratory wells drilled in each region. The regions are divided into mature regions and regions in development. When Ray Leonard showed this map at the first ASPO conference in Uppsala he pointed out that latitude 60°N marked the boundary for additional salary due to cold weather during Soviet times. That might be why region 4 has more exploratory wells drilled north of that latitude [3]

oil production came from outside Russia. The Soviet Socialist Republics contributing to oil production at that time were Kazakhstan (3%), Azerbaijan (2%), Ukraine (1%), and Turkmenistan (1%). As far as is possible we use oil statistics for Russia in the discussion below but sometimes only statistics for the Soviet Union are available.

During the twentieth century the Soviet Union was systematically explored for oil resources. Figure 14.3 shows where this exploration occurred. Every spot on the map in Fig. 14.3 represents an exploratory well. Oil production was divided up into six regions of which four are now considered mature. The most recent region to begin producing oil is Sakhalin in eastern Russia [3]. Between 1900 and 2010 Russia produced 155 Gb of oil and its reserves in 2010 as stated in the *BP Statistical Review of World Energy* [4] are close to 77 Gb (a number that we discuss later).

This means that, originally, the United States and Russia had fairly similar volumes of reserves with each possessing approximately 10% of global reserves of crude oil. The difference is that the United States developed a lifestyle requiring six times as much oil as Russia. Today, the United States is the world's largest importer of oil whereas Russia and Saudi Arabia (see Chap. 19) are the world's largest exporters of oil.

Future Oil Production and Consumption in the United States

In the spring of 2011 the Energy Information Administration (EIA) of the United States published its *Annual Energy Outlook 2011 with Projections to 2035* report. This report describes a "business as usual trend estimate" known as the "reference case" and it includes a projection of future consumption of liquid fuels in the United States up until 2035 (Fig. 14.4) [5]. From reading this document it is evident that the EIA believes the United States will never return to the levels of oil consumption seen between 2005 and 2007; that is, the United States has passed Peak Consumption of oil.

Figure 14.2 shows a declining trend of oil production from the Lower-48 but in recent years the trend has leveled off due to the introduction of new production techniques and deep water production. A number of oilfields in the United States are currently undergoing Phase III production as described in Fig. 7.4. At the beginning of this century the United States produced around 7.5 Mb/d but in 2008 this declined to 6.7 Mb/d. Some of this decline was due to Hurricane Katrina. The explosion of the Deepwater Horizon rig and the subsequent environmental catastrophe did not actually disrupt oil production from the Gulf of Mexico significantly so that in 2010 the United States managed to produce 7.5 Mb/d again, similar to its rate of production 10 years earlier.

The reserve numbers the United States has reported during the past decade have oscillated around 30 Gb. In 2010 its reserves were stated as 30.9 Gb, but this number has nothing to do with the URR for the United States. Because the United States is producing 2.74 Gb per year then with a URR of 30.9 Gb this would mean a rate of depletion of remaining recoverable resources (DRRR) of nearly 9%. In contrast, in the rapidly exploited North Sea the DRRR leveled off at a maximum near 6%. Thus the stated URR for the United States is too small and the reason for this is that the United States only reports proven reserves (1P) compared with the rest of the world which reports proven and probable reserves (2P). This also explains why US reserve values do not appear to change greatly from year

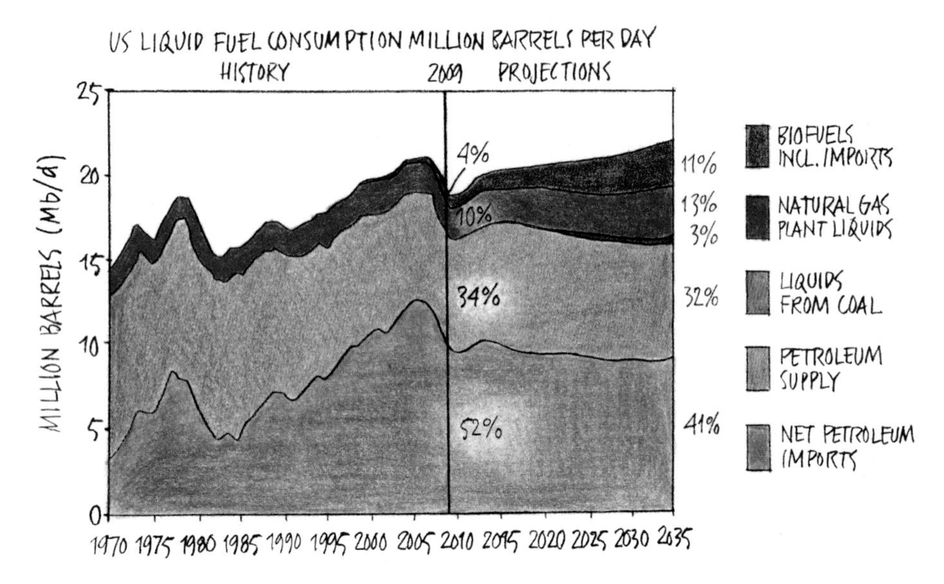

Fig. 14.4 The United States' total consumption of liquid fuels, including both fossil fuels and biofuels, is shown from 1970 to 2009, when it was 18.8 Mb/d. According to the EIA's reference case [5], consumption will increase to 21.9 Mb/d in 2035. Production from Shale Oil is included in the Petroleum Supply. The proportion of consumption provided by imports reached 60% in 2005 and 2006 before falling to 51% in 2009. The EIA expects this to fall to 42% in 2035 [5]

to year. The real question is how large are the United States' 2P reserves? We can estimate this by comparing the United States with other oil provinces that derive a large part of their production from land-based wells. A DRRR of 5% is not unreasonable for the 2P reserves of those provinces and, based on this, we can estimate that the United States' 2P reserves are approximately 55 Gb. By only reporting 1P reserves the EIA prolongs the time before they must report a decline in the United States' oil reserves.

According to the EIA the United States will produce 67 Gb of crude oil during the next 25 years which is more than our estimated 55 Gb of 2P reserves. In 2035 the EIA estimates that the United States will be producing oil at the same rate as today. This is in spite of the fact that the trend in US oil production has been downwards for 40 years, although it did increase slightly in 2010. If oil production in 2035 is to be the same as today then the United States' reserves in 2035 will need to be the same size as currently. In other words, during the next 25 years the United States must find and put into production 67 Gb of reserves which is more than twice the volume of their total current 1P reserves. This is also ten times the current 1P reserve volume in the Gulf of Mexico (see Chap. 12). The new areas where exploration might discover significant oil reserves are mainly coastal areas that are currently protected and also the Arctic National Wildlife Refuge (ANWR). ANWR is estimated to contain less than 10 Gb of reserves. The United States will never be able to produce all the oil that exists within its oilfields but new technology will allow the recovery factor to increase somewhat. However, it is clear that the EIA's expectation that the United States will experience an oil production plateau over the next 25 years is unrealistic.

Liquid fuel derived from coal is estimated to constitute 3% of the United States' production in 2035 and this is equivalent to 660,000 b/d. Compared with the 5.5 Mb/d that the National Petroleum Council suggested was possible in 2007 (see the section "Oil Production from Coal by CTL (Coal-to-Liquids)", Chap. 10) this is a very modest amount. Nevertheless, it is four times what South Africa currently produces. According to the Hirsch Report four new production facilities would be needed for production of 660,000 b/d and the minimum required planning time is 5 years [6]. At most it would be possible to build one facility per year. Full-scale operation yielding 660,000 b/d with a 50% increase in the efficiency of the Fischer–Tropsch process (see Fig. 10.6) would require an increase of coal production in the United States on the order of 20%. This is also the increase in coal production that the EIA proposes in the reference case [5].

The EIA estimates that production of natural gas liquids (NGL) will contribute 2.9 Mb/d to the United States' oil production in 2035 which is an increase over the current NGL production level of 50% [5]. In our article, "The Peak of the Oil Age," [7] we described that 15% of conventional gas production is NGL, a proportion that has been more or less constant for the past 30 years. According to the EIA [5], conventional natural gas production in the United States will decline by 30% between 2010 and 2035. However, during the same period, unconventional forms of natural gas production such as shale gas, tight gas, and coalbed methane are expected to increase by 65%. When this unconventional production is added to conventional gas production, the EIA estimates that the United States' total natural gas production will increase by 25% by 2035. Some unconventional natural gas production has been found to include significant NGL production, as in the production from the Marcellus Shale [8]. Other unconventional gas discoveries have been found to lack NGL co-production. For this reason, we conclude that expectations of a 50% increase in NGL production by 2035 are unrealistic.

There are also serious doubts over the EIA's estimate that production within the United States of biofuels for vehicles will increase from 4% to 12% of total fuel use. These doubts are supported by an analysis by Bob Hirsch and his colleagues in their book, *The Impending World Energy Mess* [6]. Increased consumption of biofuels for vehicles in the United States will need to be supported by importation of biofuels.

The conclusion of our analysis is that the EIA's reference case (its business as usual trend estimate) is based on several unrealistic assumptions. The increase in the United States' liquid fuel consumption that the EIA has projected in the period up to 2035 will require a greater level of imports than the EIA anticipates.

Future Oil Production and Consumption in Russia

After the fall of the Soviet Union, Russia was somewhat written off as an oil producer but in the early years of this century its oil production activity had recovered and by 2010 it had become the world's largest producer at 10.15 Mb/d or 3.7 Gb per annum [4]. The main reasons for this have been access to new technology and the opening of production from the island of Sakhalin in eastern Siberia. Russia's oil reserves at the end of 2009 were reported in the *BP Statistical Review of World Energy* [4] as 76.7 Gb and this means that they produced 4.8% of these reported reserves during 2010. At the end of 2010 Russia's reserves were reported as 77.4 Gb meaning that they had succeeded in replacing the produced volume and had even increased their reserves by 0.7 Gb. In Fig. 14.3 one can see that Russia possesses developed areas where new technology can increase total production and it also has areas requiring further exploration.

Since 2005 Russia has exported 7 Mb/d and so is the world's second largest oil exporter after Saudi Arabia. The largest parts of the exports go to Europe and Europe's future oil supply is dependent on Russia's future ability to export oil. Europe's dependence on Russian oil inspired the Global Energy Systems research group at Uppsala University to study future Russian oil production. Russia's system for reporting oil reserves differs from that of other nations and the actual reserve numbers are regarded as state secrets. In 2007 Aram Mäkivierikko produced a Diploma thesis on future Russian oil production in which he described how various authorities on oil reported Russian reserves to be anywhere between 70 and 170 Gb [9]. Russia has produced 15 Gb of oil since 2007 and BP estimated Russia's reserves to be 77.4 Gb in 2010. Thus Russia's reserves must be much larger than 70 Gb reported in 2007. However, because 70 Gb is a number that is openly discussed, it was important to include it in analyses of Russia's future oil export capacity. For this reason, Aram Mäkivierikko chose to project possible future Russian oil production using a Hubbert analysis based on reserve sizes of 70, 120, and 170 Gb (see Fig. 14.5). Although predicting Russia's future oil production based on the Hubbert model is, in this case, unrealistic, it nevertheless conveys an impression of the future limits of Russian production.

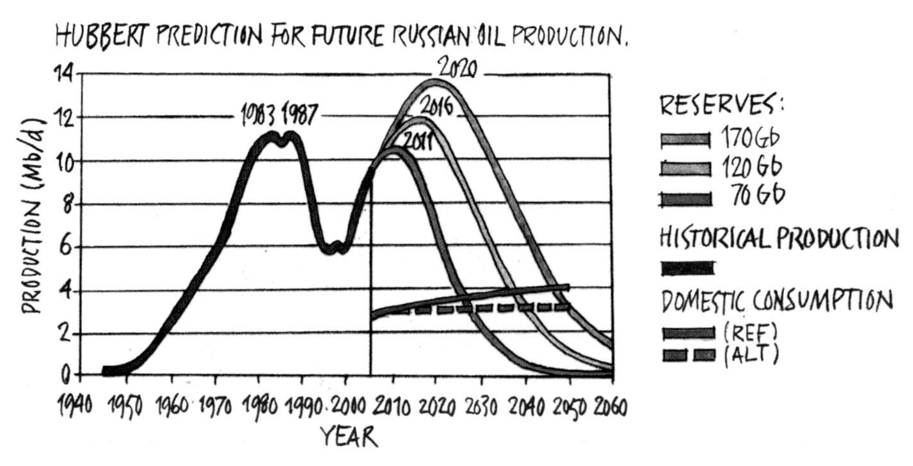

Fig. 14.5 Russia's past and future predicted oil production based on 2006 statistics. During the 1980s Russian oil production was on a plateau at around 11 Mb/d but with the collapse of the Soviet Union oil production fell dramatically. Between 2000 until 2010 it grew to 10 Mb/d and in 2006 it was 9.5 Mb/d. The area under a Hubbert curve represents the total volume of produced oil and the three curves shown for future production have been fitted according to reserve volumes of 70, 120, and 170 Gb. The figure also includes two curves for predicted consumption. The reference scenario curve (REF) shows increasing consumption and the alternative curve (ALT) assumes increasing efficiency of fuel use and lower consumption. The point at which the consumption and production curves intersect forms an estimate for the moment when Russia will no longer be an oil exporting nation. The latest date for this would appear to be around 2050 [9]

As described above, Russia produces about 5% of its reserves every year. Aram used this figure and the different possible reserve numbers to produce several predictions of future Russian production. These projections show that it is not inconceivable that Russian oil production might rise from today's level of 10 Mb/d to its previous record level of 11 Mb/d seen in the 1980s. However, it is more realistic to expect that we will now see a plateau of production at 10 Mb/d for some years. Details of Aram Mäkivierikko's research can be found in his Diploma thesis which is available online [8].

In Fig. 14.5 we show two assumptions regarding future domestic consumption of oil in Russia. In the "reference" scenario domestic oil consumption continues to increase in line with current trends. In the "alternative" scenario gains in the efficiency of oil use lead to a slower increase in domestic consumption allowing a greater capacity for oil exports. These two domestic consumption scenarios allow us to make estimates of the earliest

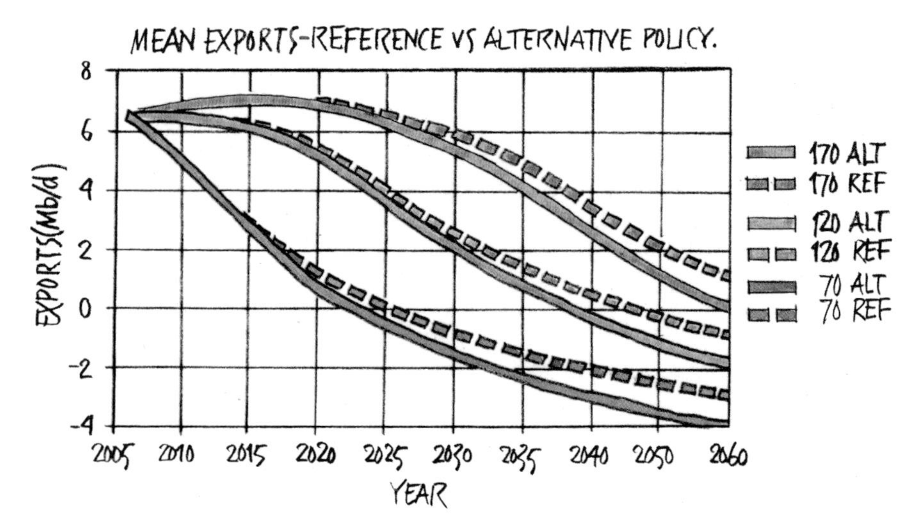

Fig. 14.6 Predictions for future exports of oil from Russia assuming highest possible production rates (highest DRRR). The two possible rates of domestic consumption shown in Fig. 14.5 have been subtracted from total oil production to give the export volumes. The predictions of export volumes are based on total reserve sizes of 70, 120, and 170 Gb in 2006 [9]. The *World Energy Outlook 2010* report (*WEO 2010*) by the International Energy Agency (IEA) shows a predicted constant rate of crude oil production until 2035 (see Fig. 11.7). At that time we see that Russia, today the largest producer of oil in the world, will be exporting oil at a rate somewhere between zero and 4 Mb/d. Maintaining today's export level of 7 Mb/d is considered impossible

and latest times when Russia will no longer be able to export oil (i.e., when the predicted production and consumption curves intersect). In examining Fig. 14.6 one should realize that the projection based on the lowest reserve estimate is clearly unrealistic because, as stated above, reserves were certainly greater than 70 Gb in 2006. For example, if we add the oil produced between 2006 and the end of 2010 to the 2010 reserve volume reported in the *BP Statistical Review of World Energy* we find that the reserves in 2006 must have been at least 93 Gb. In Mäkivierikko's analysis of 2007 we judged that 120 Gb was a realistic estimate of Russia's 2006 reserves and, if this is so, Russia will be exporting less than 2 Mb/d in 2035. Even at the higher, less realistic 170 Gb estimate of 2006 reserves, Russia's exports in 2035 would be less than today.

Russia consumes approximately 30% of the oil it produces and we estimate its consumption will increase. Clearly, Russia is unlikely ever to be an oil importer because, when their production falls below their domestic requirements, there will probably be no other land from which to import oil. We can expect that Russia has a strategy to guarantee domestic needs in the future. If they continue to export large volumes of oil this will reduce their oil reserves relatively rapidly and will hasten the day when they can only meet domestic needs. During the next 10 years we can expect them to export more than 6 Mb/d if no serious disruptions occur. After that, Russia's exports will decline. This is a signal to Russia that it needs to develop an economy that is less dependent on oil revenues. Likewise, the world, and especially Europe, should demand that Russia declare its strategy for future oil exports to, at least, 2035.

It seems very likely that Vladimir Putin will continue to lead Russia during the coming decade. In 1997 Putin actually submitted a doctoral thesis in economics titled, "The Strategic Planning of Regional Resources Under the Formation of Market Relations" at the St. Petersburg State Mining University [10]. The thesis discusses how, in a developing nation such as Russia, the local area around St. Petersburg can use its natural resource base for strategic planning and management. The thesis mainly addresses local strategy in the mining sector, a strategy that now appears to have been applied at the federal level. Experts who have read the thesis also assert that Vladimir Putin foresaw the coming "natural gas war" with Ukraine. We must presume that Russia has a strategy for natural resource development for the coming decade. The question is how this strategy will affect the European Union and the rest of the world.

Summary

In this chapter we followed the development of oil production and consumption in two rival nations that have pioneered the oil industry, Russia and the United States. Initially, they both had similar volumes of oil reserves but the United States is now living beyond its means and must import oil whereas Russia is self-sufficient in oil. Peak Oil occurred for the United States decades ago and global Peak Oil means that the United States is now entering an era where it must reduce its consumption of oil whereas Russia's internal consumption can continue to grow. However, Russian Peak Oil means that Russia's exports of oil will decrease. In Chap. 18 we examine how Peak Oil has affected the political relationship between the Soviet Union/Russia and the United States. In Chap. 19 we look more closely at the export/import endgame.

At the beginning of this chapter I described the drilling beside Oil Creek in Pennsylvania that led to the discovery of Drake's first productive oil well. Today the site of Drake's well has become an historical museum [11].

When I visited this museum at Titusville in 2005 I also bought some souvenir bottles of oil produced from the McClintock No. 1 well. They cost $2 each. When I measured the volume in each bottle I realized that I had probably purchased the world's most expensive oil, equivalent to a barrel price of $10,000.

References

1. King Hubbert, M.: Nuclear energy and the fossil fuels. Shell Development Company, Exploration and Production Research Division, Houston, TX, Publication No. 95. http://www.energybulletin.net/node/13630 (1956)
2. Laherrère, J.: Private communications (2011)
3. Leonard, R.C.: Russian oil and gas: a realistic assessment. International Workshop on Oil Depletion, Uppsala, Sweden, May 2002. http://www4.tsl.uu.se/isv/IWOOD2002/ppt/pptLR.ppt; http://www4.tsl.uu.se/isv/IWOOD2002/ppt/UppsalaRCL.doc (2002)
4. BP: BP Statistical Review of World Energy, June 2011. http://bp.com/statisticalreview (historical data; http://www.bp.com/assets/bp_internet/globalbp/globalbp_uk_english/reports_and_publications/statistical)
5. EIA: Annual Energy Outlook 2011. Energy Information Administration, DOE/EIA-0383, April 2011, Figure 1; http://www.eia.gov/forecasts/aeo/pdf/0383(2011).pdf (2011)
6. Hirsch, R.L., Bezdek, R.H., Wendling, R.M.: The Impending World Energy Mess. Apogee Prime, ISBN 978-1926837-11-6 (2010)
7. Aleklett, K., Höök, M., Jakobsson, K., Lardelli, M., Snowden, S., Söderbergh, B.: The peak of the oil age—analyzing the world oil production reference scenario in World Energy Outlook 2008. Energy Policy **38**(3), 1398–1414 (accepted 9 Nov 2009, Available online 1 Dec 2009, http://www.tsl.uu.se/uhdsg/Publications/PeakOilAge.pdf) (2010)
8. Marcellus Shale—Appalachian Basin Natural Gas Play, geology.com. http://geology.com/articles/marcellus-shale.shtml (2011)
9. Mäkivierikko, A.: Russian oil—a depletion rate model estimate of the future Russian oil production and export. Diploma thesis, October 2007. http://www.tsl.uu.se/uhdsg/Publications/Aram_Thesis.pdf (2007)
10. Putin's PhD thesis essential reading, BBC Monitoring Service, UK, 5 Feb 2006. http://slavija.proboards.com/index.cgi?board=general&action=print&thread=2343, http://en.wikipedia.org/wiki/Vladimir_Putin (2006)
11. Drake well Museum: http://www.drakewell.org/ (2011)

Chapter 15

China and Peak Oil

In the mid-1950s there was a severe oil shortage in China. Fighter jets and tanks stood still and the buses on Beijing's streets were fueled from large bags of gas on their roofs. Several drilling teams traveled northeast of Beijing to look for oil. One of those teams was led by the legendary "Iron Man" Wang Jinxi and was to drill in the eastern area of Heilongjiang province. The temperature in that area can drop to below −30°C during winter and just getting the equipment to the drilling site required heroic effort. However, they eventually succeeded and began to drill. The first two holes were dry but on their third attempt they succeeded in finding an oil-bearing layer. On September 26, 1959 they had discovered one of the world's largest oilfields, Daqing (meaning "Great Celebration") [1].

When the news of the discovery of Daqing reached Beijing the government decided to make a huge effort to develop the oilfield. Approximately 40,000 workers were sent to Daqing with 30,000 of these contributed by the Red Army. They worked around the clock in sometimes severe conditions so that the first train pulling oil tanks could leave Daqing one year later. For China it was enormously important to become self-sufficient in oil production and in December 1963 Premier Zhou Enlai could announce that China had achieved this goal.

The illustration on the cover of this book shows a woman from China hugging her nation's oil barrel. There is a special reason why this character is a woman. Her name is Wu Tan and she was 20 years old when the Swedish journalist Gunnar Lindstedt met her at Daqing in 1977. At that time Wu Tan was the leader of a drilling crew of 82 young women called "the iron girls" who worked in Daqing between 1974 and 1979. They became role models for young women in Mao's China. Twenty-eight years

K. Aleklett *Peeking at Peak Oil*, DOI 10.1007/978-1-4614-3424-5_15,

later, in 2007, Gunnar Lindstedt returned to Daqing and again met Wu Tan who now held a position in the Daqing city government. (The city sits on top of the oilfield.) An article written by Lindstedt, "Back to Daqing," quotes Wu Tan as saying, "The work as a driller was hard for a woman but it was good to contribute to the development of my nation. We worked under a blue sky and were surrounded by the green grass of the steppe."

In 1977 Daqing was still a barrack town but since then it has grown into a city of nearly three-million people. In 2005 Wu Tan had responsibility for, among other things, the city's parks. At that time a museum to honor the Ironman Wang Jinxi was under construction [2].

Since reading Lindstedt's article in 2005 a visit to Daqing has been high on my wishlist. In 2008, nearly 50 years after the discovery of the oilfield Daqing, my wish was fulfilled and the path that took me there is quite interesting.

It was at the ASPO Conference in Lisbon in 2005 that I first met Professor Pang Xiangqi, the deputy vice-chancellor of the China University of Petroleum in Beijing (CUPB) [3]. He had been invited to speak on the "Impact of Oil Depletion in China." In November 2006 I was invited to the *China Shijiazhuang Energy and Environmental Conference* as a plenary speaker. (The conference was sponsored by, among others, the city of Shijiazhuang's Swedish sister city Falkenberg, which supported my attendance and presentation.) When Professor Pang heard that I had been invited to speak at that conference he contacted me to ask whether I could visit CUPB to deliver a lecture. While in Beijing I was also able to visit my good friend John Liu (whom I mention later in the section "The Climate and Peak Oil", Chap. 19).

In September 2007 ASPO's annual conference was held in Cork, Ireland, and Professor Pang was again invited to speak. I asked Professor Pang and his colleague Professor Lianyong Feng (from CUPB's School of Business Administration) to visit me in Uppsala on their way to Cork and to speak at a symposium titled, *China and Sweden—Future Energy Perspectives.* Then, in October of that year we met again when, as the president of ASPO, I was invited to a conference in Beijing held to mark the establishment of ASPO China. During discussions in Beijing it became clear that CUPB wanted to collaborate with Uppsala University (UU) in research on oil and other fossil fuel resources. I was also invited to the Swedish Embassy in Beijing and during a meeting there I discussed this matter with Sweden's ambassador to China, Mikael Lindström. He encouraged us to collaborate and believed it was very important for Swedish research. During the evening (and night) before I was to travel back to Sweden, I worked on a draft collaborative agreement. (I received help with the wording via the Internet from Simon Snowden, a collaborator at the University of Liverpool.) In the morning I met Professor Pang and we then worked on the final wording before we both signed it. The agreement became official when the leadership of both universities had ratified it in early 2008.

As one might expect, the agreement between UU and CUPB states that, "CUPB and UU should collaborate in research related to Peak Oil, Peak Gas and Peak Coal, and also in studies of the economic and social consequences of 'peak' events." However, what are of greater interest are the statements of mutual agreement that form the foundation for the collaboration. These represent a very concise summary of the Peak Oil issue and its significance:

> Whereas research groups at CUPB and UU have recognised that oil, natural gas and coal are finite natural resources subject to depletion, and that future production of these resources will reach a maximum rate of production.
>
> Whereas research groups at CUPB and UU are aware that these products are of eminent importance to China and Sweden.
>
> Whereas research groups at CUPB and UU accept that production and consumption of fossil resources are affecting future global climate.
>
> Whereas research groups at CUPB and UU understand that consumption of oil, natural gas and coal is correlated to national figures for GDP.
>
> Whereas research groups at CUPB and UU are aware of the fact that the import and export of oil, natural gas and coal constrained by limited supply can lead to problems in the future.

The research collaboration between Uppsala Global Energy Systems (UGES) and CUPB has now been underway for over 2 years and has already resulted in publication of a number of joint scientific papers [4].

In the spring of 2008 my research group received an official invitation (which we accepted) to visit China after the Olympic Games in August 2008. During our tour of Chinese oil institutions and authorities (described in detail later) we had the opportunity to visit Daqing. As we said before, the city of Daqing is built on top of the oilfield of the same name and it was a little bizarre to see pumpjacks (also called "nodding donkeys") working away between shops, schools, and guesthouses. When we were being shown around the city I asked whether we could stop somewhere suitable for photographing the oil pumps and they chose a spot where six pumpjacks were working rhythmically lifting oil to the surface. Figure 15.1 shows an instant that I captured from the dance of the pumpjacks.

Something else I saw during our visit also struck me with its significance. Most people are not aware of the importance of oil to our society and civilization but not so the residents of Daqing. Of course we visited the Iron Man Museum (that Lindsted referred to in his 2005 article and for which construction is now complete) but, more important, we also visited the Daqing Oil Museum where we saw at the entrance the following declaration

Fig. 15.1 Six of the thousands of pumpjacks of Daqing move in a rhythmical ballet as they lift oil to the surface from the oilfield. At the moment I captured this "ballet step" I decided that if I was ever to write a book about oil then this image would be in it (Photograph by Kjell Aleklett)

in both Chinese and English text: "Petroleum has a compact relationship with a country's political, economic and military strength".

There is no doubt that the giant Daqing oilfield has been crucial for China's development and will be so in the future. The fact that it made China self-sufficient in oil production in 1963 also made China politically independent. In 1964, 5 years after Daqing was discovered and one year after it became self-sufficient in oil, Mao Zedong formulated the political slogan/policy, "Learn the lesson of Daqing." At that time the slogan was directed at the Chinese people but in the section "China's Future and Peak Oil" we show that it also has consequences for our future as we encounter Peak Oil.

Daqing and China's Other Giant Oilfields

On the list of the world's giant oilfields there are some names that resonate more deeply than others (Table 8.1). These have special significance either due to their size or political and economic impact. In a class of their own (despite uncertainty over their ultimately recoverable resources, URR) are the world's two largest crude oil fields, Ghawar in Saudi Arabia (discovered in 1948, URR 60–150 Gb) and Greater Burgan in Kuwait (discovered in 1938, URR 32–74 Gb). Further down the list we have Rumaila in Iraq (discovered in 1953, URR 19–30 Gb). A small corner of this field extends into Kuwait and it was Kuwait's overproduction from this area that was one of the sparks that ignited the First Gulf War in 1990. The list includes Samotlor, the largest oilfield in Russia (discovered 1964, URR 28 Gb). Exports from this field during the high oil prices brought on by the oil crises of the 1970s and early 1980s were an important support for the Soviet economy. When the oil price subsequently collapsed, the Soviet Union followed. The discovery of the giant oilfield in Kirkuk (URR 15–25 Gb) in 1927 was the start of the fantastic finds that made the Middle East the world's most important oil region. One of the world's largest offshore oilfields, Cantarell in Mexico (discovered in 1976, URR 11–20 Gb), is, like Samotlor, an example of how a giant oilfield can rescue a nation's economy. Cantarell's production is now falling dramatically causing drastic changes in Mexico's economic wellbeing. Finally, we come to Daqing, discovered in 1959. In Table 8.1 (which shows data from Robelius' thesis [5]) Daqing's URR is reported as 13–18 Gb. Using the information that we now have access to from China we can revise this URR up to 24.2 Gb which places Daqing among the ten largest oilfields in the world. However, it can be argued that no oilfield has had a greater political impact than Daqing.

China's oil reserves were reported as 14.8 Gb in 2010 and they produced 1.5 Gb in that year (4.07 Mb/d) which is equivalent to around 5% of global production [6]. It is obvious that the reserve numbers for China reported in the *BP Statistical Review of World Energy* (derived from data in the *Oil & Gas Journal*) are underestimates because, if they were true, this would mean that China produced 10% of its reserves during 1 year (i.e., a rate of depletion of remaining recoverable resources, DRRR, of 10%). The mysterious nature of China's oil reserve numbers is well illustrated by BP's statement of these as 13.3 Gb in 1980 followed by reserve numbers around 15 Gb in the 30 years since then. In 2007 Colin Campbell estimated China's URR as 65 Gb with 2P reserves at 25 Gb [7].

The collaborative agreement between UU and CUPB has given UGES the opportunity to make a detailed analysis of China's giant fields. In total, China possesses nine fields in the giant category with Daqing (URR 24.2 Gb) and Shengli (URR 15.8 Gb) being the largest [8]. The aggregate URR for China's giant fields is estimated as 60 Gb with a little more than half of this volume already having been produced. If one adds the past production of smaller fields to this then the total oil production from Chinese fields up to and including 2010 has been 40 Gb. BP stated China's reserves as being

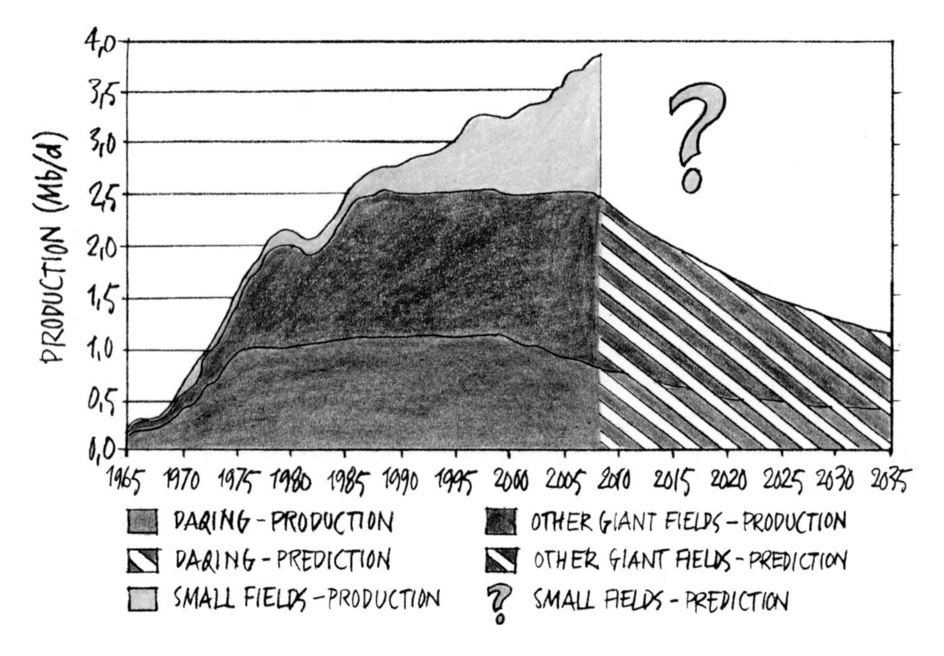

Fig. 15.2 China's historical and projected oil production based on data from our collaborative paper, Höök et al. [8]. For the period 1965–2007 we show the production from Daqing, China's other eight giant fields and 13 smaller fields. For the period 2008–2035 we give a production forecast for Daqing and the other eight giant oilfields based on our field-by-field analysis [8]. At the time our paper was published it was not possible to forecast the future production of the 13 smaller oilfields

14.8 Gb in 2010, which may be a statement of the nation's 1P reserves (proven reserves). From the data we have obtained access to, China's 2P reserves (proven and probable) would appear to be closer to 35 Gb which is somewhat higher than Campbell estimated. The URR for China should ultimately be between 90 and 100 Gb. This means that China, with approximately 20% of the world's population, originally possessed 4–5% of the world's recoverable oil (see Figs. 4.3 and 4.4).

Our collaboration with CUPB has also given us access to detailed production data from China's oilfields to the end of 2007. At that time China's total oil production was 3.74 Mb/d of which 2.5 Mb/d came from China's 9 giant fields. The remaining production came from 13 smaller fields (see Fig. 15.2). Using the detailed production data we have been able to project future production from the giant oilfields. We have calculated that, by 2035, their production will have declined from 2.5 to 1.1 Mb/d [8]. Over the same

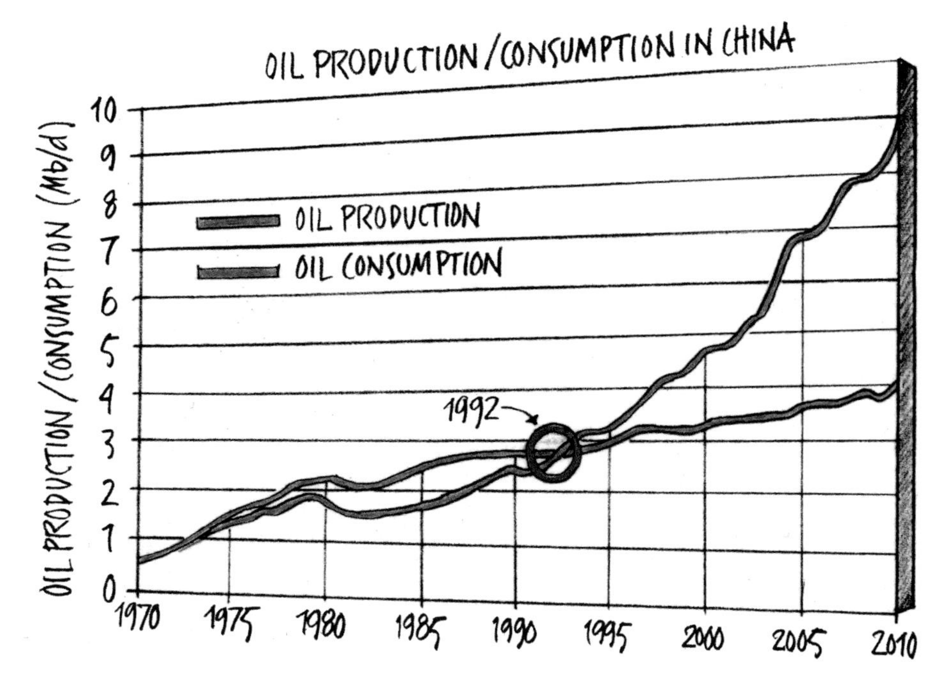

Fig. 15.3 China's annual domestic production and consumption of oil (Data from BP Statistical Review of World Energy [6])

period production from smaller fields will continue to grow for some time, meaning that China has not yet reached Peak Oil. Nevertheless, as seen for other oil-producing nations, once production from the giant fields begins to decline, Peak Oil soon follows [8] so China is expected to see its peak of oil production within a decade. Thus, the oil that China needs for its economic expansion does not exist within its own borders.

China as an Exporter and Importer of Oil

If we examine China's production and consumption of oil we see that China was an exporter of oil until 1992 after which China's consumption exceeded its domestic production. The growth of China's oil consumption has been very strong over the past 20 years averaging 7% per annum. At the start of this 20-year period (1990) China was self-sufficient in oil but by 2006 its domestic production only supplied 50% of its needs. China's dependence on oil imports has increased since then (see Fig. 15.3) [6].

In Professor Pang's presentation at the *China and Sweden—Future Energy Perspectives* symposium he described some of China's oil projects outside its national borders and he also mentioned that China's aim was to exert some degree of control over 50% of the oil production required for its consumption. In recent years we have seen Chinese oil companies becoming increasingly active in obtaining rights to production of oil in other nations. Oil industry insiders have also described how Chinese oil companies are interested in financing production from deep water fields in Africa. The companies are willing to pay the market price for the oil but they insist on the condition that the oil will be exported to China, not elsewhere. Because Chinese domestic oil production will reach Peak Oil within a decade we can expect increased international activity by Chinese oil companies. In light of this it is also worth realizing that China is currently the only nation making large investments in the education of oil industry workers at all levels. CUPB alone has 10,000 students and if we include all other oil industry education there are more than 50,000 students in China. There is a shortage of oil industry education in other oil-producing nations. It is estimated that, in 20 years, 80% of all oil industry workers will be Chinese which may give China a remarkable degree of control over the future of this industry internationally.

A Tour of Chinese Oil Institutions and Authorities

How aware are the Chinese people of Peak Oil and its implications? We found some answers to this question on a tour of Chinese institutions and authorities that we made in late August and early September 2008. Professor Lianyong Feng of CUPB was our host in Beijing and our tour guide in Daqing.

Our tour began with a working dinner with the vice chancellor and the deputy vice chancellor of CUPB and some of that institution's leading professors. I was not surprised that the head of their economics section was skeptical about Peak Oil but there were other professors present who knew more about the issue and were more supportive. In regard to this it is interesting to note that the vice chancellor of CUPB was, in 2008, also the chairperson for ASPO China! Professor Pang, who is a co-author of our article on giant fields in China, is also deputy vice chancellor of CUPB. Professor Pang has also been awarded status as a national hero for his discovery of large natural gas fields in western China.

The following day CUPB had organized a workshop on Peak Oil in which the participants were representatives of many sectors of society.

My then Ph.D. students Mikael Höök, Kristofer Jakobsson, and Bengt Söderbergh presented their research while I gave an overview of Peak Oil and Peak Gas. UGES collaborates with Simon Snowden from the University of Liverpool and he had also been invited to China with us. He presented his research on the dependence on oil of milk production in the United Kingdom. One month before our visit to China and this workshop the price of oil had reached the record level of $147 per barrel. It was notable that the participants were very keen to discuss this price but not as interested in discussing the idea that it was the rising demand from China that was driving up the price.

During the following 2 days we attended a number of symposia and a workshop on the economic and socioeconomic consequences of Peak Oil. These meetings were in central Beijing and because CUPB lies on Beijing's outskirts we needed to depart early in a car bearing the correct final digit on its registration plate. This was because limitations on traffic in Beijing imposed for the Olympic Games were still in force.

The first of the 2 days in central Beijing involved seminars for various departments of the National Development and Reform Commission, NDRC [9]. In the morning we met the vice director and various assistants of the Bureau of Energy, the department that draws up the guidelines for China's 5-year plan for energy and oil policy. They listened to our arguments regarding Peak Oil and we gave them copies of our publications.

That afternoon we had a very lively and rewarding meeting with people from the energy research institute at the NDRC. We (i.e., researchers, Ph.D. students and undergraduates from Uppsala University, the University of Liverpool, and CUPB) were met by an equivalent group. Together there were about 30 people involved. The deputy director of the institute, an economist, opened the discussion by stating that he did not believe in Peak Oil. Rather, economic forces would create the energy resources needed in the future. After my presentation on the global energy situation and the presentation by Simon Snowden from the University of Liverpool about how the costs of milk production in the United Kingdom are affected by the high oil price, we had a very rewarding discussion. We asserted that, in the future, China was relying upon being able to increase its oil consumption but the nations that would increase their oil exports were not known. The meeting concluded in a positive spirit and we agreed to have continued discussions in future.

On the second of the 2 days in central Beijing we had another very important meeting. The Chinese Academy of Social Sciences (CASS) [10] had organized a workshop on the socioeconomic consequences of Peak Oil. Once again there were approximately 30 people involved. Our hosts presented some of their research on the influence on the Chinese economy of

Fig. 15.4 Welcome representatives from Uppsala University of Sweden to the Advanced Personnel Training Center of Daqing Oilfield

the price of oil and we (the same group as on the previous day) presented our research. After the meeting we made the journey back to CUPB by public transport: underground railway and express bus. This went very well even though the queues at first seemed unending.

After our days in Beijing we then made the visit to Daqing that I described earlier. It took a day of travel by air and car to get there. The only thing I would like to add to my earlier description is the wonderful reception that we received when we arrived at the Advanced Personnel Training Center of Daqing Oilfield. A large banner had been suspended over the entry gate (Fig. 15.4). I could read "Uppsala" among the Chinese characters and I was told that the banner declared a greeting to us. We were allowed to keep the banner when we left and I still have it but I have not yet found a space large enough to display it. We also gave presentations at the Training Center and, on our second day in Daqing, we visited the university there so that oil industry students could hear my presentation, "Peak Oil and the Export–Import Oil Endgame." I discuss this "endgame" further in the section "Export and Import of Oil", Chap. 19.

During the day it took to travel back to Beijing, Professor Feng, our guide, received a call on his mobile phone. He seemed very surprised and stood up to speak. Of course, I understood nothing of what he was saying. When the call ended he explained that he had just spoken to one of the highest-ranking leaders in the China Petrochemical Corporation (CPC), the company responsible for much of China's oil refinery capacity. He had read my blog on *Aleklett's Energy Mix* [11] and realized that I was currently in China. He had inquired if I might be able to visit their head office in Beijing on the following day to participate in a meeting and discuss oil's future. That was to be the day before I traveled home and we had planned, among other things, to visit the Great Wall. My Ph.D. students got to see it but Simon Snowden and I were, of course, very willing to visit CPC. We had very interesting discussions at the company and learned that, although the company was planning new oil refineries for China, it was concerned about the future oil supply. The solution that they found after our visit was to invite Saudi Arabia to become a 49% co-owner of the new refineries. (This was not one of the solutions that we discussed with CPC but it is a very smart move).

During the afternoon of our last day in China we were, in fact, invited to visit the Great Wall and even though it was the fifth time I have seen it I never cease to be amazed by how it snakes its way over the mountains. During its long history China has engaged in a number of mammoth projects although the ultimate wisdom of some of them is debatable.

China's Future and Peak Oil

The answer to our question of how aware the Chinese are of Peak Oil and its implications is that Peak Oil is not yet part of the general debate on China's future. A great many people still harbor dreams of owning their own car and new motorways are being planned to handle the anticipated extra traffic. That the people of Daqing are conscious of Peak Oil is obvious because the oilfield's plateau production phase ended in 1999 and production then began declining by around 3.4% per annum [8]. Daqing will still produce oil for many years to come and the Chinese will certainly do everything possible to squeeze out every last drop, but slowly and surely Daqing's importance for China will wane. The fact that China is planning to increase its rate of oil use can be explained by the attitude that, inasmuch as China possesses 20% of the world's population, it has the right to use 20% of the world's oil reserves. The world is currently on a production plateau and its oil production may fall to 75 Mb/d by 2030. However, even at that level, 20% of

75 Mb/d equals 15 Mb/d which is 6 Mb/d more than China is currently consuming. At today's rate of growth, China will be using 15 Mb/d within 10 years.

Many are surprised by the enormous investments China is making in alternative energy use and to increase energy efficiency. A factor encouraging this investment must be awareness of the Peak Oil issue among Chinese energy experts. In the future we can also expect to see large investments in electricity-powered transport for personal and public transport. China will develop its own electric vehicles and will adapt them to its own needs and standards because its internal market is sufficiently large to support this.

If we apply Mao's slogan, "Learn the lesson of Daqing," to today's situation then, with Peak Oil underway, we can expect China to make enormous efforts to secure its imports of oil. China has just launched its first aircraft carrier which indicates China's intention to defend its supply lines. China's investment in construction of an oil pipeline from Burma to China is another example of its strategic thinking. However, "Learn the lesson of Daqing," also means that China will make massive investments to incorporate renewable resources into its energy system. In the section "The Climate and Peak Oil", Chap. 19, I discuss China's Loess Plateau Watershed Rehabilitation Project as an example of its current mega projects.

On our last evening in Daqing when the official program was over we gathered in one of the hotel's karaoke rooms to thank Professor Feng for all he had done for us. We also tried to do justice to some of our favorite songs. Feng was revealed as having a very fine singing voice. A memory I will always hold dear is his rendition of the song of the Iron Man Wang Jinxi, Daqing's hero, that he had learned as a boy in school.

References

1. History of Daqing Oilfield. Petro China Daqing Oilfield. http://www.cnpc.com.cn/dq/eng/qyjj/hhls/ (2009)
2. Lindstedt, G.: Tillbaka till Daqing (Back to Daqing). Veckans Affärer, #38, 19 Sep 2005
3. China University of Petroleum in Beijing (CUPB). http://department1.cup.edu.cn/~waisb/ (2003)
4. Uppsala Global Energy System Group (UGES), Uppsala University, Sweden, Publications. http://www.physics.uu.se/ges/en/publications (2011)
5. Robelius, F.: Giant oil fields—the highway to oil: giant oil fields and their importance for future oil production. Uppsala Dissertations from the Faculty of Science and Technology, ISSN 1104–2516; 69. http://uu.diva-portal.org/smash/record.jsf?pid=diva2:169774 (2007)

6. BP: BP Statistical Review of World Energy, June 2011. http://bp.com/statisticalreview (historical data; http://www.bp.com/assets/bp_internet/globalbp/globalbp_uk_english/reports_and_publications/statistical) (2011)
7. Campbell, C.J., Heaps, S.: An Atlas of Oil and Gas Depletion. Jeremy Mills Publishing Limited, Lindley (2008). ISBN-1-906600-42-6
8. Höök, M., Tang, X., Pang, X., Aleklett, K.: Development journey and outlook of Chinese giant oilfields. Pet. Explor.Dev. **37**, 237–249. http://www.tsl.uu.se/uhdsg/Publications/China_GOF.pdf (2010)
9. NDRC: Main Functions of the NDRC, National Development and Reform Commission, People's Republic of China. http://en.ndrc.gov.cn/mfndrc/default.htm (2011)
10. The Chinese Academy of Social Sciences (CASS). http://bic.cass.cn/english/infoShow/Arcitle_Show_Cass.asp?BigClassID=1&Title=CASS (2011)
11. Aleklett, K.: Meetings with energy authorities and the Academy of Social Sciences, Aleklett's Energy Mix, 31 Aug 2008. http://aleklett.wordpress.com/2008/08/30/meetings-with-energy-authorities-and-the-academy-of-social-sciences-mote-med-energimyndigheter-och-akademi/ (2008)

Chapter 16

Peak Transportation

The historical path of petroleum, from an unappreciated and mostly inaccessible underground resource to its current status of lifeblood of our technological civilization, began in 1859 in the United States, when (as described in Chap. 14) "Colonel" Edwin Drake drilled his well at Titusville, Pennsylvania. Students of oil history can begin their studies at the Drake Well Museum that features a reconstruction of Drake's original well and is sited near "Oil Creek" in Cherrytree Township a short distance from Titusville (see Fig. 14.1). As they continue their path of discovery, these students will probably want to visit the oil refinery of Engelsbergs Oljefabiks AB (literally "Engelsberg's Oilfactory Share Company"). This is the world's oldest preserved oil refinery that stands on the island of Oljeön (literally, "Oil Island") in the middle of a lake named Åmänningen in Sweden. To visit this museum one can take the train from Stockholm to the city of Västerås and then change rail lines to travel to the village of Ängelsberg. The rail platform in Ängelsberg lies right near the shore of Åmänningen and if one follows the signs giving directions to Oljeön one comes to a small ferry that transports people to the island in summer. Why is the world's oldest remaining refinery to be found deep in the Swedish forest on an island in a lake?

The refinery was built in 1875 and was placed on an island due to its inherent fire risk. Some years earlier Pehr August Ålund had built a previous refinery on the shore of the same lake but it had caught fire. The fire then spread, causing a great deal of damage. With his "new" refinery built on an island in the middle of the lake the damage from a second, uncontrolled fire would be limited to the island itself [1].

The oil for the refinery was freighted all the way from Pennsylvania. In the refinery museum is an example of one of the oil barrels in which the oil

K. Aleklett *Peeking at Peak Oil*, DOI 10.1007/978-1-4614-3424-5_16,

was transported across the Atlantic Ocean to Stockholm before it was loaded onto a barge to be taken to Ängelsberg. Ängelsberg was the site of one of Sweden's world-famous iron foundries. Iron ingots from the foundry were loaded onto barges for transport to Stockholm and then the barges returned to Ängelsberg loaded with barrels of crude oil. The refinery processed a maximum of 1,500 barrels per year and was active until 1902. Today the foundry at Ängelsberg has status as a UNESCO World Heritage Site.

The main product of the refinery on the island of Oljeön, was paraffin for the increasingly popular paraffin lamps of that time. The paraffin was required to provide lighting in the iron mines of Ängelsberg but the refinery's production was so great that it also provided for much of Sweden's needs. Fat and tallow were also imported from Russia and tar was imported from Galicia north of the Carpathians. These raw materials were blended with the heaviest fractions from the refined crude oil to produce important products such as gun oil and grease. A by-product that they had no use for at that time was gasoline. Today the refinery museum on Oljeön is owned by the oil company Preem.

A use for gasoline was found in the Otto engine developed by Nikolaus August Otto who, together with Eugen Langen, founded the first internal combustion engine production company, NA Otto and Cie [2], in 1864. The Otto engine was the first internal combustion engine to use compression of the fuel successfully to increase the engine's power output. The Otto engine was initially developed to run on lighter grades of fuel but was eventually adapted to run on gasoline. It was Rudolf Diesel [3] who extended the idea of gas compression to develop highly efficient Diesel engines that could burn even heavier fuel grades (see the section "Transport on Land" below).

The use of paraffin for lighting was largely abandoned as electricity began to provide the energy for this role but it then found use as a fuel for aviation. Today, 70% of the volume reported as oil production finds use in the transport sector but transport does not consume crude oil. Instead, transport uses various refined products of the oil. These refined products are essential to today's mobile civilization.

Oil itself must also be transported from where it is produced to where it is refined and then further distributed. The most important oil transport routes are from the Middle East to various parts of the world. Currently, nine supertankers carrying around two million barrels each leave the Middle East every day (Fig. 16.1). Six of these steer towards Asia and three navigate their way towards Europe and the United States [4]. Today's world would not function without this oil transport and an important issue is how Peak Oil will affect this vital flow of oil.

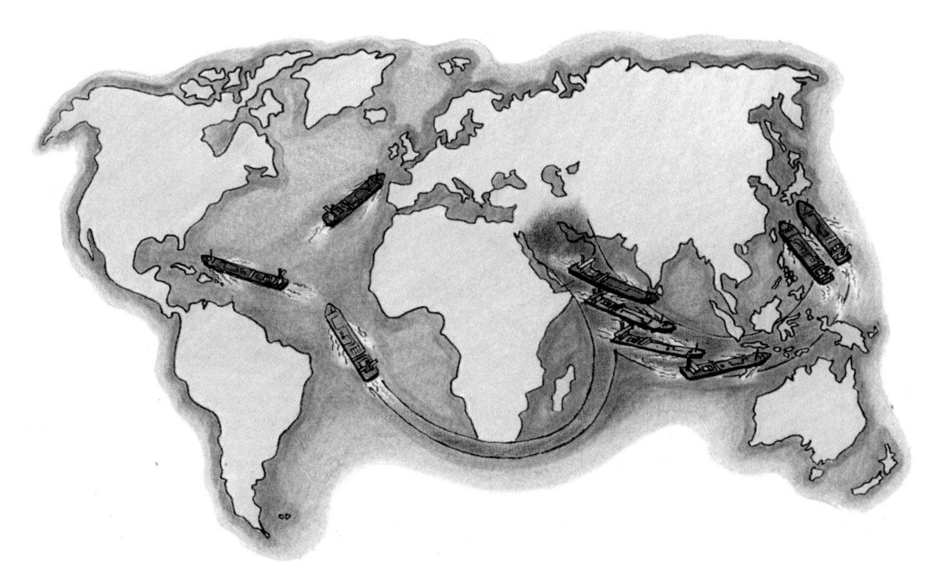

Fig. 16.1 Exports of oil from the Middle East are approximately 18 million barrels per day and can be represented by nine supertankers each carrying two million barrels of oil. Currently, six of these ships steer towards Asia and three steer towards Europe and the United States [4]. The world could not function without this transport of oil

From Crude Oil to Transport Fuel

The most commonly used transport fuels are gasoline, aviation fuel, diesel, and bunker oil. I have named them in order from the lightest fuel with the lowest boiling point (gasoline) to the heaviest with the highest boiling point (bunker oil). Crude oil is a mixture of all the possible refined products (see Fig. 4.1). By boiling crude oil and passing the vapors though a distillation apparatus these products can be separated. The distillation products are usually grouped into three categories: light distillates (LPG, gasoline, naphtha), middle distillates (kerosene, diesel), and heavy distillates and residuum (heavy fuel oil, lubricating oils, wax, asphalt).

A somewhat more detailed description of the distillation process is as follows. The crude oil is heated to approximately 350–400°C while holding it under pressure so that it remains in liquid form. It is then released into the lower end of a distillation column where the lower pressure means that it boils instantly. The distillation column is divided into a number of levels each separated by approximately 60 cm. Each level has a different temperature. The lowest levels are the hottest and the temperature

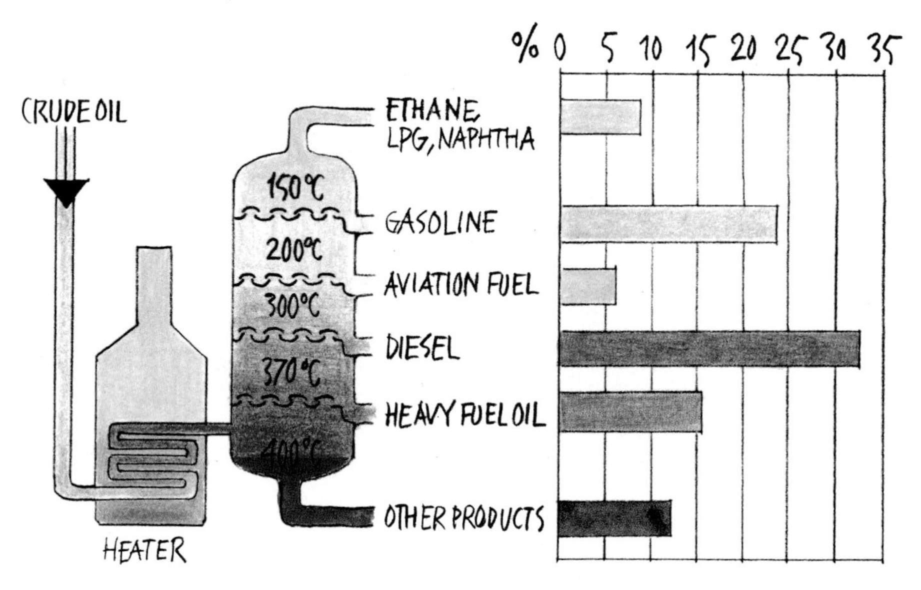

Fig. 16.2 The distillation of crude oil. The lightest products from distillation are normally subsequently burned to provide heat to boil additional crude oil. On the *right* are shown the relative amounts in percent of the various distillation products that constituted global refinery production in 2007 [5]. Individual refineries would show different relative amounts of each product depending on the grade of crude oil used and other factors

gradually decreases up the column so that the highest levels are the coolest. The boiling point of the various hydrocarbon molecules in the crude oil determines at what level in the column these molecules condense into liquids. In our article, "Aviation Fuel and Future Oil Production Scenarios" [5], we used the relative percentage volumes of the various refinery products as reported for 2007 as the basis for our discussion of possible future aviation fuel production (see also the section "Fuel for the Aviation Industry" below). In Fig. 16.2 these relative percentage volumes are used to illustrate the different amounts of products being produced by an average refinery. Boiling crude oil requires energy and this is usually provided by burning the lightest products from the distillation process. Among the heaviest products from the refining of crude oil is the raw material for asphalt. Asphalt is used to construct the new roads that facilitate automobile use thus encouraging more traffic that then requires more gasoline and diesel fuel.

In order to refine crude oil that contains a large amount of sulphur, critical components of the refinery must be built from stainless steel. During the first half of the Oil Age when we mainly used the highest-quality crude oil containing little sulphur it was possible to use cheaper refinery components. However, as we enter the second half of the Oil Age the proportion of oil production containing higher sulphur levels is increasing and this means that old refineries must be updated with stainless steel components or new refineries must be built if this oil is to be used. One example of this is oil from the oilfield Manifa in Saudi Arabia. The oil has such a high sulphur content that no existing refinery will buy it. Therefore, Saudi Arabia is building its own new refinery to process this oil (and thus enable the oil from Manifa to be produced).

The sulphur in crude oil follows it through the distillation process and segregates mainly with the lighter hydrocarbon products. It is separated from these after distillation. In the heavy oil used to power international shipping (out in international waters) some sulphur usually remains and creates highly polluting exhaust fumes. However, in coastal waters where nations can set pollution standards, shipping is now being forced to use sulphur-free fuel. Among other things these restrictions will increase the demand for diesel fuel.

The octane rating of gasoline is a measure of its ability to tolerate high pressure without premature ignition. After its production by distillation, the octane rating of gasoline is usually increased using additives, and other substances may be added to improve its characteristics under particular circumstances. Additives are also blended into aviation fuel that must tolerate very cold environments. All these additions mean that the volume of the refined fuel is increased. In statistics describing oil for consumption these volume increases are usually reported as "processing gains." (Processing gains are part of what are called "refinery throughputs.") When one realizes the complexity involved in extracting crude oil from deep underground and then converting it into the finished products for purchase it is astonishing that, at gas stations, one gallon of gasoline is still sold for less than one gallon of Coca-Cola!

The crude oil being produced today is, overall, becoming heavier as the more profitable, lighter crude oil reserves have preferentially been produced first (see the cover illustration of this book). The lighter distillation products of oil are more valuable than the heavier products, therefore the oil refining industry is introducing more advanced methods for "cracking" the larger molecules in heavier crude oil to produce greater volumes of the lighter fractions. Also, around 15% of conventional natural gas production consists of heavier molecules that are liquids (or nearly liquid) when separated from the methane that makes up the bulk of natural gas. These are termed natural gas liquids (NGL; see Chap. 10). NGL are lighter than crude oil and can be mixed into heavy oil to make it flow more easily. However, NGL are also an important feedstock for the petrochemical industry.

In the section titled, "Oil: Refinery Throughputs" of the 2011 edition of the *BP Statistical Review of World Energy* [4] they show the volumes of input to the primary distillation units of the world's oil refineries. (Data for Slovenia and Lithuania prior to 1992 are excluded.) The total input volume was highest in 2007 at 74.3 million barrels per day (Mb/d). In 2009 the volume had fallen to 73.5 Mb/d. When BP reports global oil production they do not separate this into its various fractions. However, in the *World Energy Outlook 2010* report, the International Energy Agency (IEA) states that crude oil production was 67.9 Mb/d in 2009 and that unconventional oil production was 2.3 Mb/d. Therefore, if all this crude and unconventional oil production passes through refineries then the total volume of input to refineries in 2009 would be 70.2 Mb/d. "Additives" then contributed an additional 2.3 Mb/d and then 1.0 Mb/d of NGL would be needed to get the volume to BP's stated 73.5 Mb/d.

In 2010 we have seen an increased volume of oil being processed by refineries but the level has still not returned to the peak volume seen in 2007. In Chap. 11 we calculated that crude oil production will decrease by approximately 2% per year and in Chap. 10 we showed that production of unconventional oil and NGL may increase by 8 Mb/d between 2010 and 2030. Some of this production need not be processed by refineries. However, for the calculations below we assume that 7 Mb/d of unconventional oil is subjected to refinery processing in 2030. Using this assumption we can calculate (albeit crudely) future volume flows through refineries. This shows that it is impossible to increase production of transport fuels from oil in the period to 2035. Instead, we will see a decrease so that, by 2035, the volumes of available fuel will be similar to what was available in the 1980s. It is interesting to note in Fig. 16.3 that the strong growth in transport fuel volumes seen from the end of the 1990s until the economic crash of 2008 did not result in an increase in input to the refineries of the OECD. We conclude that the maximal refinery capacity for the OECD has been reached and that any increase in future consumption will require importation of refined hydrocarbon products. Another conclusion is that, in the future, we will not see any investment in new refineries in the OECD because such investments would require guaranteed volumes of oil imports during the next 30 years and such guarantees do not exist.

When most other nations have ceased exporting oil, Saudi Arabia will be one of the last with the capacity to do so. Therefore, although China is currently

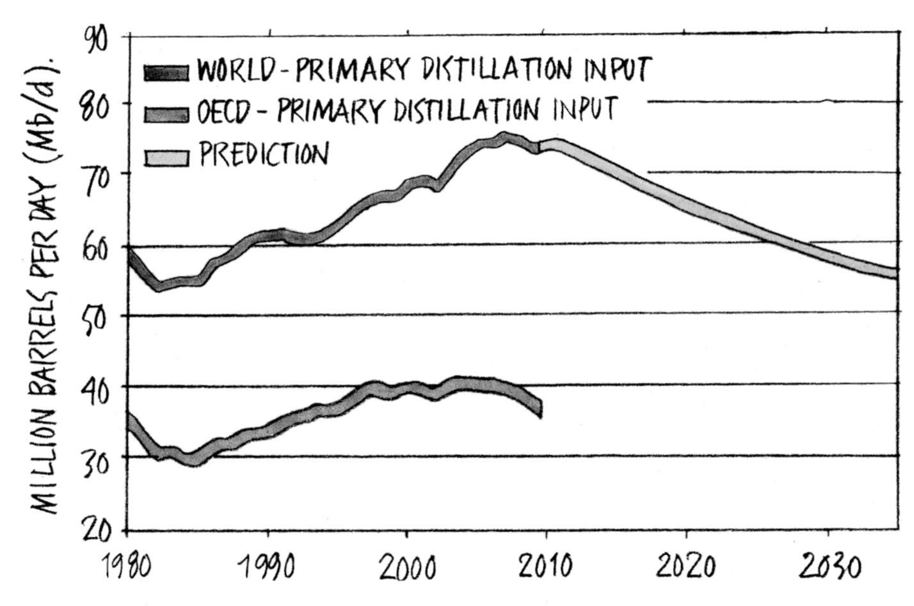

Fig. 16.3 The volume of oil and additives entering the world's refineries. This is primarily conventional crude oil, Orinoco heavy oil, and oil from Canada's oil sands. The volume of refinery input reached a maximum of 74.3 Mb/d in 2007 and then declined in 2008 and 2009. In contrast, in 2010 the volume of oil production reached the highest-ever level of 82.1 Mb/d (exceeding the 2008 production of 82.0 Mb/d). Although the volume of oil entering refineries increased in 2010, it did not reach the level of 2008 [4]. This recent difference between oil production and oil entering refineries can be explained by the fact that the increase in oil production was, in reality, an increase in NGL production and only a small fraction of the NGL is passing through refineries. The predicted declining volume of oil entering refineries from 2012 is based upon a combination of the expected decrease in conventional crude oil production by 2% per year (see Fig. 11.1) and production of unconventional crude oil rising by 9 Mb/d by 2035 (see Fig. 10.7). Refinery inputs in the OECD nations were fairly constant for 10 years and then began to decline in recent years

building new refineries it has also invited Saudi Aramco (Saudi Arabia's national oil company) to own 49% of one (and possibly more) of these. In this way China hopes to ensure that supertankers continue to deliver oil to it from Saudi Arabia.

Fuel for the Aviation Industry

The fuel used in the aviation industry is named Jet-A1. Very stringent regulations exist for what this fuel can contain. In December 2008 I was invited by the International Air Transport Association, IATA, to be one of the speakers at their *Aviation Fuel Forum* in Shanghai. At that time the hottest topic of discussion was contamination of Jet-A1 fuel by FAME (fatty acid methyl ester). FAME is a collective name for what is also called biodiesel. The maximum permissible concentration of FAME in Jet-A1 fuel is five parts per thousand (0.5%). If the FAME concentration in Jet-A1 fuel exceeds this value at an airport then that airport must be closed. The reason is that today's aircraft engines have not been tested for fuel blends containing FAME. This shows that it is much simpler to blend biodiesel into fuel for road transport (as they do in France) than to do so for aviation.

To illustrate how FAME can come to contaminate Jet-A1 fuel we can examine fuel transport in France where gasoline, Jet-A1, diesel fuel, and heating oil are transported though pipelines from Le Havre and the refineries in Normandy to depots in the Paris area. In France, diesel fuel is blended with 7% FAME. If they transport diesel by pipeline and then, immediately afterwards, transport Jet-A1 in the same pipe then the first portion of the transported Jet-A1 can become contaminated with the FAME in the diesel. The French authorities are aware of this problem and are currently working to find a solution [6].

One of the reasons I was invited to Shanghai was to present Uppsala Global Energy System's research report on future production of aviation fuel titled, "Aviation Fuel and Future Oil Production Scenarios" [5]. Air transport and Jet-A1 fuel are crucial for the functioning of our globalized world and, naturally, it is very interesting to study what the consequences of Peak Oil will be for Jet-A1 production.

Figure 4.12 in the IEA's *WEO 2010* report shows that the aviation industry's fuel consumption during 2009 was 250 million tons of oil equivalents (Mtoe), and that demand for aviation fuel is expected to grow to 380 Mtoe per year by 2035 [7]. Airbus and Boeing have announced future scenarios in which they expect transport by air to grow by 5% per year until 2026. Aircraft orders fell marginally in 2009 but now orders are increasing and are thought to have returned to their earlier growth trend. If we assume that the trends of recent years in increasing aviation fuel efficiency continue, then by 2026 it is calculated that air transport will require approximately 440 Mtoe of fuel. This is an increase of 75% over 15 years. The IEA's estimated increase in aviation fuel use by 2026 is 32% but in *WEO 2010* it does not see an equivalent increase in crude oil production by that year.

If the inflow of raw material to refineries follows the prediction shown in Fig. 16.3 and if the proportion of refinery output that is Jet-A1 fuel continues to be 6.3% then a 30–75% increase in fuel production for aviation would seem unrealistic. Instead, we can expect the aviation industry to struggle for every drop of fuel in the future.

Transport by Sea

The heaviest loads are freighted by sea and the ships that carry them also use the heaviest refinery products. As the demand for diesel increases in the future we will see an increasing number of refineries installing equipment for converting the heavier fractions from oil into diesel. Sea transport has, historically, used the dirtiest fuel. However, increasing environmental standards for coastal shipping have forced a change to cleaner varieties. Sea transport is the most efficient form of transport in terms of fuel use per freighted kilogram. Indeed, it is so efficient that lamb meat transported from New Zealand to Europe can have less environmental impact than lamb produced in Europe itself (due to differences in husbandry methods). As air freight becomes more expensive we can expect that shipping will retain or even increase its share of total transport. It is also possible that governments will seek to protect sea transport in the face of declining oil availability through regulations or subsidies.

Can Agriculture Provide Both Food and Transport Fuel?

Before discussing the fuel requirements for land transport we must examine the potential of agriculture to contribute to the fuel supply through production of biofuels. The IEA now includes production of ethanol in its "oil" production statistics. It states ethanol production in 2010 as 1.8 Mb/d (see the section "Peak Oil and Energy Demand", Chap. 2). In our article, "Agriculture as Provider of Both Food and Fuel" [8], we studied the energy content of the world's agricultural products and we asserted that there is sufficient food to feed the world's population. We also asserted that agriculture produces large quantities of residues that could be converted into transport fuels but that agricultural production is insufficient for provision of both food and fuel [8].

The energy needs of transport in 2006 were estimated to be 25,000 terrawatthours (TWh). Primarily, one can use corn or sugar cane to produce ethanol and soya beans or palm oil to produce biodiesel. In terms of energy, the 1.8 Mb/d of ethanol production in 2010 amounts to 1.5% of transport's needs [7]. Using the production numbers for 2006 we can calculate that if the world's entire corn production were converted to ethanol then it would cover 6.4% of transport's needs and conversion of the world's sugar cane to ethanol would provide a further 2.5%. If the world's entire soya bean and palm oil crops were used to produce biodiesel then an additional 2.6% of transport's energy requirements would be met. In other words, if the world's population stopped eating corn and sugar from sugar cane and if soya beans and palm oil were removed as sources of raw materials for the food industry

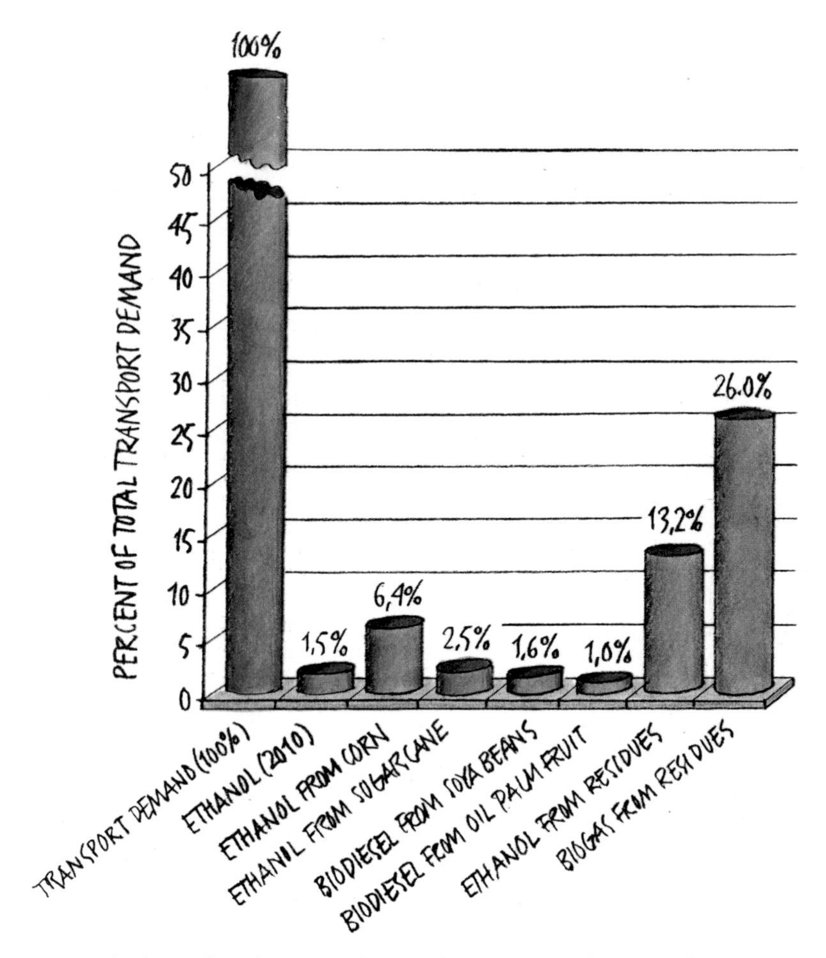

Fig. 16.4 Biofuels' real and potential contribution to total energy for transport. In 2006 transport's total energy demand was 25,000 TWh. The contribution of ethanol to transport energy is calculated from the production reported for 2009 [7]. If the world's entire crops of corn, sugarcane, soya beans, and palm fruit were used to produce biofuel they could provide the percentages of the total transport energy requirement given in the figure. Residues from agriculture could contribute more energy to transport than all of these crops combined either in the form of ethanol or, even more significantly, if converted to biogas (according to 2006 statistics) [8]

then these agricultural products could, instead, provide 11.5% of transport's energy needs. If new technology in the future allows us to produce ethanol from agricultural residues then this could provide an additional 13.2% of transport's needs. However, if the same residues were used to produce biogas then the energy contribution to transport could be nearly double that of ethanol! (Fig. 16.4).

There is a great need for further research into production of biofuels. In particular, we must investigate the possibility of agriculture itself providing the fuel required by agricultural machinery. In the calculations made by the Uppsala Global Energy Systems group we have not considered economic profitability but have only focused on maximal possible production. Even under these unrealistically favorable conditions we see that it is definitely not possible for agriculture to provide us with both food and all the fuel we need.

An excellent examination of the United States' ability to provide ethanol for its own transport needs is given in the book, *The Impending World Energy Mess* [9]. As one might expect, it is not possible for the United States to substitute the oil it uses for transport using ethanol. In fact, the total energy provided for transport by ethanol is so small that one can regard the United States' current ethanol production more as a national-scale research project than as a significant contributor to that nation's transport energy needs. In the United States, the demand for gasoline is greater than the demand for diesel but the trend of world oil production towards heavier crude oils favors production of diesel over gasoline. This imbalance is corrected somewhat by using diesel fuel (in agriculture) for ethanol production and then blending the ethanol into gasoline to increase gasoline volumes.

Transport on Land

Liquefied petroleum gas (LPG) and natural gas can be used in internal combustion engines to power transport. However, it is gasoline and diesel that completely dominate transport fuel use. Gasoline is used in Otto engines where a compressed mixture of fuel and air is ignited by an electric spark. In contrast, diesel fuel is used in Diesel engines in which air is highly compressed and so becomes very hot before diesel fuel is sprayed into the combustion chamber and self-ignition of the fuel occurs. The mixture of fuel and air before compression in an Otto engine means that there is an upper limit to how highly the mixture may be compressed lest it prematurely self-ignite due to the high temperature. The temperature limit is determined by the characteristics of the fuel–air mixture. Diesel engines can function at higher pressures and temperatures because only air is initially compressed before the fuel is introduced. Higher operating temperatures allow more efficient fuel combustion and this means that diesel-powered vehicles can travel farther than gasoline-powered vehicles for the same volume of fuel.

Most land-based goods transport is powered by diesel fuel. Also, it is becoming more popular to purchase automobiles powered by Diesel engines. Diesel is also the fuel used to power all earthmoving and mining machinery, agricultural machinery, and all other heavy vehicles. Diesel is also used for rail transport because far too few of the world's railways are electrified. Aside from its use in land transport, diesel is also seeing increased use by shipping, and diesel is commonly used to power generators in nations where the electricity supply is irregular. Therefore, Peak Oil combined with increased use of diesel fuel will lead to shortages of this fuel in the future.

In Fig. 16.2 we see that approximately 35% of the crude oil entering refineries leaves them as diesel fuel and approximately 25% leaves as gasoline. For many years diesel fuel was significantly cheaper than gasoline but recently this situation has reversed and we can expect diesel to be more expensive than gasoline in the future.

Another factor increasing the price of diesel is (somewhat paradoxically) the use of ethanol to decrease gasoline volumes as ethanol is replacing gasoline. During 2010 ethanol production was 1.8 Mb/d. However, volume for volume, ethanol only contains about two thirds of the energy of gasoline. Therefore, 1.8 Mb/d of ethanol can only replace 1.2 Mb/d of gasoline. To produce 1.2 Mb/d of gasoline would require processing of nearly 5 Mb/d of crude oil by refineries. Therefore, to have produced gasoline equivalent to 2010's ethanol production would have required world oil production to have increased from 82.1 to 87 Mb/d. If world oil production had been 87 Mb/d this would also have provided an additional 1.7 Mb/d of diesel fuel and so the price of diesel would, presumably, have been lower. Increased gasoline availability through ethanol addition can allow for increased economic activity but does not provide a commensurate increase in diesel availability. Thus, a political decision to mix ethanol into gasoline to a level of 10% leads to higher diesel prices relative to gasoline and may increase the cost of goods transport and agricultural production including those agricultural products used for ethanol production.

In Chap. 10 we showed that only limited volumes of liquid fuels can be produced from coal. However, an advantage of this form of fuel production is that it can be used directly as aviation fuel or blended into diesel. The US Army has plans to produce 300,000 barrels of fuel per day from coal which is equivalent to their daily fuel use within the United States. This shows that the US Army is expecting future difficulties in securing sufficient fuel for its transport needs.

Modern Production of Vehicle Fuel

At the beginning of this chapter I described how an oil company active in Sweden, Preem, owns the world's oldest preserved oil refinery on Oljeön. Preem also owns one of the world's most modern oil refineries, Preemraff, in the city of Lysekil in Sweden. Preemraff has recently been upgraded and its capacity is so great that it can produce more products than Sweden requires. However, one product that it refrains from producing is the aviation fuel Jet-A1. Instead, this fraction is blended into diesel to produce a more environmentally friendly form of that fuel.

There are uses for all the products produced by Preemraff, including the sulphur it extracts during refining of crude oil. The future supply of crude oil is the most important consideration for the refinery's activities but another issue is Sweden's political ambition to be free of fossil fuel use by 2050. Is Preemraff fated to become like the oil refinery on Oljeön, an industrial museum from oil's golden age?

References

1. Our sites and facilities: Oljeön, Ecomuseum Bergslagen. http://www.ekomuseum.se/english/besoksmal/oljeon.html (2011)
2. Wikipedia: NA Otto and Cie. http://en.wikipedia.org/wiki/Deutz_AG (2011)
3. Wikipedia: Diesel engine. http://en.wikipedia.org/wiki/Diesel_engine (2011)
4. BP: BP Statistical Review of World Energy, June 2011. http://bp.com/statisticalreview (historical data; http://www.bp.com/assets/bp_internet/globalbp/globalbp_uk_english/reports_and_publications/statistical) (2011)
5. Nygren, E., Aleklett, K., Höök, M.: Aviation fuel and future oil production scenarios. Energy Policy **37**, 4003–4010 (2009). http://www.tsl.uu.se/uhdsg/Publications/Aviation_EP.pdf
6. Viltart, P.: Transfer of Jet-A1 following biodiesel at Charles De Gaulle and Orly airports (Paris-France) in MPPs. Seminar on Implications of FAME/diesel Containing FAME for Jet Fuel Quality, London, January 2009. http://www.energyinst.org.uk/content/files/EI_FAME_Seminar_3.pdf (2009)
7. WEO: World Energy Outlook 2010 (Chapter 4). International Energy Agency, November 2010. http://www.worldenergyoutlook.org/2010.asp (2010)
8. Johansson, K., Liljequist, K., Ohlander, L., Aleklett, K.: Agriculture as provider of both food and fuel. Ambio – Roy Swedish Acad Sci **39**, 91 (2010). http://ambio.allenpress.com/perlserv/?request=index-html
9. Hirsch, R.L., Bezdek, R.H., Wendling, R.M.: The Impending World Energy Mess, Apogee Prime, Washington DC, USA ISBN 978-1926837-11-6 (2010)

Chapter 17

Peak Oil and Climate Change

For most people "climate change" is synonymous with the "greenhouse effect." A critical factor in climate change is emissions of carbon dioxide, CO_2. In this chapter we restrict our discussion primarily to the question of the volume of reserves of fossil fuels that can generate future CO_2 emissions.

If we regard the world as one large system enclosed within an outer boundary then it is easy to understand that the world will become warmer if the solar energy crossing the boundary into the system (mainly as visible light) is greater than the heat energy crossing the boundary to leave the system (as invisible infrared radiation; see Fig. 17.1). The atmosphere plays a decisive role in this. If the Earth had formed without an atmosphere then the solar energy striking it would have been radiated back into space at the same rate in the form of infrared radiation. The temperature at its surface would be around −15°C. Clearly, this would have restricted the evolution of life.

However, the Earth does have an atmosphere made up of various gases including water vapor. The molecular structure of the gases can make them vibrate at different frequencies and some of these frequencies are within the range of the infrared radiation that is radiating back into space. This means that the infrared radiation leaving the world system can be absorbed by these gases which causes their molecules to vibrate. Within a fraction of a second the vibrating gas molecules re-emit the radiation. However, when they do so, they re-emit it in all possible directions including back towards the ground (see Fig. 17.1). In this way the Earth's atmosphere acts like a blanket or the window of a glasshouse/greenhouse that traps heat inside. Apart from CO_2, water vapor and methane are the most important "greenhouse" gases.

K. Aleklett *Peeking at Peak Oil*, DOI 10.1007/978-1-4614-3424-5_17,

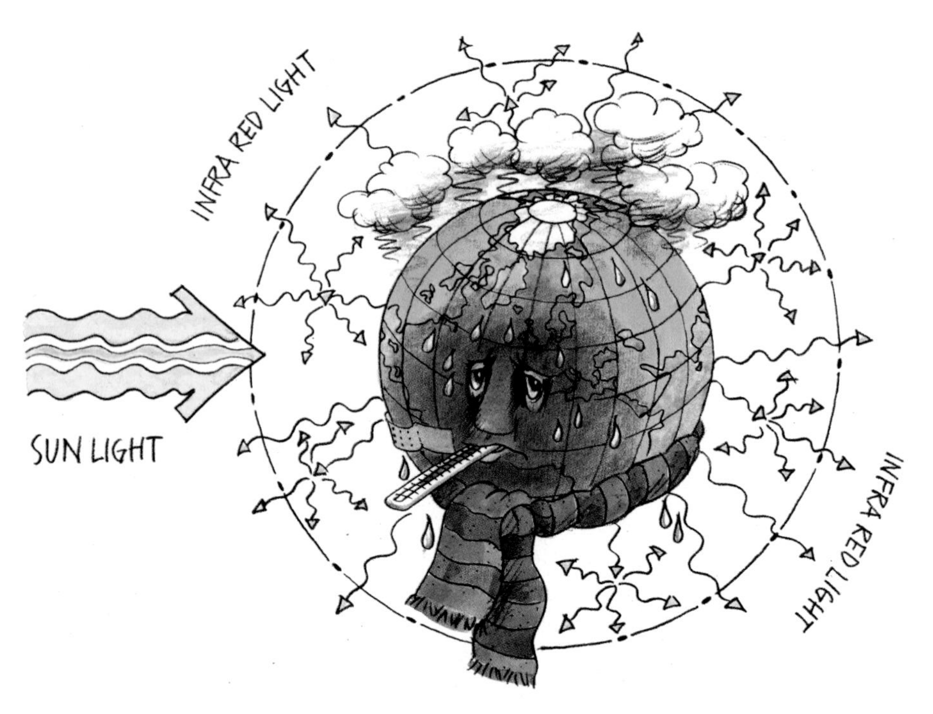

Fig. 17.1 Electromagnetic energy arrives from the sun mainly as visible light. The energy warms the Earth's surface and this heat then radiates away from the Earth as infrared radiation. Parts of this infrared radiation have the same frequency as particular vibration states of CO_2 and can be absorbed and then re-emitted in any direction. Some of the infrared radiation from the carbon dioxide is emitted back towards the ground and this means that CO_2 acts like a thick blanket retaining heat around the Earth. It is the effectiveness of this blanket that is debated when global warming is discussed

The amount of CO_2 in the atmosphere is not great enough to absorb all of the infrared radiation leaving our planet, otherwise the average temperature of the Earth would be higher. If the amount of CO_2 increases then more infrared radiation will be captured and redirected back into the world system. We will have an increased greenhouse effect and it is this that has been named global warming.

CO_2 is a gas formed when carbon or hydrocarbon molecules are burned. It consists of one carbon (C) atom and two oxygen (O) atoms. To gain some comprehension of the amount of CO_2 that human activities release, imagine that we have a large woodpile outside our house and that we burn all this wood in our fireplace during the winter. As we burn the wood, we take the ash that is left and place it beside the woodpile. In spring when the wood has all been burned we can compare the size of the ash pile with the original woodpile. The difference in weight and size between the woodpile and the ash is mainly the carbon in the CO_2 that was emitted through the house's chimney. (Depending on how dry the wood is, some of the weight is also water that leaves the chimney as steam.) And because oxygen atoms are heavier than carbon atoms, the weight of CO_2 emitted through the chimney is actually more than three times the weight of the wood! (The molecular mass of CO_2 is 3.67 times greater than that of pure carbon, C.)

There are many factors that influence increasing temperatures on our Earth but it has been calculated that around 65% of the increase is caused by increasing levels of CO_2 in the atmosphere. Factors other than CO_2 may have a greater influence than previously estimated but we do not discuss them in this chapter.

I would also like to emphasize that the Uppsala Global Energy Systems research group does not question the validity of the research reports showing, for example, that polar ice is decreasing, that the average global temperature is rising, or that the sea level has risen during the previous century. The level of CO_2 in the atmosphere has been measured on Mauna Loa on Hawaii since 1960 and it is clear that this is steadily increasing. In 1960 the amount of CO_2 in the atmosphere was less than 320 parts per million (ppm) and by April 2011 this had reached 393.18 ppm which is the highest level ever observed [1]. The fact that the amount of oxygen in the atmosphere is decreasing at the same rate as CO_2 is increasing shows that the burning of carbon (coal) and hydrocarbons (oil and natural gas) can be the main cause of the increase in CO_2 levels [2].

In Chap. 11 I described how, in 2007, the OECD gave me the task of writing a report on the world's future oil supplies [3]. At the same time I was tasked with writing an additional report on how limited production of oil, natural gas, and coal (Peak Oil, Peak Gas, and Peak Coal) would affect emissions of CO_2. That report was titled *Reserve Driven Forecasts for Oil, Gas & Coal and Limits in Carbon Dioxide Emissions; Peak Oil, Peak Gas, Peak Coal and Peak* CO_2 [2]. This chapter is mainly a description of the research in that report.

The aim of our research in this area was not to evaluate whether climate is changing or what tipping points might exist leading to dramatic changes in climate [4]. The aim was simply to evaluate the scenarios for CO_2 emissions that researchers around the world use in their calculations of climate change. The IPCC, the Intergovernmental Panel on Climate Change of the United Nations, usually asserts that it only assesses the results that other climate researchers have published in peer-reviewed journals. However, this is not the case for the IPCC's *Special Report on Emission Scenarios (SRES)* which is work for which the IPCC itself is responsible [5]. A large part of the responsibility for the production of the SRES can also be laid at the door of IIASA, the International Institute for Applied Systems Analysis in Laxenburg, Austria [6]. IIASA had primary responsibility for organizing the report. The discussion that follows below might have been my "peer-review" assessment of the SRES if I had been given the chance to review it before it was published.

Resources and Reserves Available for Emission as CO_2

The oil that exists in geological formations of sandstone or limestone is a resource and constitutes the original oil in place, OOIP. The oil that can be raised from underground economically is regarded as reserves and there are rules for how these reserves should be classified (see Chap. 6). Some of these reserves have already been exploited and CO_2 has been produced as a result.

In addition to oil we also exploit the fossil fuels, natural gas and coal. Natural gas (that some prefer to call "fossil gas") is methane and has the chemical formula CH_4. When methane is burned it reacts with the oxygen in the atmosphere to form one molecule of CO_2 and two molecules of water (H_2O). Coal is basically carbon (C) and when it is burned it forms mostly CO_2. This means that burning coal, oil, or natural gas releases different amounts of CO_2 for the same amount of energy production (see Table 17.1).

The easiest way to begin to appreciate the scale of our reserves of coal, oil, and natural gas is to examine the numbers in British Petroleum's freely available *BP Statistical Review of World Energy 2011* [7]. It states coal reserves

Table 17.1 The world's reserves of fossil fuel from the BP Statistical Review of World Energy 2010 [7] (in *bold text*) and their energy content and production of CO_2 when burned[a]

	(Mton)	(Tcf)	(Gboe)	Conv. factor ([b])	CO_2 (Mton C)	CO_2 (Mton)	CO_2 (Mton/Gboe)
Coal	**860,938**		*3,300*	*0.7326*	*630,700*	*2,315,000*	*700*
Gas		**6,609**	*1,172*[c]	*14.56*	*96,200*	*353,200*	*300*
Oil			**1,383**	*106.1*	*146,700*	*538,500*	*390*

[a]Values calculated using BP's numbers are shown italicized. The stated conversion factors are calculated from the values given by the CDIAC [8]. A molecular mass ratio of 3.67 was used to calculate CO_2 mass from Mton C (Note: The standard scientific unit for measuring energy is the Joule and one barrel of oil releases 6,117,863,200 J when burned)
[b]The conversion factors are, respectively, from coal (in Mton), natural gas (in Tcf), and oil (in Gb) to Mton carbon in the molecule CO_2
[c]The conversion factor 5.64 is used [9]

as 860,938 million metric tons (Mton), oil reserves as 1,383 billion barrels (Gb), and gas reserves as 6,609 trillion cubic feet (Tcf) or 187.1 trillion cubic meters (Tcm). The fact that these different forms of fossil fuel are quantified in different ways makes it difficult to compare them in terms of their energy content and how much CO_2 they produce when burned. However, if we compare the energy content of a given amount of coal or gas to the energy content of a barrel of oil (as "barrels of oil equivalent," BOE or boe) then these calculations are easier.

In Table 17.1 we show the world's reserves of fossil fuel as described by BP and we calculate their energy content and production of CO_2 when burned. To compare oil, coal, and natural gas we have used the conversion factors that the US Energy Information Administration (EIA) uses when making such comparisons [10]. The weight of CO_2 released by burning each fuel is calculated according to the guidelines published by CDIAC, the Carbon Dioxide Information Analysis Center [8].

The conversion factors for comparing oil and natural gas vary depending upon who is using them. According to the EIA one boe equals 5,640 cubic feet of natural gas. However, the USGS and others give one boe as 6,000 cubic feet. Where possible we try to use the same values as the International Energy Agency (IEA) and so, in this case, we assume that one boe equals 6,000 cubic feet of natural gas.

In the final column of Table 17.1 we compare the CO_2 produced by burning the different forms of fossil fuel when each fossil fuel is measured in a unified unit of energy production, gigabarrels of oil equivalent (Gboe). Coal and natural gas are both used for generation of electricity and it is obvious that coal is much more damaging (in terms of CO_2 production) when used for this purpose than natural gas. We discuss this in more detail at the end of this chapter.

The Emission Scenarios of the IPCC

In November 2000 the United Nations' Intergovernmental Panel on Climate Change (IPCC) presented its report on possible scenarios for emission of CO_2, the *Special Report on Emission Scenarios (SRES)*. In the report the IPCC presented 40 different scenarios in which future use of oil, natural gas, and coal would cause increased emissions of CO_2. The scenarios were grouped into four different families or "storylines" that were given the names A1, A2, B1, and B2. The person with lead responsibility for assembly of the report was Nebojsa Nakicenovic from the International Institute for Applied Systems Analysis (IIASA) in Austria [5].

The report showed that future emissions of CO_2 (even in the absence of policies addressing climate change) depend greatly on the choices we make as individuals. Crucial factors are how the economy is structured, the forms of energy we prefer and how we use available resources. However, the report also stated that all the different scenarios have the same likelihood of reflecting future events. In reality there are limits to how much of our energy "resources" can be transformed into "reserves" and to the rate at which these reserves can be exploited, but this is not reflected in the SRES calculations. In Peak Oil circles we refer to these rate limits as "the size of the tap" (in contrast to the "size of the tank" which is the size of the reserves and resources available). It is the "size of the tap" that will determine when the world experiences Peak Oil, Peak Gas, and Peak Coal, and the emission level of CO_2.

The IPCC states that the "Scenarios are images of the future, or alternative futures." However, "They are neither predictions nor forecasts." In the scenarios, different assumptions about the future affect the course of future history so that a number of alternative future narratives are depicted. There is a computer game named SimCity in which different cities can evolve through time depending on the conditions that a player sets and one can think of the IPCC's scenarios as a kind of "IPCC-SimCity game."

The four families of scenarios in the SRES; A1, A2, B1, and B2 each generate different amounts of CO_2 emissions depending on how much oil, natural gas, and coal they use. It is the descriptions of CO_2 emissions from these different scenarios that climate change researchers use when attempting to calculate how global temperatures might change in the future. Some broad details about the different families are shown in Fig. 17.2.

In the past decade the climate debate has revolved around the SRES families. The families A1, A2, B1, and B2 each possess, respectively, 17, 6, 9, and 8 scenarios. Most people engaged in the debate have never seen a detailed description of the CO_2 emissions that each scenario generates. However, they have probably seen diagrams of future changes in temperature over time where each scenario or family is named. In Fig. 17.3 we show how much oil is consumed in the different scenarios. We have chosen to show this in units of millions of barrels per day (Mb/d) which is commonly used in industrial circles and the media. For most people, Mb/d is more easily understood than the "zetajoule" (ZJ) unit that the IPCC uses.

In our article, "Validity of the Fossil Fuel Production Outlooks in the IPCC Emission Scenarios," [11] there is a detailed analysis of the different scenarios. The scenario that consumes the most oil in the year 2100 is "A1G AIM." It consumes 350 Mb/d which is the highest red curve in Fig. 17.3. This is more than fourfold greater than our current production of 82.1 Mb/d. The IPCC's emission scenarios were constructed from an

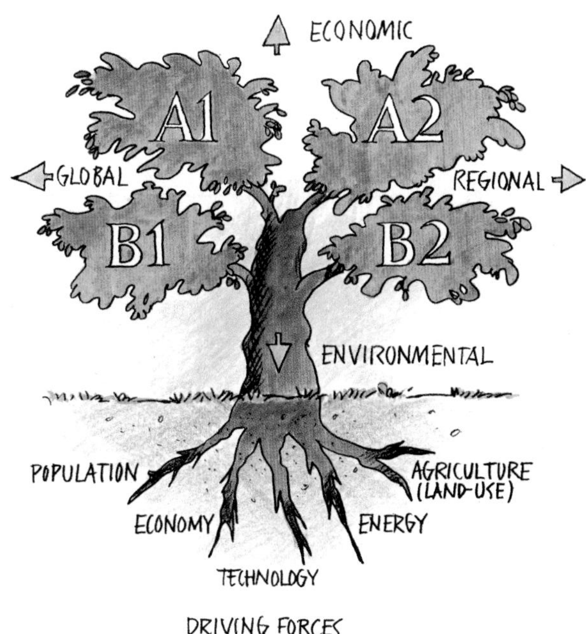

Fig. 17.2 In calculations of predicted climate change different CO_2 emission scenarios are used. The emissions are generated by the activities of the world's population. If we portray the future as a growing tree then the tree's root system is crucial for that growth. In the IPCC's SRES (Special Report on Emission Scenarios) the most important "roots" are population, the economy, technology, energy, and how we use land for agriculture. The tree has four main boughs that differ in the extent to which future societies and economies are characterized by global versus regional development and economic versus environmental development. The lesser branches of each bough differ according to other particular assumptions. The different boughs and their various branches are described as "families" and are named A1, A2, B1, and B2 [5]. The CO_2 emitted in each scenario comes mainly from use of fossil fuels but no limits have been placed on the total quantities of oil, natural gas, and coal that it is possible to burn. That the supply of these fuels is limited such that Peak Oil, Peak Gas, and Peak Coal are inevitable is nowhere to be found in the calculations

energy economics perspective. To assert that a production rate of 350 Mb/d in 2100 is in any way consistent with reality shows that we cannot leave predictions of the future to energy economists. In Chap. 9 we described how oil is found within the pores of sandstone or limestone reservoirs and that this fact means that individual reservoirs cannot be emptied faster than by approximately 10% per year. For the North Sea—so far the world's most rapidly produced region—the rate of depletion of remaining

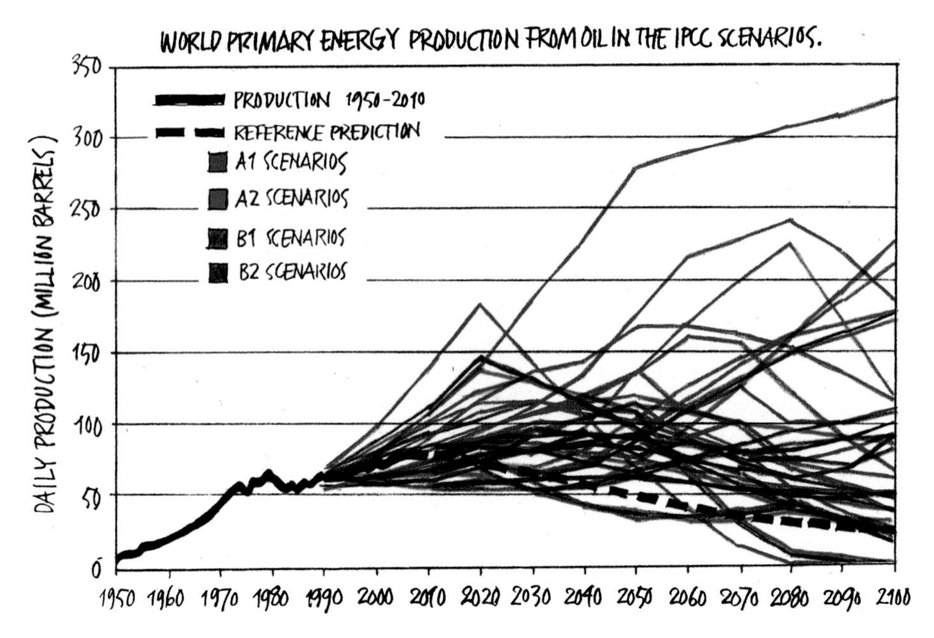

Fig. 17.3 The oil consumed by the 40 CO_2 emission scenarios published in 2000 by the IPCC that extend from 1990 until the year 2100. Families *A1, A2, B1, and B2* possess, respectively, 17, 6, 9, and 8 scenarios and are marked in *red, blue, green, and brown*. Oil production up to 2010 is shown by the *black curve* and our reference prediction for future production is shown by the *dashed black curve*

recoverable resources (DRRR) is only 6% per year. A rate of 6% means that a production rate of 350 Mb/d (128 Gb/year, Gb/y) would require a reserve size of over 2,100 Gb which is far larger than today's existing reserves. If we examine the total amount of oil this production rate would require during the next 90 years (assuming a linear increase to 350 Mb/d by 2100) then, in Gb, this is 7,112 Gb [(83 + 350)/2 × 90 × 365/1,000 = 7,112]. If current crude oil reserves are 900 Gb, unconventional oil reserves are 300 Gb and if we anticipate finding an additional 300 Gb of reserves, then total current and future reserves are only 1,500 Gb compared with the IPCC's requirement of about 7,100 Gb. If, in addition, we also consider the reserve size (2,100 Gb) that would be required for a production rate of 350 Mb/d by year 2100 then the A1G AIM scenario's requirement exceeds 9,000 Gb. To assert that this is possible shows just how inconsistent with reality is the A1G AIM scenario. It is this and other unrealistic emission scenarios that the IPCC has provided to climate change researchers for use in their calculations.

A Study of the World's Oil Resources and a Comparison with the IPCC's Emission Scenarios

Jean Laherrère is one of the founders of ASPO, the Association for the Study of Peak Oil and Gas. He has served as head of exploration for the French oil company Total, participated in the discovery of large oil and gas fields, and has worked with oil statistics for many years. He is truly an "insider" when it comes to oil and gas. In June 2001, 7 months after the IPCC presented its *Special Report on Emission Scenarios*, Jean was invited to a workshop on resources organized by the IIASA of Austria. He prepared a paper for presentation at the workshop titled, "Estimates of Oil Reserves" [12]. Jean has described how Nebojsa Nakicenovic, who was responsible for assembly of the SRES, was present when he delivered his paper. By the end of the presentation, Jean had shown that the world's oil and gas reserves were insufficient for the scenarios of the SRES to be realized. I asked Jean how this information was received by the workshop and I was told, "The only reaction was to ignore my paper because Peak Oil and the end of cheap oil were regarded as a myth since the price of oil in 2001 was $25 per barrel" [12]. It is apparent that energy economist Nebojsa Nakicenovic did not want to listen to what an experienced oil geologist had to say; there was no room for reality among the IPCC's emission scenarios!

In January 2003 I had the opportunity to recruit Anders Sivertsson as a Master degree student to my newly formed research group at Uppsala University. Of course, Anders' research project was to be on the topic of Peak Oil. However, inspired by the paper Jean had prepared for the IIASA's workshop, Anders' project was expanded. He would now compare the results of his Peak Oil analysis with the scenarios of the SRES. As mentioned above, a fact that is not discussed in the SRES is that those scenarios envisaging high levels of oil production in 2100 also require large remaining reserves for them to be plausible. Therefore, Anders also made estimates of the reserves required in 2100 [13].

A Possible Scenario for Emissions of CO_2 from Oil

In Chap. 2 I defined "oil" as including crude oil, oil from oil sands, liquid products synthesized from coal and natural gas, oil from oil shale, and, lastly, "natural gas liquids" including condensate. However, in calculating our reference predictions for emissions of CO_2 used for comparison with the IPCC's scenarios, liquid products synthesized from coal and natural gas are considered part of coal and natural gas production, respectively.

Using the data we have on conventional and unconventional oil reserves we can now make a realistic prediction for oil production from these sources up until 2100.

In Chap. 11 we discussed how much crude oil has been discovered and how much we can expect the oil industry to find in the future. In total, there will be about 1,200 Gb of conventional crude oil that can be used between now and 2100. So far we have consumed around 1,100 Gb. In Chap. 10 we discussed forms of unconventional oil. The oil sands in Canada and Venezuela are estimated to see a production increase of 4.5 Mb/d by 2035 and together they reach a total production of 2.5 Gb/y in 2037. If we make the assumption that this level of production will continue until 2100 then the total production from oil sands will be 200 Gb in the period between 2010 and 2100. There will then exist an additional 200 Gb left to produce but our judgment is that environmental concerns and other factors will constrain yearly production so that it does not exceed 2.5 Gb/y during the remainder of this century.

In *WEO 2010* the IEA estimated that production of oil from oil shale will reach 0.1 Mb/d by 2020 and will then increase to 0.3 Mb/d by 2035. If we assume that this rate of production increase would continue for the rest of the century then the total production from oil shale between 2010 and 2100 would amount to 20 Gb.

Natural gas liquids (NGL) are produced in association with natural gas production. In Chap. 10 we calculated that NGL production will increase until 2035 but that it will then decrease at the same rate that natural gas production decreases. (See the section "A Possible Scenario for Emissions of CO_2 from Oil".) Total production of NGL by 2100 will amount to 220 Gboe. Oil production between 1990 and 2010 amounted to 560 Gb. If we add to this the predicted production from conventional and unconventional sources by 2100 then the total oil produced from 1990 until 2100 according to our reference prediction is 2,100 Gb. During the same period the IPCC estimates that total oil consumption may be as high as 8,300 Gb and its least thirsty scenario requires 1,800 Gb. In Fig. 17.3 we compare our future estimate for the rate of oil production with the rates required by the 40 scenarios of the SRES. If, instead, we examine the total oil consumed by the 40 scenarios over the next 90 years then we see the data in Fig. 17.4. The upper boundary is formed by scenario A1G AIM that requires production of 8,300 Gb of oil between 1990 and 2100. The lower boundary is formed by scenario A2 MINICAM that requires the least oil at 1,800 Gb. We now know that the actual production/consumption of oil that occurred

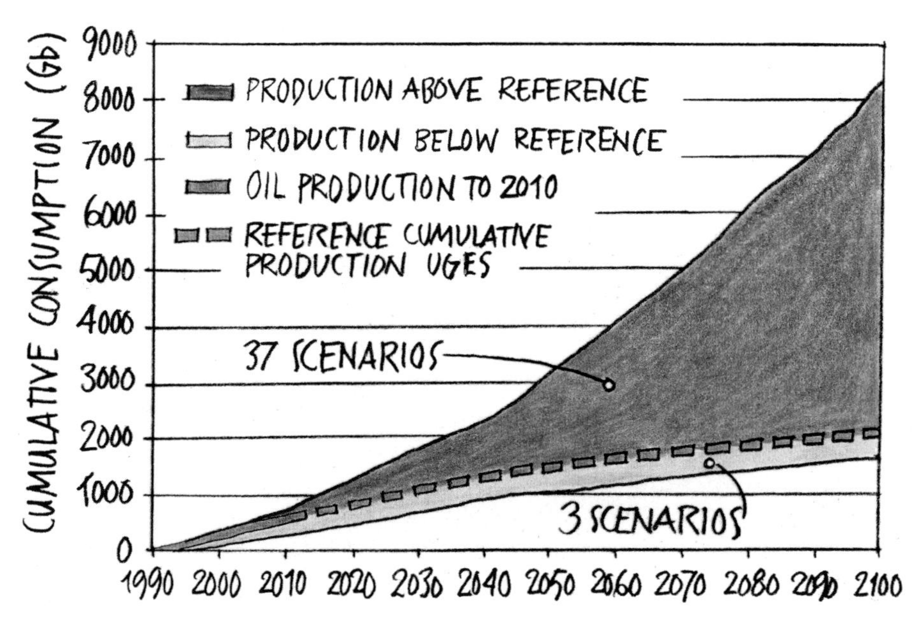

Fig. 17.4 The total oil consumption until 2100 of the 40 scenarios in the SRES is compared with our realistic estimate of possible oil production. By 2100 three of the scenarios lie under (*within*) our estimate and 37 are above (*outside*) it. The various scenarios of the SRES also have set volumes of associated natural gas and coal production and, clearly, it is total emissions from all fossil fuels that must be considered

between 1990 and 2010 was 560 Gb and the total consumption required under the various scenarios during the same period ranged between 408 and 703 Gb.

Our calculations show that from 2010 until 2100 it is possible to produce/consume 1540 Gb of oil (1100 + 200 + 20 + 220 Gb, see above). This would cause release of around 600 billion tons of CO_2. For the period 2010–2050 the amount is 380 billion tons CO_2.

Natural Gas and Climate Change

In Chap. 4 we described how conventional natural gas and oil have the same origin. Therefore, we can expect that the history of natural gas discovery will follow a similar pattern to that of oil. The Uppsala Global Energy Systems research group has not studied the world's natural gas resources and natural gas production in detail. However, we have made detailed

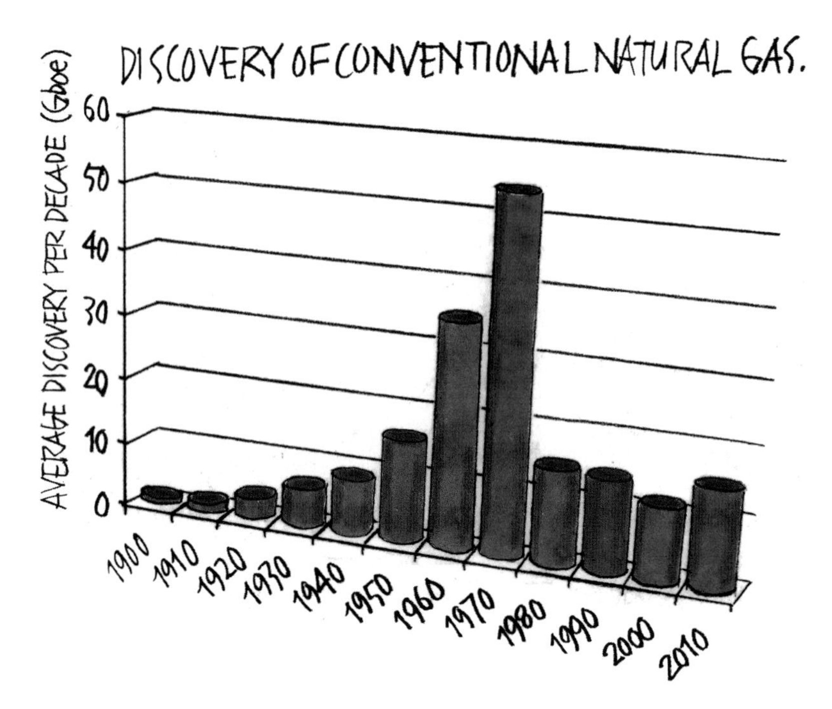

Fig. 17.5 Discoveries of natural gas reached a maximum during the 1970s when the world's largest natural gas field was found on the border between Iran and Qatar. Normally volumes of natural gas are described in Tcf but we have shown them as barrels of oil equivalent to compare them more easily with oil discoveries

studies of this fossil fuel for the North Sea and for Russia and a summary of these studies can be found in Bengt Söderbergh's Ph.D. thesis [14]. The research in the thesis shows global discoveries of natural gas in giant natural gas fields reached a maximum (peaked) during the 1970s and then declined. In Fig. 17.5 we show the total discoveries of natural gas per decade. Usually, the volume of natural gas discoveries is described using the unit Tcf (trillions of cubic feet) or Tcm (trillions of cubic meters). However, in order to compare the energy content of natural gas discoveries with those of oil we describe them in barrel of oil equivalents (boe). One billion boe (1 Gboe) = 5.64 Tcf [9]. Until 1990 the total natural gas that had been consumed was about 100 Gboe and from 1990 until 2010 an additional 310 Gboe was consumed. Table 17.1 shows natural gas reserves as 1,172 Gboe and if we add the gas that has already been consumed to these reserves we find that the total volume of natural gas discovered is approximately 1,600 Gboe. That is 300 Gboe less than the total volume of conventional oil discoveries of 1,900 Gb. From a climate change perspective this means that conventional natural gas has had, and likely will have, less influence on the climate than oil (Table 17.1).

The extrapolation of previous natural gas discoveries to predict the total of future gas discoveries involves more uncertainty than for discoveries of oil where a downward trend is clear. For a region such as the North Sea we see that there are approximately as many natural gas fields as oilfields. We have calculated that in 2100 we will have produced 2,100 Gb of crude oil. If we assume that global natural gas production (in boe) will be similar in magnitude then this will be equivalent to 11,850 Tcf. In the most recent estimates made by Jean Laherrère he sees a volume of 12,500 Tcf produced/consumed by 2100 [15]. The natural gas production estimate that Colin Campbell and I published in 2003 was based on nation-by-nation estimates [16]. It was subsequently refined by Anders Sivertsson [13] and has continued to be updated by Colin Campbell in subsequent years [17]. Total natural gas production is currently estimated at 10,250 Tcf by 2100 which is marginally less than the volume calculated by Jean. For our discussion below we use 2,100 Gboe as the amount of natural gas produced by 2100. Because we had already consumed 410 Gboe by 2010 this means that we have the potential to consume an additional 1,690 Gboe of natural gas by 2100.

The base for unconventional production of natural gas is coal discoveries and oil shale. By far the largest quantities of these resources exist in the United States and the United States dominates world production of unconventional natural gas. According to the IEA's *World Energy Outlook 2010* report (*WEO 2010*) [18] total natural gas production in 2008 was 18.6 Gboe and unconventional natural gas accounted for 12% of that (2.2 Gboe). By 2035 unconventional natural gas production is expected to have grown to 5 Gboe per year. During the period from 2008 until 2035 the total amount of unconventional natural gas produced is projected to be 97 Gboe. It is unrealistic to expect that unconventional natural gas production will increase after 2035 but let us assume that production remains constant until 2100. This would mean total production of 325 Gboe of unconventional gas between 2035 and 2100 and total production of this gas between 2010 and 2100 would be around 400 Gboe. Total production of all natural gas would therefore be around 2,500 Gboe by 2100. When combined with the total oil production calculated in the previous section we can make the following comparison with the consumption required by the various scenarios of the SRES (see Fig. 17.6).

A detailed description of the differing patterns of natural gas consumption for the 40 scenarios of the SRES is given in our article, "Validity of Fossil Fuel Outlooks in the IPCC Emission Scenarios" [11]. If we examine

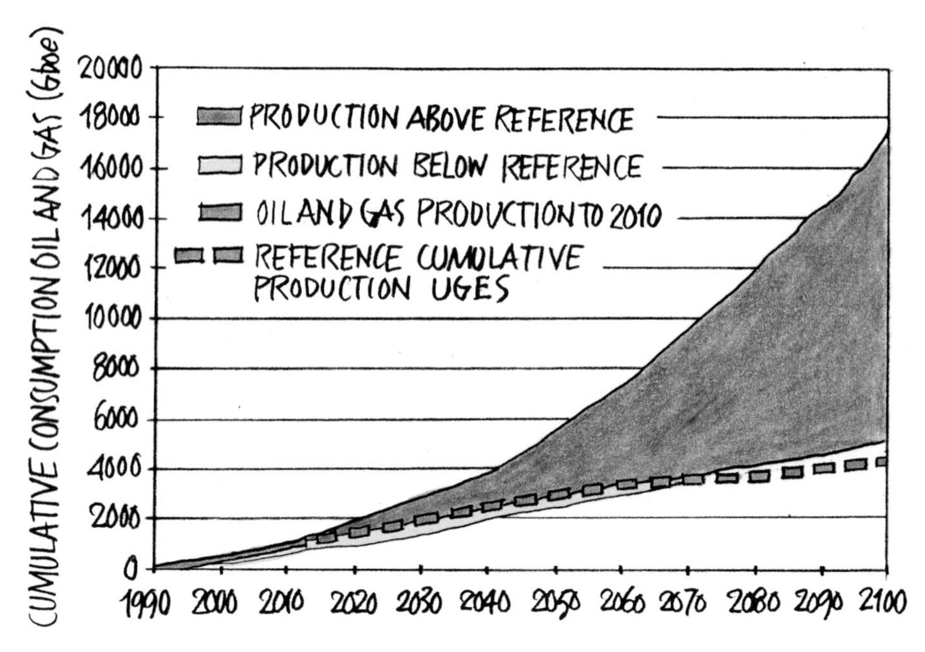

Fig. 17.6 Total oil and natural gas consumption for the 40 scenarios in the SRES compared with that from our reference prediction. In around 2030 half of the scenarios for oil and gas lie under our calculated production but from 2070 all of the scenarios lie above this

their required cumulative consumption of gas in the same way as we did in Fig. 17.4 then we find that, by 2100, the scenarios have consumed between 9,585 Gboe (A1 GAIM) and 2,699 Gboe (B1 MINICAB). This means that all the scenarios consume more natural gas than is realistically possible (2,090 Gboe, see below). In Fig. 17.6 we have shown the cumulative oil and natural gas production of the different scenarios and compared this with the reference prediction by Uppsala Global Energy Systems, UGES.

Around the year 2030 approximately half of the 40 scenarios of the SRES show a combined rate of oil and natural gas consumption that is just under our estimated possible production. However, by 2100 all the scenarios are consuming more than our predicted production. We conclude that all of the 40 scenarios published by the IPCC in 2000 use more oil and natural gas by 2100 than is realistically possible.

Between 1900 and 2010 the world consumed 410 Gboe of the 2,500 Gboe of natural gas that it is feasible to produce and consume by 2100. This means that we can consume an additional 2,090 Gboe by 2100. Using the information in Table 17.1 we can calculate that 2,090 Gboe of natural gas contains approximately 630 billion tons of CO_2. The amount of CO_2 we predict to be emitted between 2010 and 2050 is around 370 billion tons.

Reactions to Our Research from IIASA and IPCC

Earlier I mentioned my conversation with Jean Laherrère where he described how the person co-ordinating production of the 40 scenarios of the SRES, Nebojsa Nakicenovic, did not react to Jean's demonstration that none of the scenarios were possible. In September 2003 Anders Sivertsson completed his Master thesis. He was due to make a public defense of his thesis in the last week of September after which, if successful, he would have earned his degree from Uppsala University. However, some weeks before this a writer for the magazine *New Scientist*, Andy Coghlan, rang and wanted to ask a few questions about Peak Oil. During our conversation I mentioned Anders' work and Andy was intrigued. I sent him a copy of Anders' Master thesis on the condition that nothing could be published until Anders' had made his defense. On October 5, 2003 an article appeared in print in *New Scientist* titled, "Too Little Oil for Global Warming" [19].

Andy Coghlan also contacted Nebojsa Nakicenovic and then wrote the following text:

> Nebojsa Nakicenovic, an energy economist at the University of Vienna, Austria who headed the 80-strong IPCC team that produced the forecasts, says the panel's work still stands. He says they factored in a much broader and internationally accepted range of oil and gas estimates than the "conservative" Swedes. Even if oil and gas run out, "there's a huge amount of coal underground that could be exploited," he says.

The electronic version of the *New Scientist* article was released at midnight (Greenwich time) on October 2nd and that morning CNN's London office rang and wanted to do an interview for broadcast around noon. They also published an article on their website regarding Anders' work [20]. Of course, I had expected that there would be considerable interest in this research and that we would be invited to various climate conferences to describe it. However, there was almost no reaction from the scientific community. In fact, the only reaction I received came before the data were published. One person connected to the IPCC (who attended ASPO's 2003 conference in Paris in May 2003 [21]) thought we should not publish anything that might harm "the cause." It was not this statement that led to a delay in further research on this topic but, rather, Nakicenovic's words that "there's a huge amount of coal underground that could be exploited." I began planning for a new Ph.D. project focused on examining coal reserves and production. Unfortunately, 3 years passed before I could appoint Mikael Höök as a Ph.D. student in 2006 but then the work began in earnest.

Coal and Climate Change

Coal and natural gas are used to generate electricity. According to the information in Table 17.1, using coal for electricity generation causes release of a little more than twice as much CO_2 as when natural gas is used. When Nebojsa Nakicenovic says, "There's a huge amount of coal underground that could be exploited," he is expressing a belief held by many. The ongoing discussion regarding the use of "carbon capture and storage" (CCS) methods to prevent CO_2 emissions from coal is largely driven by the idea of the availability of almost unlimited coal resources.

If we study where the world's coal resources are located we find that 90% of these lie in six nations: the United States, China, Russia, India, Australia, and South Africa. These nations also consume 80% of the world's coal production. According to our research, the 90% of future coal-derived CO_2 emissions that these six nations are responsible for will amount to 65% of all future CO_2 emissions. Therefore, we can conclude that, although CO_2 emissions are a problem for all the nations of the world, it is the six nations above that will be responsible for the bulk of those emissions. (This fact is discussed in Chap. 19.)

The first issue that our coal research addressed was whether coal reserves are, in practical terms, "unlimited." Coal production already appears to have peaked in some nations such as the United Kingdom, Germany, and Japan. Hubbert curves can be fitted quite well to the production histories of these nations (see Fig. 17.7). Therefore, it is valid to believe that a peak in world coal production will occur at some point in time as more and more nations experience declining internal coal production. The world's largest reserves of coal exist in the United States so we began by making a state-by-state examination of its production. We concluded that the possibility exists to increase US coal production by 40% by the year 2050 but afterwards coal production would level off until 2100. The bulk of any future increase in US coal production would come from the states of Wyoming and Montana. However, there is public opposition to coal mining, particularly in Montana. The fact that the nation with the world's largest coal reserves can only increase its rate of coal production by 40% is

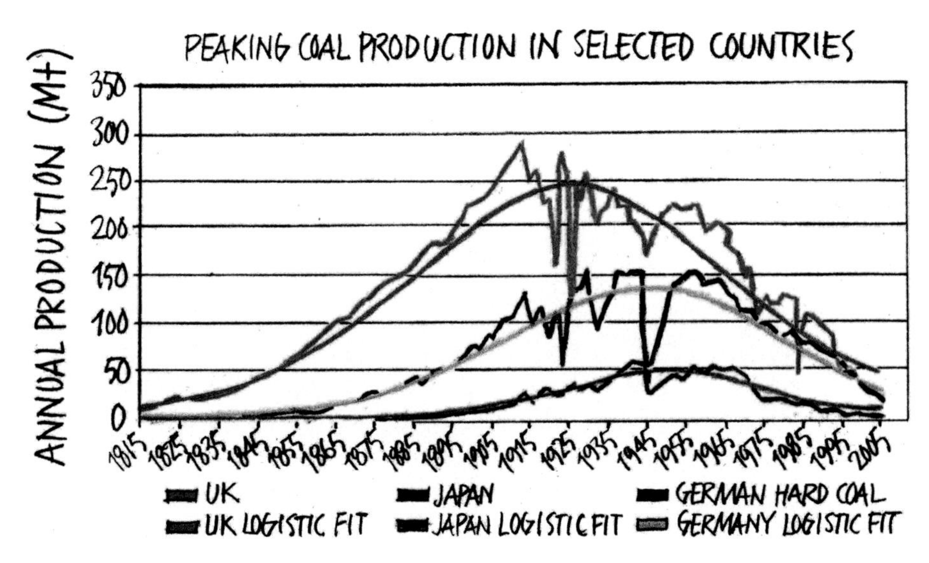

Fig. 17.7 The histories of coal production in the United Kingdom, Germany, and Japan all show that peaks in production occurred followed by declines. A Hubbert curve can be fitted to each nation's production history showing that all three nations have passed Peak Coal. The peaking of production in individual nations before occurrence of a world production peak has been observed for conventional oil so discussions of a peak in world coal production are also valid [22]

particularly relevant for the rest of our discussion below. For details of our calculations please refer to our published paper, "Historical Trends in American Coal Production and a Possible Future Outlook" [22].

Analyzing future global coal production was a massive task and we did it in collaboration with W. Zittel and J. Schindler of Germany. By the autumn of 2007 we had made a first estimate and this was included in my report to the OECD [2]. To prepare a scientific paper describing our research and have it published in a peer-reviewed journal took several attempts. We sent it to the journal *Fuel* in April 2009 and then had to write a revised version that was accepted for publication in June 2010. The paper then became available to the public on the Internet. The conclusion we had arrived at—and that proved so controversial—was that global production of coal would reach a maximum around 2040 and then decline. This was very different from the widely held belief that future coal production is essentially unlimited [23]. The overall result of our analyses is shown in Fig. 17.8. According to our calculations maximal coal production (Peak Coal) for the world should occur around the year 2030. For China, Peak Coal should occur around 2020. The fact that China is currently buying into Australian coal mines shows that they are preparing for a peak in Chinese coal production.

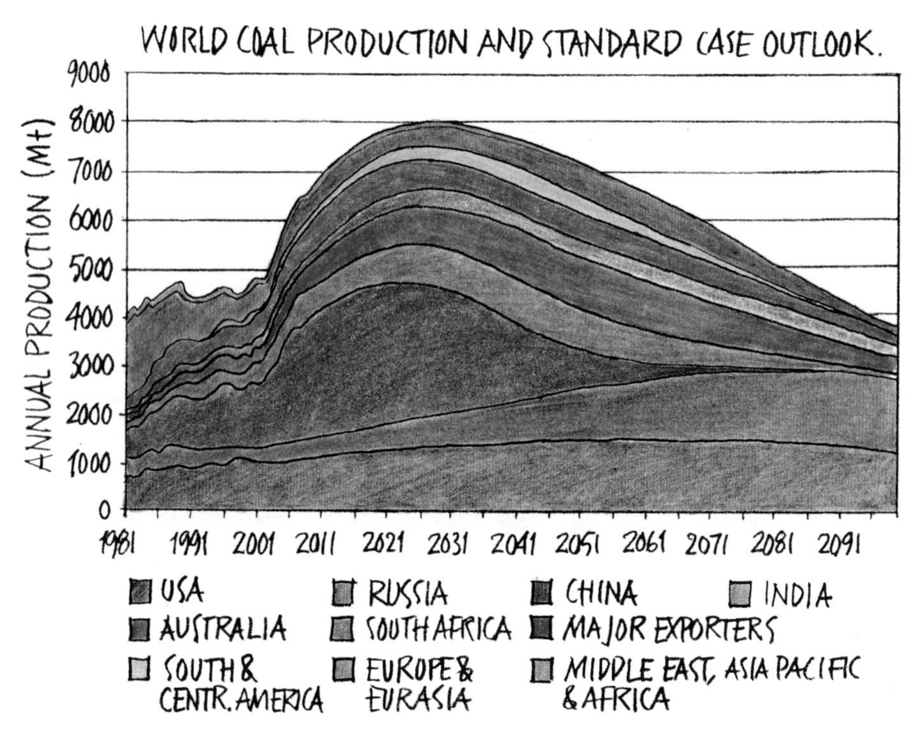

Fig. 17.8 Production of coal as shown in our publication, "Global Coal Production Outlooks Based on a Logistic Model" [23]. Around 80% of all coal production occurs in six nations: the United States, Russia, China, India, Australia, and South Africa. Note that Chinese coal production is expected to reach a maximum around 2020 and the subsequent decline in Chinese production means that global maximal coal production will occur around 2030

Coal production in Russia after 2050 will come principally from discoveries in Siberia but there is currently limited infrastructure for mining or transport of coal in that area. During the autumn of 2010 I had the opportunity to study the coal production and export infrastructure in Newcastle in Australia. I could conclude that making the infrastructure investments necessary to produce Siberia's coal will require a coal price significantly higher than currently prevails.

If we compare our expectations for world coal production with that in the scenarios of the SRES then we see that some scenarios use less coal. However, it is remarkable that some scenarios involve production of coal in 2100 that is 500–600% greater than today's production level. The United States with the world's largest coal reserves can only increase its production by a maximum of 40%. Russia can increase its production by a larger percentage but it is starting from a significantly lower production level than the

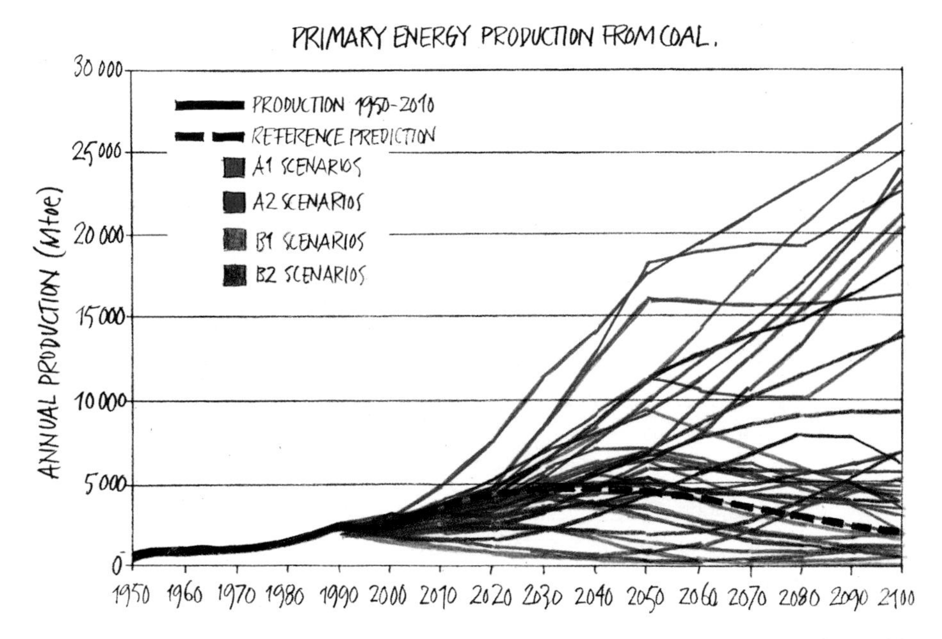

Fig. 17.9 Comparison of UGES's assessment (the "reference prediction") of possible primary energy production from coal with that of the 40 scenarios published in 2000 by the IPCC that extend from 1990 until the year 2100. Families *A1, A2, B1, and B2* include, respectively, 17, 6, 9, and 8 scenarios and are marked in *red, blue, green, and brown*. Coal production up to 2010 is shown by the *black curve* and the reference prediction is shown as a *dashed black curve* (The original figure is shown in Ref. [22] and describes energy production from coal in millions of metric tons of oil equivalents, Mtoe)

United States. The world coal production maximum that we predict in 2030 is only 25% higher than the current production rate and by 2100 coal production will be below today's rate. The scenarios of the SRES that expect an increase of 500–600% are commonly described as "business-as-usual." It is these scenarios that are expected to increase global temperatures by 6°C. We calculate that consumption of coal from 2010 until 2100 can equal 1,550 Gboe and is equivalent to about 1,100 billion tons of CO_2. By 2050 emissions of CO_2 from coal consumption can be about 600 billion tons (Fig. 17.9).

Future Emissions of CO_2 from Coal, Oil, and Natural Gas

Several of the scenarios of the SRES that have relatively low emissions of CO_2 from coal have relatively large emissions from oil and natural gas. In Fig. 17.10 we have added together the consumption of oil, natural gas, and

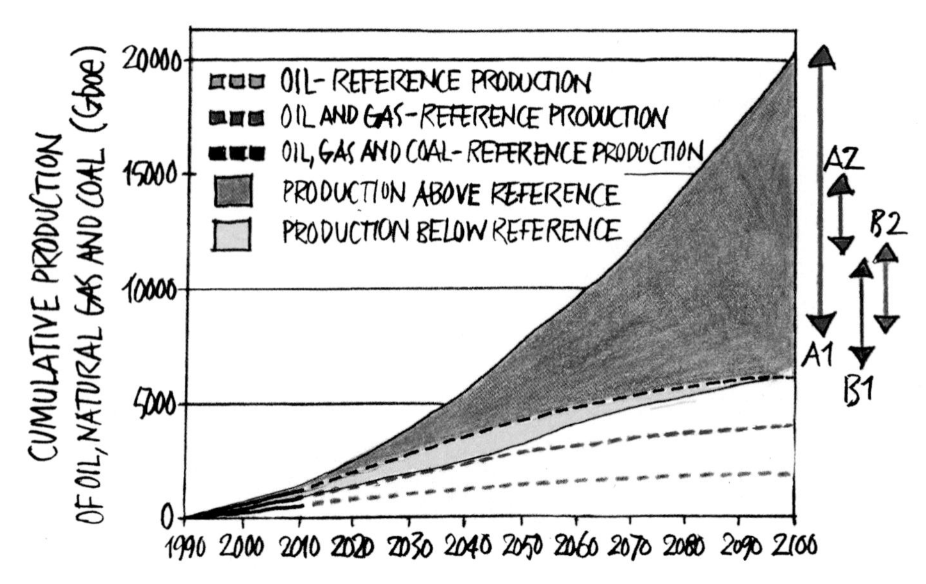

Fig. 17.10 The total consumption of oil, natural gas, and coal of the 40 scenarios in the SRES until 2100 is compared with the sum of our reference predictions for oil, natural gas, and coal of 5,740 Gboe. By 2100 none of the scenarios lie within our estimate and by 2050 a total of 30 are above it. The ranges of consumption variation in 2100 for the different members within the A1, A2, B1, and B2 families are indicated by arrows on the right. If we consider the uncertainty of our reference production predictions to be ±30% (see above) then predicted cumulative production might be as high as 7,500 Gboe and this means that families A1, A2, and B2 are outside the upper limit of possible production

coal for the various scenarios. We have also summarized the maximum production of these fossil fuels that we have discussed as reference predictions in this book. Together these amount to 5,740 Gboe (oil of 2,100 Gboe plus natural gas of 2,090 Gboe and coal of 1,550 Gboe). We see that, in 2010, most of the scenarios would have consumed less fossil fuel in total than was actually observed. However, by 2050 only 10 of the 40 scenarios have a total fossil fuel consumption that is lower than the maximal possible production we predict. By 2100 all of the 40 scenarios are consuming more fossil fuel than we consider realistic according to our reference predictions.

To make a stringent estimate of the possible degree of error in our predictions is not possible but we can discuss some of the factors that may have an impact on them. History shows that the world economy cannot tolerate excessive oil prices (see Chap. 19). Also, we are now seeing investment costs for deep water oil exploration increasing as are the costs of maintaining production from older fields. These factors may mean that we have overestimated future possible fossil fuel production. In contrast, new technology can increase the amount of oil we can produce from a field.

If the world's total crude oil reserves of 2,000 Gb (i.e., remaining reserves and those already produced) are based on a global recovery factor of 40% then the world's OOIP must be approximately 5,000 Gb. If we could increase the recovery factor by 10% then we would have an additional 500 Gb to consume. Similar arguments can be made for production of natural gas and coal and these show that we must assume around 20–30% uncertainty in our production predictions. Therefore, if we calculate maximal cumulative fossil fuel production from 1990 to 2100 as equaling 5,740 Gboe then a ±30% margin of error means that the actual production may be as high as 7,500 Gboe. If we examine the scenario families of the SRES in this light then we see that consumption of 7,500 Gboe by 2100 approximately coincides with the lower limit of fossil fuel consumption for B2 and that those B1-scenarios consuming the least fossil fuels may be realistic. This means that families A1, A2, and B2 are outside the upper limit of possible production.

We have also calculated the CO_2 emissions from the use of fossil energy from 2010 to 2100. Burning of the 5,740 Gboe of fossil fuels available by 2100 produces 2,330 billion tons of CO_2 with the emission of 1,350 billion tons of CO_2 between 2010 and 2050. From 2000 to 2010 we saw emissions of around 300 billion tons of CO_2 and in total we calculate that over 1,600 billion tons of CO_2 can be released between 2000 and 2050. This is more than the 1,000 billion tons of CO_2 that Malte Meinshausen et al. [24] find as the level of emissions between 2000 and 2050 that gives a 25% probability of producing more than 2°C of global warming. Therefore, even though Peak Oil, Peak Gas, and Peak Coal will force us to live in an environmentally conscious world of low CO_2 emissions, this may not be enough to prevent a temperature increase of 2°C.

What will happen with global temperatures? I hope that many of the research groups that make estimates of climate change will make use of the resource quantities that we have calculated to investigate this question. However, a crude but immediate answer to the question of future global temperatures can be found by referring to the temperature changes for the scenarios of the SRES that the IPCC presented in 2008 (Fig. 17.11). If the lower half of the B1 family is representative of our future then we will see a temperature rise of 1.5–2.0°C. The large temperature increase of 6°C that some of the IPCC scenarios predict would seem to be completely unrealistic.

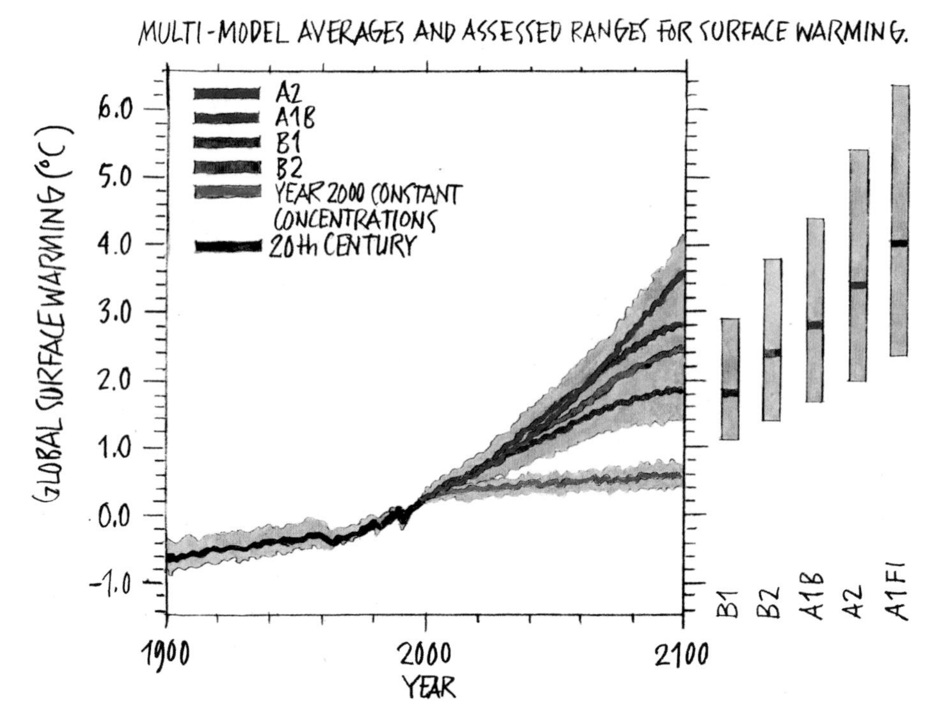

Fig. 17.11 Projections of future global average surface temperature for various IPCC scenarios. The graph shows temperature changes (as compared with the 1980–1999 average, which is used as the baseline) for three scenarios (A2, A1B, and B1). *Solid colored lines* represent the "most likely" trends presented by the IPCC [25] and the *solid green line* is an extrapolation of the B2 scenario. The *gray shaded* region represents the "probable ranges" for each family. The *gray bars* on the right represent temperatures in 2100 for all six scenarios; the colored stripe represents the "best estimate," and the *shaded gray* region represents "likely ranges." The different scenarios and models predict temperature changes between one and slightly more than 6°C (The source for the graph is the Fourth Assessment Report: Climate Change 2007 [25])

Why Are Peak Oil, Peak Gas, and Peak Coal Ignored?

In April 2011 the leading scientific journal *Nature* published an article with the title, "Opening the Future" [26]. It discusses new scenarios for emission of CO_2 and this leads us back to where we began this chapter:

> Some resource experts—such as Jean Laherrère, a petroleum geologist retired from French oil company Total—say this is unrealistic, because people won't be able to produce enough oil, coal, and gas to produce that much carbon dioxide. Nebojsa Nakicenovic agrees, but for different reasons: "The high end is impossible," he says, because the impacts would be so severe that it's inconceivable that the world would not take some kind of action.

Previously we described how Nebojsa Nakicenovic disregarded the information that Jean Laherrère presented in 2001. A decade later, it seems he is still ignoring the work of people researching Peak Oil, Peak Gas, and Peak Coal, even though these research results can now be found in peer-reviewed publications. The Uppsala Global Energy Systems group has now produced more than 20 peer-reviewed publications that are relevant to this issue [27]. The IPCC is now considering using four scenarios for future calculations of climate change (IMAGE–RCP3–PD 2.6, MiniCAM–RCP 4.5, AIM–RCP 6.0, and MESSAGE–RCP 8.5). However, these are based on old scenarios from the SRES. In light of the information presented in this chapter we can already say that the RCP 6.0 and RPC 8.5 scenarios are not realistic.

In the summer of 2009 I was invited once again to participate in the *Tällberg Forum*, in a symposium that was to discuss future energy scenarios. The leader of the symposium was Gerald Davis from IIASA. I was keen to describe the research that the UGES group had conducted at Uppsala University. I had hoped to be able to comment at the start of the symposium but it was only in the last 10 min of a late morning session that I was finally able to mention the research I have described in this chapter. I was quite upset by this as were a number of the other participants.

On the last evening before the end of the *Tällberg Forum* I met Gerald Davis (or "Ged" as he is commonly called) at one of the hotels in Tällberg. We discussed the symposium over a beer and, naturally, I couldn't resist criticizing the SRES and explaining once again that we had shown the scenarios to be erroneous. To my great surprise he then explained that it was he himself who had constructed the scenarios. I then understood why I was allowed so little chance to participate in the symposium. For many years Ged has also been responsible for the "Shell energy scenarios" released by that oil company. (Today we know that those scenarios are also not realistic.) Ged and I discussed the use of scenarios and I was led to understand that it is not regarded as wrong to use an unrealistic scenario if it can help one comprehend how an unrealistic future might appear. For me that sounded very strange. Using this redefinition of the purpose of the IPCC's scenarios we can now state that scenario families A1 and A2 can be used as examples of unrealistic futures. The problem then is that various research groups have considered the scenarios to be realistic including the temperature changes relating to these scenarios. The consequences of this are discussed in Chap. 19.

In this chapter I have tried to highlight the significant contribution of Jean Laherrère to our understanding of fossil fuel resources. Unfortunately, Jean is also an excellent example of how a successful industry professional and analyst can be ignored. This naturally leads us to ask in whom does the

IIASA place its faith? Most people conducting research into energy resources would not have heard of that particular person but for most climate change scientists he is the foundation upon which their world rests.

In our article "Validity of the Fossil Fuel Production Outlooks in the IPCC Emission Scenarios," [11] we have analyzed in detail the significance of Han-Holger Rogner for the production of the IPCC's emission scenarios. I conclude this chapter with edited excerpts from the section of that article subtitled, "SRES Dependence on Rogner":

Resource availability in the SRES [5] is built around the works of Rogner [28] and Gregory and Rogner [29], and relies on them for detailed discussion of the estimated hydrocarbon amounts. . . . The message of Rogner is that vast reserves of unconventional hydrocarbons together with previously observed rates of technology change imply that fossil energy will be available for hundreds of years at low cost, that is, at market prices not significantly higher than those of the 1990s. Rogner also states that additional forms of hydrocarbon outside what are commonly considered part of the possible resource base mean that fossil fuels can serve as an almost unlimited source of energy. As Rogner states, "The sheer size of the fossil resource base makes fossil sources an energy supply option for many centuries to come."

Rogner based his conclusions on compilations of hydrocarbon resource estimates made prior to 1997. Some of the major sources of these estimates are BP, the World Energy Council, and the German Federal Institute of Geosciences, and other estimates are taken from academic studies. For conventional crude oil, Rogner states an ultimate recoverable resource base of 2,800 Gigabarrels (Gb) and the aggregate number for various forms of unconventional oil is 16,500 Gb, where these include heavy oil, tar sands, and oil shale. Clearly, these numbers imply that unconventional oil would have to be the main source of oil in the longer term. Comparisons of the estimates in Rogner and more recent estimates of available unconventional oil can be found in Greene et al. [30].

Rogner describes a similar picture for natural gas, with only 2,900 Gboe as conventional gas (3,100 Gboe if natural gas liquids are included) but 142,000 Gboe available from unconventional sources. Over 95% of the unconventional gas is assumed to be methane hydrates, and only minor volumes come from coal-bed methane, fractured shale, tight formation, and remaining in situ. As a result, Rogner sees future gas availability as tightly connected to methane hydrates and their development.

Rogner states that drilling is yet to confirm the existence of the methane hydrates suggested by seismic surveys. In reality, few thick deposits have actually been found by drilling and generally the existence of gas hydrates is assumed from no more than uncertain seismic information [31]. It is also worth noting that estimates of methane hydrate resources differ by three orders of magnitude [32]. Some research into this potential energy source is underway but any commercial production is very far off and methane hydrates are unlikely to contribute to world gas production for the next 30–50 years [33].

For coal, Rogner highlights the considerable variation in assessments of world reserves and resources. His overview of coal availability is very brief in comparison with oil and gas and relies mostly on BGR (1980) as a source of information. The total coal resource is placed at 45,800 Gboe, which would equal 8,744 Gt of coal (assuming 30 GJ/t coal). Nearly 60% of this coal is classed in the most uncertain category.

In their paper from 1998, Gregory and Rogner largely rely on the resource estimates previously published by Rogner but they also include resource estimates for renewable and nonfossil fuels. A significant portion of the article is devoted to speculation and envisioning the feasibility of future conversion technologies, ranging from fuel cells and hydrogen to unconventional hydrocarbons such as oil shale or methane hydrates.

Interestingly, Gregory and Rogner also mention the "pessimistic" view on ultimate recoverable resources as promoted by geologists and including Laherrère and Campbell. They contrast this with the "optimistic" view, led by economists Adelman and Lynch. As an energy economist, Rogner also sides with the optimists and emphasizes the importance of unconventional hydrocarbon resources.

References

1. Earth System Research Laboratory: Trends in Atmospheric Carbon Dioxide. NOOA, US Department of Commerce. http://www.esrl.noaa.gov/gmd/ccgg/trends/ (2011)
2. Aleklett, K.: Reserve driven forecasts for oil, gas and coal and limits in carbon dioxide emissions. Discussion Paper No. 2007-18, December 2007. Joint Transport Research Center, Paris, France. http://www.internationaltransportforum.org/jtrc/DiscussionPapers/DiscussionPaper18.pdf (2007)
3. Aleklett, K.: Peak oil and the evolving strategies of oil importing and exporting countries. Discussion Paper No. 2007-17, December 2007. Joint Transport Research Center, Paris, France. http://www.internationaltransportforum.org/jtrc/DiscussionPapers/DiscussionPaper17.pdf (2007)
4. Rockström, et al.: A safe operating space for humanity. Nature **461**, 472–475 (2009). http://www.nature.com/nature/journal/v461/n7263/full/461472a.html
5. Nakicenovic, N., Swart, R. (eds.): Special Report on Emissions Scenarios (SRES). Intergovernmental Panel on Climate Change (IPCC). http://www.grida.no/publications/other/ipcc_sr/?src=/climate/ipcc/emission/ (2000)

6. International Institute for Applied Systems Analysis (IIASA). http://www.iiasa.ac.at/ (2011)
7. BP: BP Statistical Review Of World Energy. http://bp.com/statisticalreview (historical data; http://www.bp.com/assets/bp_internet/globalbp/globalbp_uk_english/reports_and_publications/statistical) (2011)
8. Marland, G., Rotty, R.M.: Carbon dioxide information analysis center—conversion tables, Table 6. Factors and units for calculating annual CO_2 emissions using global fuel production data. Carbon dioxide emissions from fossil fuels: a procedure for estimation and results for 1950–1981, DOE/NBB-0036, TR003, U.S. Department of Energy, Washington, DC. http://cdiac.ornl.gov/pns/convert.html#6 (1983)
9. Numerous conversion factors are in use for comparing the energy content of dry natural gas and oil. When examining world production, HIS (http://www.ihs.com) uses 1 boe = 6 kcf. However, the energy content per unit volume of natural gas can differ depending on the source of the natural gas. If we use the heat content values for volumes of oil and natural gas given by the American Physical Society (http://www.aps.org/policy/reports/popa-reports/energy/units.cfm) then we calculate a conversion factor 1 boe = 5.64 kcf. In our discussions in this chapter we have used this number since it is related to the energy content and CO2 emission potential of these fossil fuels
10. U.S. Energy Information Administration. http://www.eia.gov/ (2011)
11. Höök, M., Sivertsson, A., Aleklett, K.: Validity of the fossil fuel production outlooks in the IPCC emission scenarios. Nat. Res. Res. **19**(2), 63–81 (2010). http://www.springerlink.com/content/j6577353716vqn5h/;http://www.tsl.uu.se/uhdsg/Publications/IPCC_article.pdf
12. Laherrere, J.: Estimates of oil reserves. Paper Presented at the EMF/IEA/IEW Meeting IIASA, Laxenburg, Austria, 19 June 2001. http://www.iiasa.ac.at/Research/ECS/IEW2001/pdffiles/Papers/Laherrere-long.pdf (2001)
13. Sivertsson, A.: Study of world oil resources with a comparison to IPCC emissions scenarios. Diploma thesis, Uppsala University, 2004-01-01, ISSN 1401-5765. http://www.tsl.uu.se/uhdsg/Publications/Sivertsson_Thesis.pdf (2004)
14. Söderbergh, B.: Production from giant gas fields in Norway and Russia and subsequent implications for European Energy Security. Doctoral thesis, Uppsala University, Acta Universitatis Upsaliensis, urn:nbn:se:uu:diva-112229, ISBN: 978-91-554-7698-4; http://urn.kb.se/resolve?urn=urn:nbn:se:uu:diva-112229 (2010)
15. Laherrère, J.: Private communications (2011)
16. Aleklett, K., Campbell, C.: The peak and decline of world oil and gas production. Miner. Energy: Raw Mater. Rep. **18**, 5–20 (2003). http://www.ingentaconnect.com/content/routledg/smin/2003/00000018/00000001/art00004, http://www.peakoil.net/files/OilpeakMineralsEnergy.pdf
17. Campbell, C.: Private communications (2001)
18. WEO: World Energy Outlook 2010. International Energy Agency. http://www.worldenergyoutlook.org/2010.asp (2010)
19. Coghlan, A.: Too little oil for global warming. New Sci. 2003-10-05. http://www.newscientist.com/article/dn4216-too-little-oil-for-global-warming.html (2003)

20. Jones, G.: World oil and gas 'running out'. CNN, Thursday, 2 Oct 2003, Posted: 1245 GMT. http://edition.cnn.com/2003/WORLD/europe/10/02/global.warming/ (2003)
21. Association for the Study of Peak Oil and Gas: Second International Workshop on Oil Depletion, Paris, France, May 26–27, 2003. http://www.peakoil.net/conferences/iwood-2003-paris (2003)
22. Höök, M., Aleklett, K.: Historical trends in American coal production and a possible future outlook. Int. J. Coal Geol. **78**(3), 201–216 (2009). http://www.sciencedirect.com/science/article/pii/S0166516209000317; http://www.tsl.uu.se/uhdsg/Publications/USA_Coal.pdf
23. Höök, M., Zittel, W., Schindler, J., Aleklett, K.: Global coal production outlooks based on a logistic model. Fuel **89**(11), 3546–3558 (2010). http://www.tsl.uu.se/uhdsg/Publications/Coal_Fuel.pdf
24. Meinshausen, M., Meinshausen, N., Hare, W., Raper, S.C.B., Frieler, K., Knutti, R., Frame, D.J., Allen, M.R.: Greenhouse-gas emission targets for limiting global warming to 2°C. Nature **458**, 1158–1162 (2009). http://www.ecoequity.org/wp-content/uploads/2009/07/meinshausen_nature.pdf
25. Intergovernmental Panel on Climate Change (IPCC): IPCC Fourth Assessment Report: Climate Change 2007 (AR4). http://www.ipcc.ch/publications_and_data/publications_and_data_reports.shtml#1 (2007)
26. Inman, M.: Opening the future. Nat. Clim. Change **1** (2011). http://www.nature.com/nclimate/journal/v1/n1/pdf/nclimate1058.pdf?WT.ec_id=NCLIMATE-201104, pp. 7–9
27. Uppsala Global Energy Systems, Uppsala University, Publications. http://www.physics.uu.se/ges/en/publications (2011)
28. Rogner, H.H.: An assessment of world hydrocarbon resources. Annu. Rev. Energy Environ. **22**, 217–262 (1997)
29. Gregory, K., Rogner, H.H.: Energy resources and conversion technologies for the 21st century. Mitig. Adapt. Strat. Glob. Change **3**(2–4), 171–229 (1998)
30. Greene, D.L., Hopson, J.L., Li, J.: Have we run out of oil yet? Oil peaking analysis from an optimist's perspective. Energy Policy **34**(5), 515–531 (2006)
31. Laherrere, J.: Data shows oceanic methane hydrate resource over-estimated. Offshore **59**(9), 156–158 (1999)
32. Collett, T.S., Kuuskraa, V.A.: Hydrates contain vast store of world gas resources. Oil Gas J **96**(19), 90–95 (1998)
33. Collett, T.S.: Energy resource potential of natural gas hydrates. AAPG Bull. **86**(11), 1971–1992 (2002)

Chapter 18

Why Military and Intelligence Agencies Are "Peeking at Peak Oil"

In the spring of 2003 I received a telephone call that was, to me, astonishing. A lady introduced herself and told me that she worked for MUST. She and a colleague wanted to come to Uppsala to discuss Peak Oil with me. MUST is Sweden's Military Intelligence and Security Service (*M*ilitära *u*nderrättelse- och *s*äkerhets*t*jänsten). My only previous interaction with MUST had been to watch actors portray its agents in Swedish films. It felt a little strange that someone from this organization now politely but firmly said that they wanted to meet me. On the Swedish Armed Forces website one can read the following about MUST [1]:

> Opponents can be diffuse and often combine both military and civilian activities. Threats can arise from states, organisations or individuals acting alone. A changing threat environment also changes the demands placed on intelligence services. For most intelligence services the world after 11 September 2001 is completely different to what existed before. This also applies to the Swedish services. At Must's intelligence offices a multifaceted group of people is working to identify and map the threats of this new age. Security policy is now focussed on identifying, predicting and responding to so-called asymmetrical threats. Terrorism, distribution of weapons of mass destruction and cyber security are issues that have largely replaced those from the time of the Cold War.

As I read this paragraph again I wondered how it applied to Peak Oil, but the lady and her colleague did come to Uppsala and I gave them a 2-h run-through on the issue. I still remember their parting words, "We at MUST are very interested in Peak Oil and what you are working on."

When Colin Campbell came to Uppsala in September 2003 to participate in Anders Sivertsson's defence of his Master thesis I told him about the visit

K. Aleklett *Peeking at Peak Oil*, DOI 10.1007/978-1-4614-3424-5_18,

from MUST. He related a similar experience. The CIA had telephoned him and had paid him a visit. He retells the events as follows.

> About 6 months before the invasion of Iraq I received a call from the Office of US Naval Intelligence in London (which I later found was part of the CIA) inviting me to a conference in Washington. I was not able to go because I had to go to the hospital for an operation and so I politely declined.
>
> Later, about 6 weeks after the invasion of Iraq I received a second call saying that they insisted that I go to Washington because they were having a very high-level meeting on the topic, including people from the Pentagon. At that time I was not interested in attending and I told them that.
>
> But then a few weeks later I got a phone call in Ballydehob as follows. "Dick Haines, here, Office of US Naval Intelligence, we'd sure like to meet you." To which I replied: "Yes, with pleasure, but I live in a little village in the West of Ireland." He replied "Yes, we know, we're by the phone booth at the bottom of your hill (Fig. 18.1)."
>
> It happened that my son-in-law was with me at the time, and we walked down the hill to meet them. But there was no one by the phone box. I then looked across the road and saw a Hollywood figure hiding in a doorway with dark glasses, a hat pulled down over his face, and a raincoat over his arm. He was looking around. I eventually crossed and asked him if he was by

Fig. 18.1 Meeting with agents. The scene is from Ballydehob where Colin Campbell was contacted by agents from CIA

> any chance looking for me. His reply was "No, but those that are, are in the shop opposite."
>
> I accordingly went into the shop, a little electrical shop where I found a man, age about 50, and two women. He was effusive saying, "Hi, Hi. Can I buy you a drink?"
>
> So we walked up the road to Coughlan's Pub, trailed by the man with the dark glasses. Haines said, "We were just having a holiday in little old Ireland and thought we would look you up." I replied asking where he was staying, to which he replied in a slightly hesitant way, saying "We're just motoring round, you know."
>
> I then invited them home to see the data, etc. They walked back but declined to come into the house, but before they left Haines made the following comment, "You know that the admiral is very interested in what you guys are doing.".
>
> Later I received in an envelope, without any covering letter, a copy of the proceedings of the meeting I was supposed to attend. It was about what the US military, which used a lot of petroleum in its operations, would do for fuel in the future, and where it might be fighting. I imagine that Dick Haines sent me an unofficial copy off the record.
>
> It is a bizarre story, but shows that behind the scenes governments are awake to the impact of Peak Oil.

When Colin told me this story I was struck by the similarity in the parting comments made to each of us by the intelligence services. It was Colin who suggested we form ASPO, the Association for the Study of Peak Oil & Gas, and I had just become president of ASPO. It is hard for me to believe that it was a coincidence that representatives of the intelligence services visited us simultaneously.

The identities of the people from MUST who visited me must be known within that organization but when I searched through public data using the only names I had, only the contact person's name gave relevant information. Searching with the other person's name gave nothing relevant. Indeed, regarding its personnel, MUST states [1] that, "A confidential number of MUST employees have a so-called "shielded occupation." These individuals do not present themselves using their real names but instead use a number or alias."

The question still remains, "Why are intelligence services peeking at Peak Oil?"

The Cold War, CIA, and Peak Oil

On April 15, 1977, 3 months after his inauguration as president of the United States, Jimmy Carter held a press conference where he said that the CIA had informed him that the global energy situation was significantly more serious

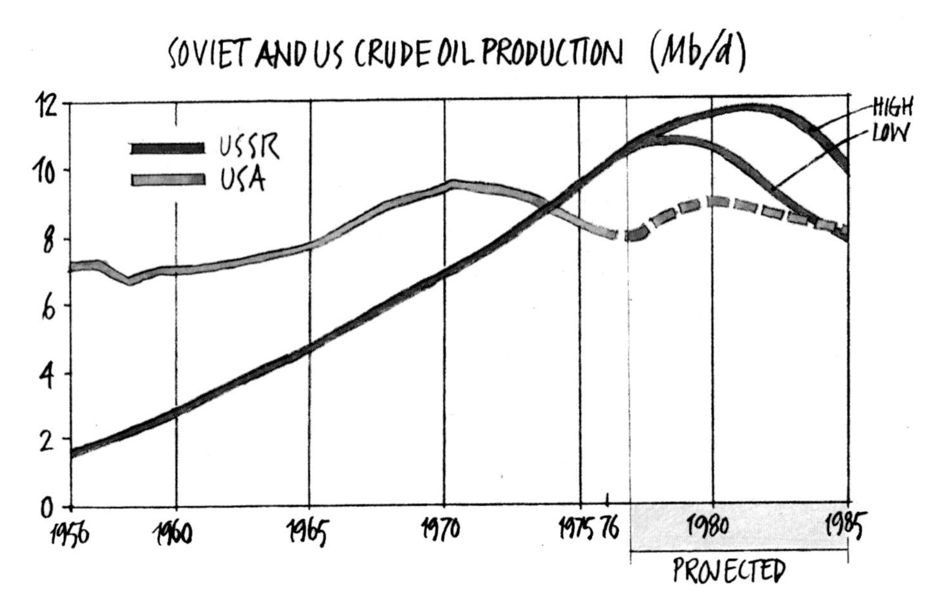

Fig. 18.2 This figure is based on one in the 1977 CIA report on Soviet oil and shows that the USSR surpassed the United States in oil production in 1974. Evidently the CIA understood as early as the 1970s that the USSR would reach Peak Oil but the point in time when it would occur was somewhat uncertain. In reality, Soviet oil production in 1985 was 11.9 Mb/d so the CIA's prediction was incorrect

than was generally accepted. He also intended to frame his energy policy in the light of this. That the CIA was engaged in energy analysis was regarded as important news and calls came immediately for classified information on energy to be made available to the public. Just 3 days later a report was released, *The International Energy Situation: Outlook to 1985* [2] and an additional report was released 1 week after that, *Prospects for Soviet Oil Production* [3]. The following day the head of the CIA was summoned to testify before a committee where he stated that the people who had performed the analyses had access to secret information that was not available to other analysts. This caused a great deal of speculation.

It was primarily the CIA's report on future Soviet oil production and its assertion that the USSR and other nations of the eastern bloc would need to import between 3.5 and 4.5 Mb/d in 1985 that caused the most disquiet. The report was filed in 1977 as *Intelligence Memorandum: The Impending Soviet Oil Crisis,* and is now available as a document in the "CIA Historical Review Program" [4]. The central figure in the report is reproduced above as Fig. 18.2. Note particularly the event in 1974 when oil production in the Soviet Union surpassed that of the United States. The report begins by analyzing problems affecting Soviet oil production and summarizes them in three points:

(a) An emphasis on development drilling over exploration, with the result that new discoveries are failing to keep pace with output growth
(b) Overproduction of existing wells and fields through rapid water injection and other methods, with the result that less of the oil in place is ultimately recovered
(c) New capacity requirements that soon will run far beyond the Soviet oil industry's capability

The issues that the CIA reported regarding Soviet oil production can be described by one term, "Peak Oil." The question is whether this was the first time that the CIA discussed a "peak" in oil production and, likewise, it would be interesting to read their current documents on Peak Oil. Based on my and Colin Campbell's experiences, we believe that such documents exist.

The conclusions of the report on Soviet oil production were regarded by many as controversial and of such importance that the Senate's Select Committee on Intelligence Interest decided to investigate how comprehensive the analytical work had been and to evaluate the quality of the CIA's energy studies. Answers were sought to the following questions.

1. Did the analytical or estimative process respond to the Administration's preferred outcome?
2. Was the manner and style of the release of the CIA information appropriate?
3. How was the study on future Soviet oil production received by the public and by other petroleum analysts?
4. What is the track record of the CIA on the subject of Soviet oil?
5. On what sources of information did the CIA base its estimate?

The report on the committee's investigations was finalized in May 1978 and printed for use by the committee. The 15-page report is now available to the public and makes very interesting reading [5].

Jimmy Carter began his presidential term in the wake of the first oil crisis. But we should not forget that his presidency ended with the second oil crisis and the invasion of Afghanistan by Soviet troops. In his third State of the Union Address on 23 January 1980 he raised the issue of the importance of oil [6].

> The region which is now threatened by Soviet troops in Afghanistan is of great strategic importance: It contains more than two-thirds of the world's exportable oil. The Soviet effort to dominate Afghanistan has brought Soviet military forces to within 300 miles of the Indian Ocean and close to the Straits of Hormuz, a waterway through which most of the world's oil must flow. The Soviet Union is now attempting to consolidate a strategic position, therefore, that poses a grave threat to the free movement of Middle East oil.
>
> This situation demands careful thought, steady nerves, and resolute action, not only for this year but for many years to come. It demands collective efforts to meet this new threat to security in the Persian Gulf and in Southwest Asia. It demands the participation of all those who rely on oil from the Middle East and who are concerned with global peace and stability. And it demands con-

> sultation and close cooperation with countries in the area which might be threatened.
>
> Meeting this challenge will take national will, diplomatic and political wisdom, economic sacrifice, and, of course, military capability. We must call on the best that is in us to preserve the security of this crucial region.
>
> Let our position be absolutely clear: An attempt by any outside force to gain control of the Persian Gulf region will be regarded as an assault on the vital interests of the United States of America, and such an assault will be repelled by any means necessary, including military force.

Today, as one reads President Carter's speech, one can wonder why he coupled together the Soviet Union's entry into Afghanistan with an attempt to take control over the Middle East. There were no indications that the Soviet Union was preparing to March into Iran. However, when one reads the CIA's memorandum, *The Impending Soviet Oil Crisis,* from 1977 together with the report from the Select Committee on Intelligence, *The Soviet Oil Situation: An Evaluation of CIA Analyses of Soviet Oil Production,* from 1978, the pieces of the puzzle seem to fall into place. What follows are some of those pieces.

In Fig. 18.2 we can see that a "low case" and a "high case" are presented for forecast Soviet Oil production. The organization responsible for this analysis was the CIA's Office of Economic Research (OER) that had studied the Soviet Union's oil production in detail for many years. During a period of 10 years from the 1960s to early 1970s the OER believed that Soviet oil production would grow slowly until 1990. However, at the beginning of the 1970s some OER analysts began to find information from various sources implying that their projection of Soviet production was too optimistic. They began to discuss a completely different future with dramatically reduced production. It was only in 1975 that this idea became more accepted within the CIA when they understood that the rapid growth of Soviet oil production was dependent on the Soviets having received permission to purchase a large number of American pumps capable of pumping both oil and water. The Soviets could then resume pumping oil from fields that had previously been pressurized with water before being abandoned. They had also received permission to buy other advanced technology for drilling wells. The low and high cases for projected Soviet oil production were simply reflections of the two opinions prevailing within the CIA at that time. The Senate's Select Committee on Intelligence Interest came to the conclusion that there was no particular reason to favor one or the other opinion. There was speculation at the time that President Carter had manipulated the analysis for political gain but that was not the case.

On the issue of analyzing the quality of the intelligence on Soviet oil production in the CIA's report, the Select Committee decided that it was more difficult to make a determination. The problem was that the report indicated the Soviet Union would become a net importer of oil—"We estimate that the Soviet Union and Eastern Europe will require a minimum of 3.5 million barrels per day of imported oil by 1985"—and the question was how the USSR would then act. The US media began to discuss this issue and one suggestion was that the oil industry had created "the low case" in order to maintain high oil prices. At the same time there were those who noted that the Soviet Union did not have the means to pay for imports of 3.5 Mb/d. In their opinion the discussion of imports was irrelevant. Other experts concluded that the Soviet Union would cope by focusing production on its own requirements. An important part of the report was the analysis made by the CIA's vice-director Walter McDonald which can be summarized by the expression, "The Soviets will do virtually anything to prevent them from becoming an oil importer of that magnitude."

Among the recommendations made by the Select Committee I find the following especially intriguing, "Finished intelligence products which are written by highly trained specialists should be reviewed by generalists with a more multidisciplinary view. This will enhance the likelihood of avoiding projections of technical outcomes which may be politically unlikely."

On December 27, 1979, 4 weeks before Jimmy Carter's final presidential address to the United States, Soviet troops occupied the central buildings in Kabul, Afghanistan. We do not know what information and analysis Jimmy Carter received from the CIA regarding this. However, the fact that he subsequently declared so clearly what would be the consequences of any Soviet entry into a nation of the Persian Gulf (such as Iran) indicates that he believed the Soviet action was a possible response to falling internal oil production. It would be interesting to know with certainty to what extent Peak Oil in the Soviet Union lay behind the words, "such an assault will be repelled by any means necessary, including military force." We could discuss this in more detail but the most important observation we can draw from the events above is that a clear connection exists between the CIA and Peak Oil from 1977 onwards.

If we consider it probable that concerns over Peak Oil in the Soviet Union lay behind President Carter's threat to declare war then it can be interesting to examine the quality of the CIA's analysis. In this regard, another report by the CIA that is now publicly available is, *The North Caspian Basin: Salvation for Soviet Oil Production?* [7]. The report was "created" on April 1, 1989 and is based on known data up until February 15, 1989. Table 1 of that report states that Soviet oil production in 1985 was

11.9 Mb/d and that this increased to 12.48 Mb/d in 1987 and was 12.45 Mb/d in 1988. No peak in Soviet oil production was yet evident in contrast to the CIA's predictions in 1977's *Prospects for Soviet Oil Production* report. The fact that the oil price at that time was high and that they were able to sell large quantities of oil to the West meant, in effect, that the Western nations were financing the Communist Soviet Union. The 1989 report analyzed how much oil exists in the area north of the Caspian Sea. The CIA assessed this as at least 30 billion barrels (Gb) but possibly over 80 Gb. The Soviet Union's invasion of Afghanistan could now be seen in a different light. They did not have a suitable export route for the Caspian oil but pipelines through Afghanistan would allow them to pump oil to the Indian Ocean. If the price of oil remained high this would give the Soviet Union the hard currency it needed for international trade.

The CIA's *North Caspian Basin* report states February 15, 1989 as the date on which the Soviet Union was forced to leave Afghanistan. A contributing factor may have been that the oil price fell dramatically in 1985 and drastically reduced the Soviet Union's income from exports. In January 1992 the Soviet Union collapsed like a house of cards and there are many theories about what contributed to this. The prevailing low oil price and the fact that they had not managed to establish a pipeline to export oil from the Caspian region may have been contributing factors.

The IEA's Attempt to Suppress Peak Oil Research

In 2004, when examining Chap. 3 of the IEA's *World Energy Outlook 2004* report (*WEO 2004*) [8], I could see that the IEA had begun to discuss crude oil resources in the same manner as Colin Campbell and I had done in our 2003 article, "The Peak and Decline of World Oil and Gas Production" [9]. What was missing was a cipher or code that could be used to translate the IEA's text into plain language. Therefore, I set aside a few weeks to write, "The Uppsala Code" and published it on ASPO's website [10].

In February 2005 I received a telephone call from Paris. It was one of the directors at the IEA. He told me that the IEA planned to make a special analysis of the energy situation in Sweden. He wondered whether I was interested in discussing the analysis with his personnel. Naturally, I was interested but then came the catch: "but they think that you should remove your analysis from the Web." I was stunned and I asked, "Who are 'they'?" There was a short silence and then he answered, "Personally, I do not have any problems with the analysis but they think you should remove it." I explained that I was a professor at Uppsala University, I would not allow myself to be pressured, and the analysis would remain on the Web.

Unfortunately, I did not make a recording of the conversation but it is one I will never forget.

By relating this story in this particular chapter I do not need to explain whom I believe "they" are. Then again, I could be wrong.

Peak Oil Symposium for the CIA and Sister Organizations Within the OECD

In March 2009 I received information that the Global Futures Forum (GFF) wanted to contact me to discuss the arrangement of a workshop on Peak Oil. I was also informed that the GFF had connections to the CIA. On the CIA's website I found the contact information that I needed to take the next step. I also found the following description of the GFF [11].

> In late 2005, the GFP [The Global Futures Partnership] launched the *Global Futures Forum (GFF)* a multinational, multidisciplinary intelligence community embracing intelligence, national security, and nongovernment experts who engage in strategic level, unclassified dialogue and research to better understand and anticipate transnational threats. GFF members from more than 35 countries have begun to work together in a number of topic-based communities of interest.

So, apparently, the GFF was an initiative of The Global Futures Partnership that, itself, is

> A strategic 'think and do tank' that undertakes unclassified global outreach for CIA and other Intelligence Community elements on the most important issues facing the intelligence community today and in coming years. It conceptualizes and implements interdisciplinary and multi-organizational projects on key intelligence issues with leading thinkers from academia, business, strategy, and intelligence consultants [11].

In light of what had happened earlier I was not surprised that the members of this forum wanted to learn more about Peak Oil. Rather, I saw it as confirmation of the importance of our research. I sent an email to the person who was named as the contact and waited with interest for the reply.

Three days later I received a reply in which the Global Futures Forum was described in more detail. In addition to the information above I was told that around 40 experts had participated in the Forum's meetings/workshops. I was also told that Sweden was a member of the GFF and had previously organized one meeting for the Forum. The GFF had begun a new energy/environmental project and had already had a first meeting in Washington. Now they wanted me to organize a workshop for them in

Sweden. I decided to take a positive attitude to the enterprise after reading the following text in their email: ". . . because GFF is an intelligence rather than policy focused initiative, we want to put the emphasis on helping policymakers anticipate what may take place rather than to come up with a precise set of policy recommendations."

I saw this as an opportunity to get the message about Peak Oil through to decision makers in a different manner than I had previously thought possible.

I personally did not have the authority to organize a workshop for the GFF. Therefore, I passed the GFF's request to the rector (vice chancellor) of Uppsala University. To my great surprise I was told that the answer was "No." Uppsala University would not have anything to do with the CIA or other intelligence services. Therefore, I got in touch with my then contact-person at MUST and asked if they were interested in organizing a workshop on Peak Oil. I also told them that Sweden was a member of the GFF. My contact-person did not know who Sweden's representative was in the GFF and so passed the issue on to his superiors. Some weeks later I received an answer that Sweden did not currently have the opportunity to organize a workshop on Peak Oil because the nation would soon take over the presidency of the European Union and that would keep the personnel of MUST fully occupied.

Peak Oil and Military Activities

The Second World War has been analyzed from all possible angles. There is no doubt that access to oil was a decisive factor both in strategic decisions that were taken during the war and in determining the war's outcome. There is also no doubt that the war's brutality was increased by access to oil. The initial aggressor nations, Germany and Japan, had no domestic oil production. In Germany the Fischer–Tropsch process was developed that could produce paraffin from coal but the Germans also attacked Stalingrad as a first step to reaching the oil wells of Baku. Japan tried to secure its oil requirements by attacking what is now Indonesia. The air warfare of WWII—including the fire-bombing of cities—would not have been possible without oil. Attacking civilians became a new warfare strategy. Factories producing weapons were bombed but cities were also bombed to break down civilian morale and reduce support for the conflict.

In the spring of 2005 Kelly Way from Hollywood began to record material for a documentary that had the working title, *Asleep in America*. The idea was that my group's research would be a central theme running through the film.

The ambition was there but, unfortunately, the budget was too small so the film was never finished. All that is left is a 7-min trailer on YouTube in which Kelly chose to focus on the military's need for oil [12].

"Petroleum has a compact relationship with a nation's political, economical and military strength." These words can be seen at the entrance to the Oil Museum in Daqing, China. The Daqing oilfield is one of the world's ten largest with initial reserves of more than 20 billion barrels of oil. When it was discovered in 1959 China's energy supply system was catastrophically bad. They even lacked fuel for the buses in Beijing. The discovery of Daqing meant that China became self-sufficient in oil and its industrial and political power grew. Without Daqing, the world today would be a different place. Modern military power requires access to oil.

The United States has the world's largest military but official figures on how much oil that military consumes are hard to find. In the United States the Department of Energy (DoE) gathers statistical information on domestic energy use by the military. These data can be recalculated and expressed in barrels of oil equivalent (boe) per day. Domestic military oil consumption has decreased in the United States since the 1980s but the use of oil by the US military outside the United States has increased. In recent years total domestic and foreign oil consumption by the US military is estimated to be between 300 and 400 thousand boe per day. This is a greater rate of oil consumption than that of many nations [13].

We may ask whether Peak Oil is influencing future planning in various military organizations. The answer is "Yes" and below are two examples that have received a great deal of attention.

The first is a report from the Engineering Research and Development Center (ERDC) in the US Army Corps of Engineers. By September 2005 the ERDC had assembled a report with the title, *Energy Trends and Implications for U.S. Army Installations*. The report was not available to the public. However, one of the readers of the Peak Oil website, the *Energy Bulletin* (*EB*) [14], had access to an electronic version of the report's summary and they sent it to the *EB*'s editorial staff. On March 11, 2006 the report's summary was then forwarded to a number of people. As the president of ASPO, I was one of those who received a copy. Because there was no indication on the document I received that it was classified (secret), I made it available on ASPO's website on March 12 [15]. A summary of the document was written and posted online by EB the same day. I know that, among others, US Congressman Roscoe Bartlett (who has been a great advocate for Peak Oil) then requested a complete copy of the report. On March 16 the complete report was made available on the Internet [16]. If one examines the section, "Conclusions about Petroleum," then one can see it contains a version of the oil production prediction (updated with 2004 data) that Colin Campbell and

I published in our 2003 paper, "The Peak and Decline of World Oil and Gas Production" in the journal *Minerals and Energy* (see Fig. 11.4 in this book). Commenting on our analysis, the report states, ". . . these are considered pessimistic projections. Others predict far higher production for the future, but discoveries to date have not born [sic] out the predictions of the optimists."

This can only be interpreted as meaning that the report's authors have greater confidence in our analysis than that of the "optimists." The next figure in the ERDC's report is a 2004 version of Fig. 3.1 combined with Fig. 11.4. It came as quite a surprise that the US military was making judgments on future oil supplies using research from Uppsala University!

In the ERDC report one can also read a discussion regarding alternative forms of energy before arriving at the section titled, "General Conclusions and Implications" [16]. This states,

> Demand now exceeds production and we are seeing that effect on prices. After the peak is reached, geopolitics and market economics will result in significant price increases above what we have seen to date. Security risks will also rise. To guess where this is all going to take us would be too speculative. Oil wars are certainly not out of the question.

The second example of military awareness of Peak Oil comes from the German armed forces. On September 1, 2010 the German newspaper *Der Spiegel* published news that spread like wildfire around the world. The headline was, "'Peak Oil' and the German Government; Military Study Warns of a Potentially Drastic Oil Crisis." A summary of the article by Mikael Höök was published by ASPO on its website on September 2 [17]:

> A study by a German military think tank has analyzed how "peak oil" might change the global economy. The internal draft document—leaked on the Internet—shows for the first time how carefully the German government has considered a potential energy crisis.
>
> The Peak Oil issue is so politically explosive that it's remarkable when an institution like the Bundeswehr, the German military, uses the term "peak oil" at all. But a military study currently circulating on the German blogosphere goes further.
>
> The study is a product of the Future Analysis department of the Bundeswehr Transformation Center, a think tank tasked with fixing a direction for the German military. The team of authors, led by Lieutenant Colonel Thomas Will, uses sometimes-dramatic language to depict the consequences of an irreversible depletion of raw materials. It warns of shifts in the global balance of power, of the formation of new relationships based on interdependency, of a decline in importance of the western industrial nations, of the "total collapse of the markets" and of serious political and economic crises.
>
> According to the German report, there was "some probability that peak oil will occur around the year 2010 and that the impact on security is expected to

be felt 15–30 years later." The Bundeswehr prediction is consistent with those of well-known scientists who assume global oil production has either already passed its peak or will do so this year.

Reflections

In this chapter I have described some events showing that military and intelligence organizations have taken an interest in Peak Oil. There are additional relevant meetings in which I participated but the people involved have expressly requested that the meetings remain confidential. When told, "I suggest that you do not write about this meeting in your blog," the message is clear and I choose to follow that advice. For me personally, every such meeting is a confirmation that a number of intelligence services believe that Peak Oil is important for our future, a future that may be just around the corner.

The information that President Carter received from the CIA in 1977 was not complete and the analysis they produced was in error. The consequence of that inaccurate analysis was a presidential threat of war. The US House of Representatives' Select Committee on Intelligence Interest recommended that the reports the CIA had submitted should be reviewed by experts inasmuch as "this will enhance the likelihood of avoiding projections of technical outcomes which may be politically unlikely." In the ERDC report *Energy Trends and Implications for U.S. Army Installations* we read that, "Oil wars are certainly not out of the question." In the German military's report on Peak Oil the long-term consequences of declining oil production are described as involving drastic economic and geopolitical changes.

For me, working to spread awareness of the inevitability and consequences of Peak Oil is a peace project. If we can write scientific articles that clearly define the future in ways that leave little room for doubt then it is my hope that this knowledge will prevent conflict. We have shown that oil supplies are very important to military forces and an important question is whether Peak Oil will mean reduced resources available to military forces or whether those forces will continue their high rates of consumption to the detriment of civilian activities. The ideal outcome would be a world without military forces engaged in conflict but in the real world in which we live it is a fact that military power plays an important role in defending human rights. For example, the United Nations has the right to use military force to defend human rights when a nation's leader violates them. The UN-sanctioned actions to defend civilians in Libya are one such example but we should not be blind to the fact that oil is also an important aspect of that conflict.

In 2005 when the ERDC's report was written there was ongoing conflict in Iraq. Without discussing the events that led to that war we can, nevertheless, assert that the future of world oil production was a very important contributing factor. I must point out that the invasion of Iraq has not led to an increase in Iraqi oil production. The infrastructure for oil production is very expensive to build. It is also easy to destroy if there is no army to protect it.

An important question for our future is who has the right to extract oil. Fifty years ago the international oil companies had this right whereas today the national oil companies are dominant. Another important question is who has the right to buy the oil that is produced. After Peak Oil we may be faced with a market where an oil-producing nation or company chooses to sell its oil only to a few select customers. We cannot be certain that the future world market for oil will be free, fair, or open.

References

1. Militära underrättelse- och säkerhetstjänsten (Must): Försvarsmakten. http://www.forsvarsmakten.se/hkv/Must/ (2011)
2. Central Intelligence Agency: The International Energy Situation: Outlook to 1985. http://catalog.hathitrust.org/Record/007418225 (1977)
3. Central Intelligence Agency: Prospects for Soviet oil production: A Supplemental Analysis. ER 77–10425. http://catalog.hathitrust.org/Record/000090124 (1977)
4. Central Intelligence Agency (CIA): Intelligence Memorandum: The Impending Soviet Oil Crisis. Secret ER 77–10147, March 1977. http://www.peakoilpolicy.com/documents/CIA%20THE%20IMPENDING%20SOVIET%20OIL%20CRISIS.pdf (1977)
5. Staff report of the Senate Select Committee on Intelligence: The Soviet Oil Situation: An Evaluation of CIA Analyses Of Soviet Oil Production, United States Senate, May 1978. http://intelligence.senate.gov/pdfs/95soviet_oil.pdf (1978)
6. Carter, J.: State of the Union Address 1980, 23 Jan 1980. http://www.jimmycarterlibrary.gov/documents/speeches/su80jec.phtml (1980)
7. The North Caspian Basin: Salvation for Soviet Oil Production? CIA/SOV/89-10028x. http://catalog.hathitrust.org/Record/000090124 (1989)
8. WEO: World Energy Outlook 2004. International Energy Agency. http://www.iea.org/textbase/nppdf/free/2004/weo2004.pdf (2004)
9. Aleklett, K., Campbell, C.: The peak and decline of world oil and gas production. Miner. Energy Raw Mater. Rep. **18**, 5–20 (2003). http://www.ingentaconnect.com/content/routledg/smin/2003/00000018/00000001/art00004, http://www.peakoil.net/files/OilpeakMineralsEnergy.pdf

10. Aleklett, K.: International Energy Agency accepts Peak Oil—An analysis of Chapter 3 of the World Energy Outlook 2004. Association for the Study of Peak Oil and Gas. http://www.peakoil.net/uhdsg/weo2004/TheUppsalaCode.html (2004)
11. The Global Futures Partnership. https://www.cia.gov/offices-of-cia/intelligence-analysis/organization-1/gfp.html (2010)
12. Way, K.: Asleep in America Promo, YouTube (7 min). http://www.youtube.com/watch?v=jeHs-RDK1b4 (2006)
13. Karbuz, S.: US military energy consumption—facts and figures. Energy Bull. http://www.energybulletin.net/node/29925 (2007)
14. Energy Bulletin. www.energybulletin.net
15. Fournier, D.F., Westervelt, E.T.: Energy trends and their implications for U.S. army installations. U.S. Army Corps of Engineers, Washington DC, September 2005. http://www.peakoil.net/Articles2005/Westervelt_EnergyTrends__TN.pdf (2005)
16. Fournier, D.F., Westervelt, E.: Energy trends and their implications for U.S. army installations. U.S. Army Corps of Engineers, Washington, DC, September 2005 (complete report). http://stinet.dtic.mil/cgi-bin/GetTRDoc?AD=A440265&Location=U2&doc=GetTRDoc.pdf (2005)
17. Höök, M.: Peak Oil—An Analysis by German Military, Höök, M. ASPO, 2010–0902. http://www.peakoil.net/peak-oil-an-analysis-by-german-military (2010)

Chapter 19

How Can We Live with Peak Oil?

Ernst Solvay was a famous Belgian chemist, industrialist, and philanthropist. Over 100 years ago in 1911, he invited a number of the leading physicists of the age to Brussels to discuss the "new physics," the physics that would ultimately revolutionize our world and our understanding of the universe. Einstein had already presented his theory of relativity but the atomic structure of nature was still a secret. We did not know the source of the sun's energy, but the new physics could explain how its energy was transported to our planet. Nobody had heard of the "Big Bang." In 1911, *Conseil Solvay* became the world's first international conference on physics.

The *Solvay Conferences* on physics are now an important part of the history of science. The most famous is the fifth conference in 1927 where the topics for discussion included Heisenberg's uncertainty principle. That idea—that one cannot determine simultaneously the exact position and speed of an object—is the foundation of quantum mechanics. This book is not the place to begin a discussion of quantum theory but we should remember Einstein's comment on that theory that "God does not play dice" to which Nils Bohr is said to have replied, "Einstein, stop telling God what to do." When one studies the photograph of the delegates to the *Fifth Solvay Conference* taken outside the Bibliothèque Solvay in Leopold Park (Leopoldspark) in Brussels in 1927, one sees that 17 of the 29 delegates eventually received the Nobel Prize [1]. As a young physicist I used to daydream about participating in those historic conferences.

Friends of Europe is a think tank with its headquarters in the Bibliothèque Solvay in Brussels. It is primarily concerned with issues affecting Europe. On April 1, 2011 I received an invitation to participate in a round table conference on "The Future of Oil—Moving to a Post-Oil Society: How Realistic?" When I received the invitation I did not immediately think it

K. Aleklett *Peeking at Peak Oil*, DOI 10.1007/978-1-4614-3424-5_19,

remarkable that the meeting was at the Bibliothèque Solvay but once there it suddenly struck me that I was on hallowed ground for physicists. Presumably there was nobody else in the room who had the same reaction. However, anyone who has worked with research in nuclear physics for 30 years should know that it was here that some of the world's greatest scientists met to discuss the physics that has changed our world. This invitation to participate in a modern *Solvay Conference* on oil's future suddenly took on a new significance for me.

There were around 50 people invited to this discussion on the future of oil including representatives of the EU Commission and the EU parliament, the International Energy Agency (IEA), oil companies, the petroleum, plastic, and chemical industries, environmental organizations, vehicle manufacturers, the nuclear power industry, the renewable energy sector, university researchers, and other EU observers from near and far. There were also about 100 registered onlookers of the debate [2].

The morning session of the round table conference examined "How much oil is there and how long will it last?" There were five of us invited to open the discussion. We each had 5 min to present our views and the time was measured using an hourglass. The first to speak was Mechtild Woersdörfer who has responsibility for preparation of the EU's Energy Road Map 2050 policy document. Then it was my turn and I concentrated on our article "Peak of the Oil Age" [3] (see Chap. 11). I explained that the IEA and the Uppsala Global Energy Systems group (UGES) agree on many points but that our research does not support the future scenario given by the IEA. John Corbin from the IEA then countered my arguments with their well-known assertion that an additional 2,000–2,500 billion barrels (Gb) of crude oil exist to be consumed and that unconventional oil of a similar volume also exists. In the next breath he said that we consumed 31 Gb of oil in 2010 but that the oil industry only discovered 14 Gb. Phillippe Lamberts represented the political view and advanced the idea that because the world is finite with finite fossil fuel resources then Peak Oil must occur. Whether the moment of Peak Oil production is now, or in 10 or 20 years' time is of lesser importance. The last of the five introductory speakers was Isabella Muller who represented the petroleum industry. She asserted that the industry has faith in the IEA and plans its future activities according to the IEA's prognoses. She concluded by stating that it is much too early to be talking about divorcing society from its use of oil.

After the introductory speeches it was time for the others around the table to pose questions and make additional contributions. We five who had introduced the session were then given another 5 min each to respond to the questions we thought most relevant. The response I wished to make required longer than the sand falling through the hourglass allowed. Partly,

Fig. 19.1 Blood streaming through arteries and veins sustains the life of a human body but for our modern human society a steady flow of oil is required. Oil is a finite resource. Although we have more than half of this resource left to consume, the rate of oil flow will decrease. The speed of this decrease will be decisive for our future

I wanted to explain how our lives will be affected by Peak Oil. Like the sand falling through the hourglass at the conference, it is the flow of oil that will be decisive for our society's future (see Fig. 19.1).

Of the 27 nations of the EU, only four produce oil (UK: 1.339 million barrels per day, Mb/d, Denmark: 0.249 Mb/d, Italy: 0.106 Mb/d, and Romania: 0.089 Mb/d). The European Union's total production of oil is 1.8 Mb/d but this is very little compared with the European Union's oil imports of 11.6 Mb/d (see Fig. 19.4). For nations that must import oil, the moment when world oil production peaks is less important than when the amount of oil available on the world's export market begins to decline. One of the questions asked during the morning session of the *Solvay Conference on The Future of Oil* was, "How will world oil exports be affected by Peak Oil?" We must examine this before looking at how Peak Oil will affect the wellbeing of humanity as a whole.

Export and Import of Oil

In 2006 I published an article in the *Oil & Gas Journal* in which I discussed the fact that the world is divided into nations that are net oil exporters and nations that are net oil importers [4]. For the calculations in that article I used data for 2005 as presented in the 2006 edition of the *BP Statistical Review of World Energy*. BP has now published data for 2010 in the 2011 edition of their *Review* and we have had the opportunity to examine how the import and export of oil by the world's nations has changed over the last 5 years [5]. The consumption numbers in the *Review* now include ethanol but BP's oil production statistics exclude it. In our analysis below we have subtracted ethanol from the oil consumption statistics of Brazil, the European Union, and the United States [6]. For other nations ethanol makes very little difference to their oil consumption volumes.

When analyzing oil consumption statistics BP's *Review* includes "Processing Gains," substances that are added to the final products from a refinery (see the section "From Crude Oil to Transport Fuel", Chap. 16). In 2010 Processing Gains made up 2.7% of total world consumption. In the analysis of oil exports and imports for 2010 the Processing Gains fraction is removed by reducing consumption by 2.7%. The difference between production and consumption is then calculated for each nation and a positive number means that a nation is net exporter of oil. BP has not reported oil consumption statistics for some nations. For those nations we use numbers from the *CIA World Fact Book* [7]. In 2010 there were 35 nations that were net exporters with Saudi Arabia topping the list at 7.31 Mb/d and Egypt at the bottom with 0.01 Mb/d. The 20 largest exporting nations are shown in Fig. 19.2. Oil exports from Russia were just 100,000 barrels less than from Saudi Arabia, and in terms of total oil production Russia has now passed Saudi Arabia as the largest producer in the world.

According to BP, world oil production during the 6 years from early 2005 until the end of 2010 was 81.5, 81.7, 81.5, 82.0, 80.3, and 82.1 Mb/d, giving an average of 81.5 Mb/d (see the section "What Is Reported as Oil?", Chap. 2, regarding production of oil). During the 6 years prior to 2005, oil production had increased by 9.1 Mb/d. When production began to plateau in 2005, the global volume of exports was 47.9 Mb/d and 5 years later in 2010 it had decreased to 43.8 Mb/d. This means that consumption of the oil produced in the oil-producing nations had increased by 4.7 Mb/d, from 33.6 Mb/d (81.5 – 47.9 Mb/d) to 38.3 Mb/d (82.1 – 43.8 Mb/d), which is an increase of 2.5% per year. From 2000 to 2005 the increase was 1.3% per year.

Since 2005, oil consumption has been increasing in the oil-producing nations and total world oil production has been flat. Therefore, the nations

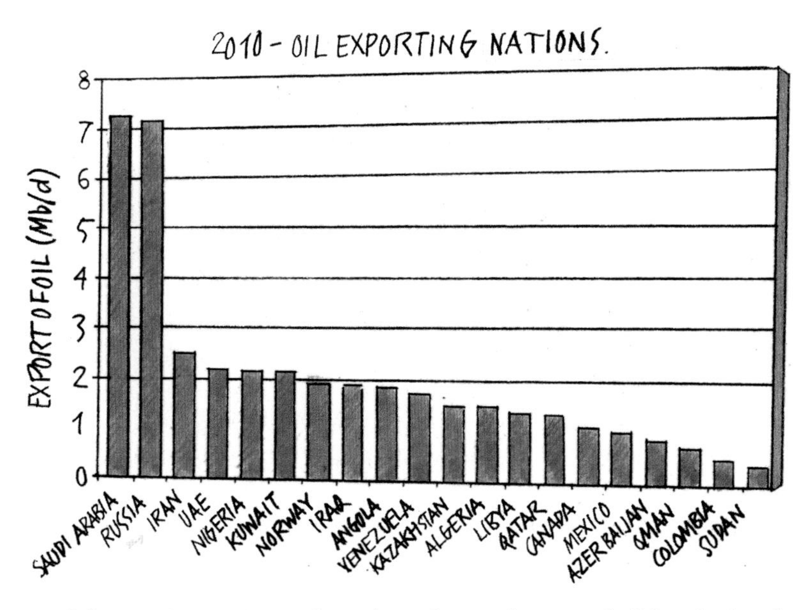

Fig. 19.2 Oil exporting nations where the volume of exported oil is calculated as the difference between total production and total consumption of oil according to data from the BP Statistical Review of World Energy 2011 [5]

that are net importers of oil have been forced to reduce their consumption by 4.1 Mb/d. The most notable exception to this is China, which during the same period increased its oil imports by 1.7 Mb/d. India, Singapore, and the rest of the non-OECD nations in SouthEast Asia (SEA, excluding South Korea and Taiwan) [8] have also increased their consumption. Together with China this amounts to an increase of 2.7 Mb/d which is an increase of 6.5% per year. The same group of nations increased their oil imports by 9.2% per year between 2000 and 2005. Therefore, the other oil importing nations reduced their consumption by 6.8 Mb/d (4.1 Mb/d + 2.7 Mb/d) between 2005 and 2010. The greatest reductions in consumption occurred in the United States (−2.2 Mb/d), the European Union (−1.2 Mb/d), and Japan (−0.9 Mb/d).

Over the next 10 years we can expect the oil exporting nations to continue to increase their oil consumption. This means that these nations will have less oil left over to export and so importing nations (as a whole) will have less oil available to import. Opinions differ on how the remaining importable oil will be distributed between the China, India, SEA group, the OECD, and the rest of the world. An economist might say that the OECD nations imported less oil in recent years because of the recession and very slow subsequent economic growth. Thus, non-OECD nations have had more oil available at prevailing prices. Against this argument is the fact that China, India, and SEA increased their share of total oil imports during the period of economic growth from 2000 to 2008 and we assume that this will continue.

The increase in consumption by the oil exporters was 1.3% per year between 2000 and 2005 and 2.5% per year between 2005 and 2010. A business-as-usual, extrapolation predicts that the oil exporters would continue to increase their consumption by 2.0% per year over the next 10 years. The increase in consumption by the China, India, and SEA group was 9.2% between 2000 and 2005 and 6.5% between 2005 and 2010. Over the next 10 years we think their rate of increase in oil consumption will slow to 5% per year. In Chap. 11 we argued that world oil production will be lower in 2020 than in 2010 but for simplicity in this discussion we assume that it stays constant at 81.5 Mb/d. In Fig. 19.3 we show the anticipated change in the flow of oil during the next 10 years (until 2020) under this assumption. We

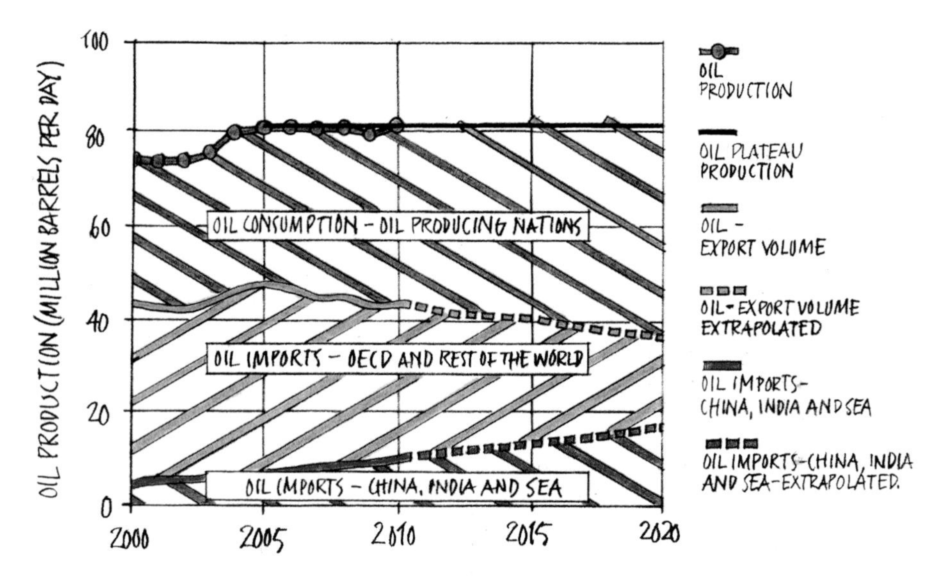

Fig. 19.3 The consumption of global oil production can be separated into three groups: the oil-producing nations, importing Asian nations that will increase consumption in the next 10 years (China, India, and other nations in SouthEast Asia [SEA] with the exception of South Korea and Taiwan), and importing nations that will, overall, decrease consumption (OECD-nations, South Korea, Taiwan, and the rest of the world). The volume of oil exported was greatest in 2005 and by 2010 it had decreased by 4.1 Mb/d. Although total oil exports during this period decreased, the volume imported by China, India, and the SEA nations increased by 2.7 Mb/d. This means that, overall, the other oil importing nations and, in particular, the OECD nations, experienced a decrease in oil imports of 5.8 Mb/d. In a business-as-usual linear extrapolation of past trends, this volume would decrease by a further 15 Mb/d by 2020. (Calculations are made using data from the BP Statistical Review of World Energy [5].) With the advent of Peak Oil in the period to 2020, the decreased availability of oil on world export markets will be much more severe

see that the volume of exported oil peaked in 2005 and is now declining. When we include the fact that China, India, and the SEA group are increasing their consumption we find that, under current trends, the importing OECD nations and the rest of the world will receive a smaller fraction of the total volume of exported oil. For this group, in 2005, total oil imports amounted to 40.5 Mb/d and in 2010 they were 33.8 Mb/d. If we assume that world oil production remains constant at around 81.5 Mb/d and we make a linear extrapolation of the declining oil importation by the importing OECD nations and the rest of the world, then by 2020 these nations will only be importing 19.5 Mb/d, a decrease of 42% from 2010 levels. If world oil production declines before 2020 then these nations will be restricted to even lower levels of oil imports.

To cope with their decreased future ability to import oil the OECD nations must introduce dramatic changes, foremost in the transport sector. The energy program that President Obama announced in the spring of 2011 is essential for the United States' future [9]. The judgment of UGES is that the United States will have difficulty competing with China over access to oil. In Fig. 15.1 we show the destinations of the supertankers that currently deliver oil from the Middle East to the rest of the world. In the figure, six supertankers currently depart for Asia for every three supertankers that depart for Europe and the United States. In 5 years one of those three ships currently departing for Europe and the United States will, instead, be heading for Asia as well.

In earlier chapters we have seen that larger oilfields usually have a longer production plateau before the inevitable decline begins. The world as a whole has now experienced a production plateau of 6 years in spite of the fact that, in some individual oil-producing nations such as Mexico, Norway, and the United Kingdom, we have seen dramatic declines in production. To maintain the world's production plateau new infrastructure must be constructed in other nations. The greatest opportunities for increased oil production exist in Iraq and expansion of infrastructure there was also a precondition for realization of the "Best Case" future oil production scenario from our research (see Fig. 9.7). The required investments in Iraq have begun but the oil production there has not yet reattained the volumes it had around the year 2000. We also see that Russia is attempting to raise its oil production back to the levels of the 1980s and there is also an increase in deep water production that is maintaining the current plateau of world oil production.

In Figs. 19.2 and 19.4 we show the world's 20 largest oil exporting nations and its 20 largest oil importing nations, respectively. In Fig. 19.4 we also show the total volume of oil imported by the 27 nations of the European Union (EU-27). Compared with 2005, exports from Saudi Arabia have decreased by nearly 2 Mb/d and Russia has increased its exports somewhat.

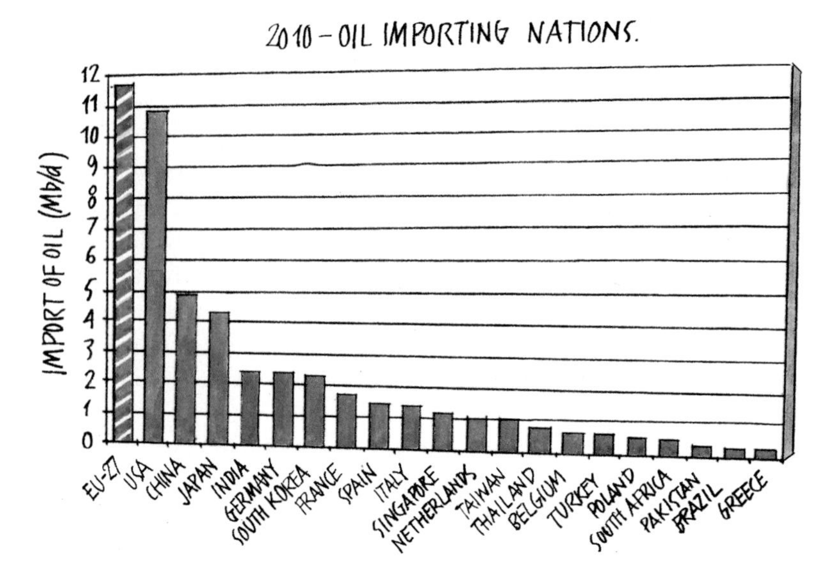

Fig. 19.4 Oil importing nations where the volume of imported oil is calculated as the difference between total consumption and total production of oil according to data from the BP Statistical Review of World Energy 2011 [5]

In terms of total production of oil, Russia now exceeds Saudi Arabia's production. Other changes since 2005 are that Norway has fallen from being the third largest oil exporter to the seventh, a decline that will continue in the future. Another nation losing its capacity to export oil is Mexico. Iran has taken over as the third largest exporter of oil and the United Arab Emirates (UAE) holds fourth position, closely followed by Nigeria and Kuwait. These six main oil exporting nations control over 50% of all the oil that is exported. We can assume that Iraq's exports will start to grow to maybe double their current level within the next 10 years as long as that nation remains relatively politically stable.

Among the oil importing nations we see imports into the United States and Japan declining whereas China is now the world's second largest importer of oil and India has taken fourth place. The question now is when India will surpass Japan. It is becoming increasingly common to see the European Union regarded as a single region when compared with the United States so it is interesting to see that both regions have oil import volumes of similar size with the European Union's oil imports being slightly larger.

Except for China and Thailand, the oil importing nations are stable democracies and over half of the oil exporting nations do not have democratic rule. Some of the remaining oil exporting nations have "democratic" political systems in name only but there are also nations with long-standing democratic systems such as Norway and Canada. The early days of the recent uprising in Libya showed that other nations with interests in the oil production of Libya found it difficult to support its democracy movement. When Saudi Arabia sent its troops over the bridge to suppress the democracy movement in Bahrain, the world's self-proclaimed leading advocate of democracy, the United States, did not protest. The United States has a naval base in Bahrain and is dependent on oil from Saudi Arabia. The constraint that an importing nation's need for oil places on its democratic behavior and ideals is what I have described as "democracy's black straightjacket."

Both BP and the IEA now include ethanol consumption as part of total oil consumption. The importance of ethanol for our economy is discussed in the section "The Economy and Peak Oil" below.

We can now return to the *Solvay Conference on The Future of Oil* where the afternoon session was given the theme, "How do we manage a smooth transition to a low-carbon economy?" The questions asked during that round table discussion together with all the previous questions on how much oil exists and how long it will last can be gathered into one issue, "How can we live with Peak Oil?"

The Human Wellbeing Equation

When I discuss the future with my students at Uppsala University we begin by discussing what will be important for them personally. At the top of the priority list comes food, shelter, and a job that will allow them to pay for those necessities. Somewhat further down the list is our climate, democratic rights, world peace, biological diversity, and so on. Usually, discussions of the future focus on economic issues although in recent years the issue of climate has become important. However, it seems the climate issue (or at least society's engagement with it) may have reached its "peak" in Copenhagen with COP15 (*UN Climate Change Conference Number 15*). The unstructured discussion among the world's leaders during the conference's final evening showed that any international agreement on measures to reduce future climate change must address the basic issues that my students placed at the top of their list of priorities.

In 2011, the tumultuous events around the democratization movements in North Africa and the Middle East were discussed by my students. They supported the demonstrators calling for the peaceful withdrawal of their dictatorial leaders to make way for democratic rule. More than half of the crude oil remaining to extract exists in that region of the world. The behavior of western nations towards these democracy movements can be understood in the light of the idea that oil is "democracy's black straightjacket" (see the section "Export and Import of Oil").

As a physicist, one tries to describe physical reality in terms of equations. In the discussion of the future in which we now engage we must hold humanity and its wellbeing at the center. I like to summarize this discussion with what I call the human wellbeing equation, HWB (see below). In principle, this equation must encompass those things that my students regard as important for their future. Energy (E) impinges on every aspect of our future so we must discuss how Peak Oil, Peak Gas, and Peak Coal will affect human wellbeing.

The individual components that affect the wellbeing of humanity can be described as Food(E), Shelter(E), Economy(E), Environment(E), and Peace(E) where each of the components has some kind of coefficient. Our wellbeing can be described as the product of these components and the different coefficients are collected into one constant.

Human Wellbeing, HWB(E) = constant • Food(E) • Shelter(E) • Economy(E) • Environment(E) • Peace(E)

To "solve" the human wellbeing equation none of these individual components can be ignored. (Factors not obviously included in the equation above can be considered to influence the coefficients for each component. For example, biological diversity could be included in the HWB in the coefficient affecting Environment.) Solutions may exist for each individual component above but only when we consider all the components simultaneously in a global system will we find a solution to human wellbeing in the face of declining fossil fuels. (Now you can understand why I call my research group at Uppsala University, "Uppsala Global Energy Systems.")

Food and Peak Oil

In 1950 the world's human population was 2.5 billion in number. Now, in 2011, it has passed 7 billion. In every year until 2000 the number of children being born increased. However, we have now reached "Peak Child" because the rate of new children being born has leveled off and reached a maximum. If we look at people in the age group of 60–65 years we have not yet reached "Peak Late Middle Age." Instead, the numbers in this age group will continue to grow. If we assume that the rate of childbirth will continue to be constant and that the other age groups will reach their maxima before 2050 then we cannot prevent the world's population from exceeding nine billion (see the presentation by Hans Rosling [36]). This means that between now and 2050 the world's population will increase by an amount equal to the world's total population in 1950. These additional billions of people will require food, the issue that was at the top of my students' list of life's priorities. The question of how population growth and development will affect the future was one of those posed at the Bibliothèque Solvay.

Without solar energy food crops cannot grow but to put food on the tables of today's seven billion people we also require energy from fossil fuels. The energy content of food is commonly measured in kilocalories. In our current world some people eat too many kilocalories and become obese whereas others cannot find enough to eat and go hungry. Our body requires a certain minimum amount of energy in order to function normally. The threshold below which one begins to starve is 1,800 kcal per day. If a person is to do anything other than simply subsist then their daily energy intake should be 2,500 kcal. That means that the world's seven billion inhabitants require 1.75×10^{13} kilocalories every day which is 20 terawatthours (TWh) and is equivalent to the energy content of around 12 million barrels of oil per day.

UGES has studied the world's food production in detail by examining the energy content of 129 different crops produced in 2006. We have also calculated the energy content of the residues generated during production of these crops. Annual global primary crop production contains 19,900 TWh of energy of which 17,560 TWh is judged to be edible. This edible production amounts to 48 TWh per day and is more than double what is needed to feed the world's current population. However, losses occur so that only a portion of this edible primary production can actually be used and find its way to our dinner tables.

Different crops suffer different forms of loss after harvest. All of these losses have been entered into a database to allow us to calculate the net energy that is ultimately available from these foods. The details of these losses have been published in a Master thesis by Kersti Johansson and Karin Liljeqvist as well as in our article, "Agriculture as Provider of Both Food and Fuel" [10]. However, the losses can be broadly divided into five classes: seed set aside for next year's crop, losses occurring when cultivating the crop, storage losses, produce removed from the supply chain due to poor quality, and crops used to feed animals. Of course, in addition to the energy content of crops one must also assess the energy in meat and fish and other products obtained from the oceans.

In our analysis we present two calculations of the energy available from agriculture, a "high case" and a "low case." In the high case we assume that all residues remaining after the initial upgrading (oil crop meals, husks, bran from cereals, and sugar crop fibers after sugar extraction) are available as food, giving a net energy of 9,265 TWh globally. However, the residues

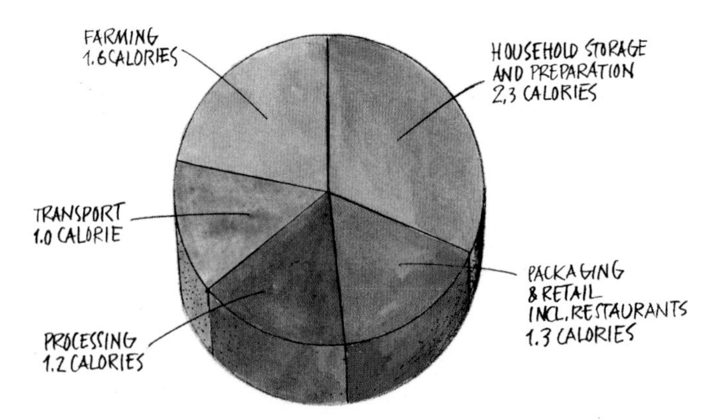

Fig. 19.5 Energy expended in producing and delivering 1 cal of food. Approximately 7.3 cal are used by the US food system to deliver each calorie of food energy. Farming accounts for less than 20% of this expenditure but still consumes more energy than the farm delivers

are preferentially not eaten, and may even be inedible or, at least, not digestible by humans. In the low case, it is assumed that all the residues in question are inedible. The low case gives a net energy of 7,225 TWh which is 20 TWh per day and equals exactly the requirements of seven billion people (see above). In our calculations we have not considered any losses in food storage, preparation, or use by households. If these are as great as 30% then the world does not produce sufficient food. However, we have also failed to consider any food production by households themselves. This may compensate for household losses so that the world's current food production is, in fact, sufficient.

As part of our research we have also calculated the energy content of the residues from agriculture. These amount to 18,200 TWh, nearly as much as the total edible primary production [10]. We discuss this later.

To discuss how Peak Oil will affect food production we need to examine this energy system from the seed grain to the final baked loaf of bread. Figure 19.5 shows how energy is used by the American food system. For every 1 cal delivered to the dinner table, many additional calories have been consumed in food production and supply: 1.6 cal is used on the farm and 1.0 cal in transport from the farm. Food processing, packaging, and retail consume an additional 2.5 cal and households themselves use 2.3 cal in food purchase, storage, and preparation. If we assume that 1 cal used on the farm is oil and the rest is fertilizer-based on natural gas, that oil powers the transport of food and that one of the calories used by households in purchase, storage, and preparation comes from oil then we can see that providing 1 cal of food requires 3 cal of oil. The remaining energy use (when fertilizer is removed)—3.7 cal—is mainly electricity. In the United States, approximately 70% of electricity is generated using fossil fuels and this means that an additional 2.6 cal of fossil energy are included in the total of 7.3 cal required to provide every 1 cal of food. Thus, 6.2 cal of fossil energy (0.6 + 3.0 + 2.6) are used to provide every 1 cal of food.

Food that is transported by air involves more oil use than average whereas food grown for local consumption requires less. If we assume that the global average for energy use in food provision is around 5 cal per calorie of food delivered to the dinner table then 30% of all fossil energy use in the world is involved in the provision of food.

Could we feed 9.5 billion people on this Earth? Our calculations show that we can produce sufficient edible food for seven billion but this requires 30% of fossil energy use. In 2050 when the world has 9.5 billion human inhabitants we will still be using fossil fuels but the supplies of these will be less than today and the proportion of fossil fuel use comprised by oil will also be less. Therefore, the methods of, and infrastructure for, food production and provision must be adapted to reduced fossil fuel availability. This change in energy use must begin with the source of the food, agriculture itself. Can we feed 9.5 billion people on this Earth? I am optimistic and say definitively YES.

In recent years we have seen strong growth in the production of liquid biofuels. In the *BP Statistical Review of World Energy 2011* liquid biofuel production for 2010 is reported as equaling 59,261 t of oil equivalents which corresponds to 0.43 Gb of oil. This is a 200% increase over the amount produced in 2005. In the first years after 2005 the annual rate of increase was around 30% but by 2010 the rate had dropped to 14%. This behavior of a rapid early percentage increase when volumes are small followed by lesser percentage increases as volumes increase is a classic pattern seen in the expansion of new technologies and the development of nations. When oil use was introduced at the beginning of the twentieth century the annual growth in its use was 30% but during the massive increase in oil use of the 1960s the annual percentage increase had dropped to 7% (which represents a doubling of use over a 10-year period).

"Can agriculture provide both food and fuel?" was the question we posed when we began our research into global agriculture [10]. The answer we arrived at was that agriculture could not. To obtain the necessary data for the various elements of our analysis we had to use statistics from 2006. In that year the global transport sector consumed 25,000 TWh of energy. The contribution of liquid biofuel to this consumption was only 276 TWh or about 1%. By 2010 biofuel production had grown to 3% of transport energy consumption.

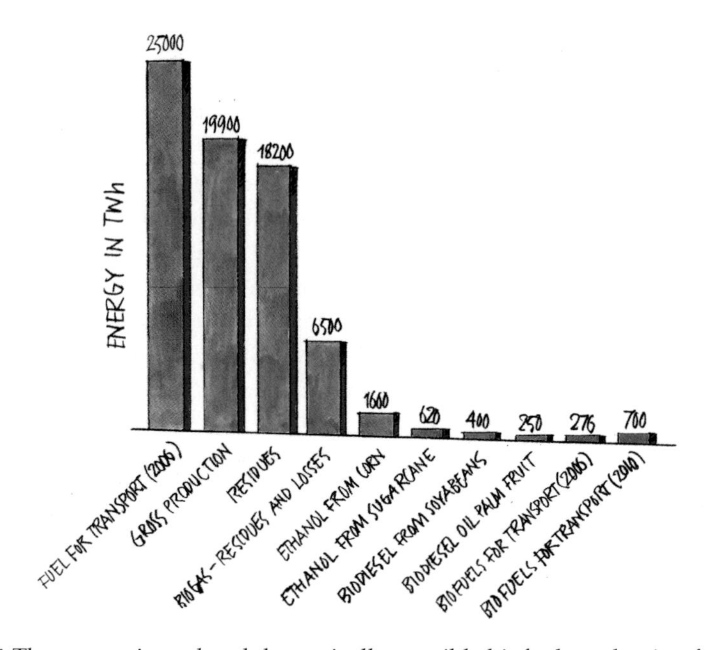

Fig. 19.6 The energy in real and theoretically possible biofuel production from various sources compared to actual fuel use by the transport sector. The data are for 2006 and are taken from the paper by Johansson et al. [10]. In that year the transport sector's energy consumption was 25,000 TWh which is equivalent to approximately 14.1 billion barrels of oil equivalents (Gboe). Energy in primary production from the world's agriculture is 19,900 TWh with residues estimated to represent an additional 18,200 TWh. If residues and losses from primary production were used for biogas production this would equate to approximately 6,500 TWh. In the unrealistic scenario that all maize, sugarcane, soya beans, and oil from oil palm could be transformed into suitable biofuel for transport this would equate to 2,870 TWh, which is far from sufficient to replace crude oil use. In 2006 production of liquid biofuel was 276 TWh and by 2010 this had increased by a factor of 2.5

If we toy with the idea of not eating any maize or sugar but of using it instead to produce ethanol, and if all soya beans and palm oil were to be transformed into biodiesel, then these products of agriculture would contribute 2,870 TWh of liquid fuel each year which is equivalent to a little more than 10% of transport's needs. Of course, this idea is unrealistic but it is interesting to note that even this unrealistic amount of biofuel production could only replace a small part of our current oil use in transport. What is far more interesting (and somewhat more feasible) is the idea that the cumulative by-products of agriculture could be transformed into 6,500 TWh of biogas.

The most important factor for our future survival is food. Peak Oil will force change in our systems of food production. The fact that global agriculture currently generates residues that (at least in theory) could substitute for oil use in agriculture means that such substitution should be a priority issue. Those engaged in agriculture should be encouraged to produce biogas and to use it as a fuel. They should be supported to do this in ways that ensure that use of biogas is not more costly than use of diesel.

The world's primary agricultural production is currently sufficient to provide food for more than nine billion people. The main problem is one of losses from primary production. Therefore, every nation should institute programs to reduce losses from the crops that they grow. As world oil production declines, locally produced food will become increasingly important. As can be seen from Fig. 19.6 the potential exists for large savings in energy use by reducing losses from primary production. If the price of food more closely reflected both the energy used to produce it and the food's energy content then this might encourage production of food in the most energy-efficient manner.

Shelter and Our Cities After Peak Oil

The large oil discoveries of the 1960s and the increasing car ownership that this supported changed the way our communities and homes were organized. In nations where homes must be heated in winter, furnaces and heaters burning coal or wood were replaced by ones burning oil or using electricity. City centers with dense residential areas where proximity to work and services had previously been prioritized were abandoned as dormitory suburbs spread outwards and family homes expanded in size. The positioning of roadways and provision of parking spaces for cars became important aspects of urban planning as most people began to use cars to travel the longer distances between home, shops, and their place of employment. Local shops disappeared as large shopping centers grew up around cities. In many cases the old city centers degenerated into slum areas. Cheap oil influenced almost every aspect of our new living arrangements.

After the oil crises of the 1970s many nations began to reconsider the use of oil in domestic heating so that heating based on electricity and natural gas became increasingly common. Insulation of houses also improved but journeys by car to work and for shopping continued to increase. Urbanization of the world's population has increased steadily and today the majority of people live in cities. Unfortunately, urban planning has not yet been influenced by Peak Oil considerations even though we can now be certain that anything constructed today will still be standing after the peak of oil production has passed. Peak Oil will change the way we live and travel.

The debate over Peak Oil in Sweden inspired the Swedish Royal Institute of Art (Kungliga Konsthögskolan, KKH) to organize a Master degree program in which, among other things, students have made a special study of the effects of Peak Oil on Shanghai, *Beyond Oil: Shanghai* [11]. The students came up with many suggestions for solutions to the problem of how an integrated large city could function using less oil. Therefore, the idea of a large city functioning on far less oil appears to be possible, at least in theory. Growing of food close to home is one important change that must occur in our city environment.

As a follow-up to the work on the project, *Beyond Oil: Shanghai,* Anton Redfors at the UGES research group has studied the possibilities for introducing a Personal Rapid Transit (PRT) system as an extension of the high-speed rail network from Shanghai's main airport to the center of the city [12]. PRT systems may be an alternative to cars as a form of transport in cities in the future.

Peak Oil is commonly misunderstood to mean the end of oil. In reality, Peak Oil means that oil production and availability will only decrease, not cease. The greatest issue then is how to prioritize use of oil in the future. In some cities, notably Bristol in the United Kingdom and Brisbane in Australia, analyses have already been made of how Peak Oil will affect their future ("Building a positive future for Bristol after Peak Oil" [13] and "Oil Vulnerability in the Australian City" [14]). Today, in many cities in industrial nations, city centers that became neglected during the 1980s and 1990s are being renovated and often only wealthier people can afford to live there. Meanwhile, the households with the lowest incomes often live farthest out on the city fringes and must commute long distances to work. In the United States before the financial crisis in 2008 it was noted that it was these poorer, fringe-dwelling households that were the first to be affected by high oil prices. The more than doubling of the oil price from 2005 to 2008 took a huge toll on the budgets of these households. One way for them to cope was to abandon their mortgage payments and give their house keys back to the banks. Thus, Peak Oil and the financial crisis were intimately linked.

Another way for people and communities to plan for, and respond to, the challenges of Peak Oil is seen in "the Transition Town movement" founded by Rob Hopkins. A Transition Initiative (which could be a town, village, university, or island, etc.) is a community-led response to the pressures of climate change, fossil fuel depletion, and, increasingly, economic contraction. Despite gradually increasing awareness in some quarters of how high oil prices and Peak Oil might affect cities in the future, one cannot yet discern any global trends in urban planning [15]. However, the challenges of providing both physical and information communication between people's residences and their workplaces should provide inspiration and motivation for many of today's architects.

The Economy and Peak Oil

The third component of the human wellbeing equation is the world economy. A crucial factor for the economy of most individuals is that someone in their family, one or several people, is employed and that person's work provides an income. In very many nations it is not possible for the poorest people to live on the incomes they earn so the state contributes to their survival. The most important issue is that a family's income must be able to cover the first two factors in the HWB: food and shelter.

The fact that the world's population is increasing every year means that, for humanity's wellbeing to increase, not only must the entire economy expand commensurately but the proportion of the economic pie that the world's poorest people receive must also increase in order to lift them out of poverty.

The US Energy Information Agency, EIA, has noted that one cannot have global economic growth without increasing world energy consumption (e.g., see part of Fig. 19.13). In 1945 at the end of the Second World War Sweden was quite a poor nation. However, in subsequent decades its gross domestic product (GDP) increased dramatically so that by 1970 it was the world's third wealthiest nation on a per capita basis. In the period between 1945 and 1970 Sweden's use of energy increased fivefold which corresponds to a 7% increase per year for 25 years. This energy came primarily from cheap oil imported from the Middle East. Thus, by 1970 Sweden's economy had become the world's most oil-dependent. It was even more oil-thirsty than the United States!

The world's poorest nations are found south of the Sahara Desert (Sub-Saharan Africa, SSA). We have studied how their poverty can be related to the use of oil. In 1980 SSA's GDP (related to purchasing power parity, PPP) per capita was the same as for India and China. In our study we followed the economic development of SSA, China, and India up until 2005 and compared it with changes in the per capita oil consumption of these nations [16]. Since 1980, China and India have seen economic progress whereas SSA has, as a whole, seen no economic development at all. There are exceptions, but the picture for the SSA region as a whole is shown in Fig. 19.7.

China's economic advancement has been very strong and we see a strong correlation between China's increased per capita GDP and its per capita oil consumption. China is now trying to tread the same developmental path that Sweden took between 1945 until 1970 but there is a large difference: Sweden's population was then 6 million whereas China's population is currently 1,331 million. The increase in oil consumption that Sweden needed for economic growth did not significantly affect global demand,

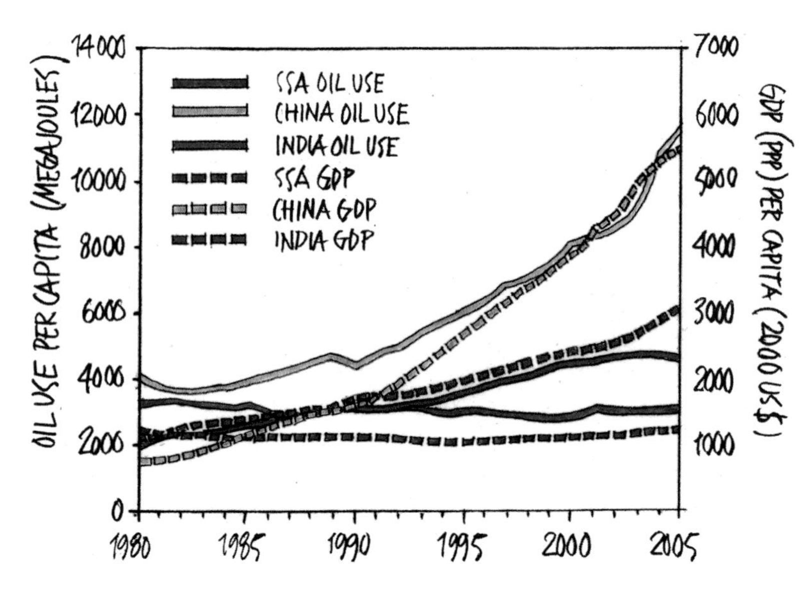

Fig. 19.7 Development of GDP (PPP) and oil use per capita in Sub-Saharan Africa (SSA), China, and India from 1980 to 2005. Prices are normalized to the value of the US dollar in 2000 [16]

whereas China's and India's needs will have an impact on us all (see the section "Export and Import of Oil" in this Chapter).

From 2005 to 2010 world oil consumption was fairly constant. If we examine global primary energy consumption (oil, coal, natural gas, biofuels, nuclear energy, hydroelectric power, wind power, and solar energy) during that period then we see an 11% increase. That increase was mainly due to the growth of primary energy consumption in the developing nations China and India. In the autumn of 2008 the global economy crashed and 2009 was a year of recession in large parts of the world. Globally, energy consumption declined by 1.5% but not in China and India where total energy consumption grew. During 2010 there was, once again, global economic growth and the world reached the highest level of energy consumption ever seen.

For many years the IEA has related increased demand for oil to global economic growth. In its *World Energy Outlook 2008* report the IEA showed the correlation between oil demand and economic growth up to 2007 [17] (in a similar manner to what is shown in Fig. 19.8) and asserted that global economic growth requires increased oil production. From 1990 up to and including 2010 oil production increased at an annual rate of 1.2% and average yearly economic growth was 3.3%. The first conclusion we can draw from this is that the one-to-one correlation between increased oil use and economic growth seen for China and India does not apply to the world as a whole. If we were to assume that, during the coming 25 years, we will experience global economic growth of 3% and the same corresponding rate

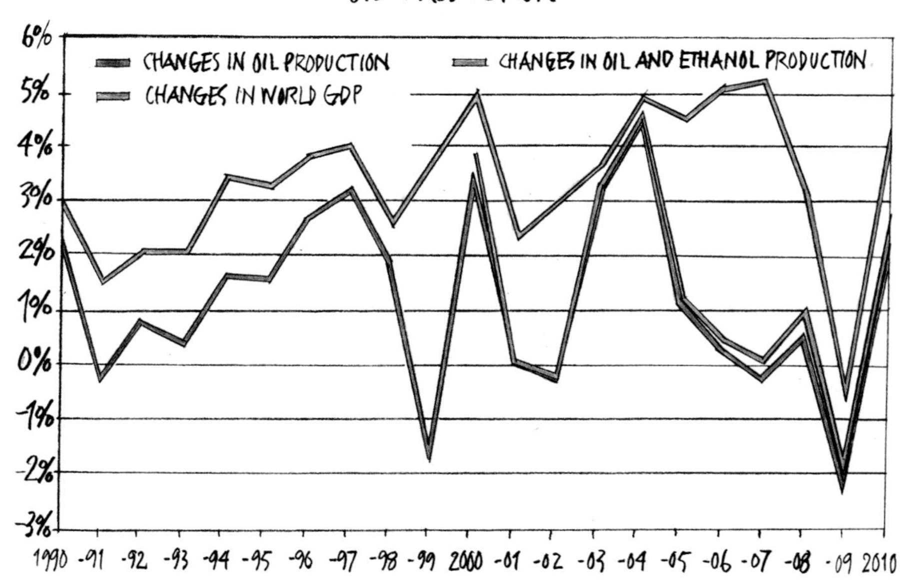

Fig. 19.8 The yearly percentage change in global economic growth and the simultaneous change in oil production. During the period from 1990 until 2004 access to oil was relatively easy (and so relatively inexpensive) and we saw the classic correlation between changes in GDP and changes in oil production. In 2005 the pattern changed as we reached the Peak Oil plateau of 81.5 Mb/d. Increased production of ethanol helped to maintain economic growth until 2008, but increasing property prices around the world also contributed to GDP during this time. When the global economy was in recession in 2009 oil production also declined. The coupling between growth and oil consumption was re-established once again in 2010

of increased oil use that we saw from 1990 to 2010 (1.2%) then this would require an increase in oil production from 82.1 Mb/d (see Fig. 9.7) to 111 Mb/d. A strong association between global economic growth and increased oil production has been central to the future scenarios presented by the IEA in the past decade.

In its *World Energy Outlook 2004* report [18] the IEA predicted an increase in oil demand to 121 Mb/d in 2030 based on the requirements of increased economic activity (see the section "Peak Oil and Energy Demand", Chap. 1, for a discussion of differences in oil production and oil demand). In subsequent reports, this demand then decreased every other year to 116 Mb/d in *WEO 2006* [19] and then to 106 Mb/d in *WEO 2008* [17].

While these reports were being published, UGES has continued to assert that these IEA prognoses are exaggerated. Our analysis of the IEA's prognosis in *WEO 2008* was published by *Energy Policy* in our paper, "The Peak of the Oil Age" [3]. In *WEO 2010* [20] the IEA continued to assert (as its "Current Policies Scenario") that the world will require 107 Mb/d of oil in 2035 which is an increase of 1% annually over current levels. However, the IEA also presented a "New Policies Scenario" that requires only 99 Mb/d and, if the world is to hold carbon dioxide (CO_2) levels in the atmosphere below 450 ppm, then the IEA sees a maximal rate of oil use at 87 Mb/d in around 2020 declining to 81 Mb/d in 2035 (i.e., Peak Oil). The IEA does not discuss how it thinks the world economy will develop under these circumstances.

During the period 1990 to the middle of 2004 the oil price was low and there was good access to oil. For this period we saw a very strong correlation between economic growth and increased oil consumption. At the end of 2004 the price passed $50 per barrel. In 2005, when we reached the Peak Oil plateau production level of around 81.50 Mb/d, the price began to rise strongly. Economic growth continued at a high level until 2007 and a significant increase in ethanol production assisted in maintaining that growth. The unrealistic growth in property values that occurred during this period can also have contributed to maintaining growth. However, in 2008 the price of oil spiked up to reach US$147 per barrel and the global economy crashed. The economic growth and oil production statistics of 2009 and 2010 once again showed the strong connection between these numbers (see Fig. 19.8).

When comparing the price of crude oil at different times one usually determines an index price and then adjusts this for inflation. We have chosen an index price that we calculate as the average of the Dated Brent, West Texas Intermediate and Dubai Fateh benchmark prices. An oil price that most people remember is the record high price of US$147 per barrel that was achieved on the New York Mercantile Exchange (NYMEX) on July 11, 2008. We have chosen this price as the reference against which to calculate inflation-adjusted prices for other dates in Fig. 19.9. The oil prices discussed below have been adjusted for inflation relative to this.

When Colin Campbell wrote his first ASPO newsletter in January 2001 the price of crude oil was $28.50 per barrel. By December of that year it had fallen to $20.10 per barrel. In May 2002 when ASPO organized the world's first Peak Oil conference in Uppsala the price was once again up at the same level as January 2001. At that time this was a price that OPEC regarded as reasonable. At the ASPO conference in Berlin in 2004 the main topic of conversation was whether the barrel price of oil would exceed $50 which it subsequently did in September that year. During the meetings ASPO 2005 in Lisbon, ASPO 2006 in Pizza and ASPO 2007 in Cork the price varied between $60 and $80 per barrel but by the end of 2007 it had passed $100. During the spring of 2008 there was fevered speculation that the price

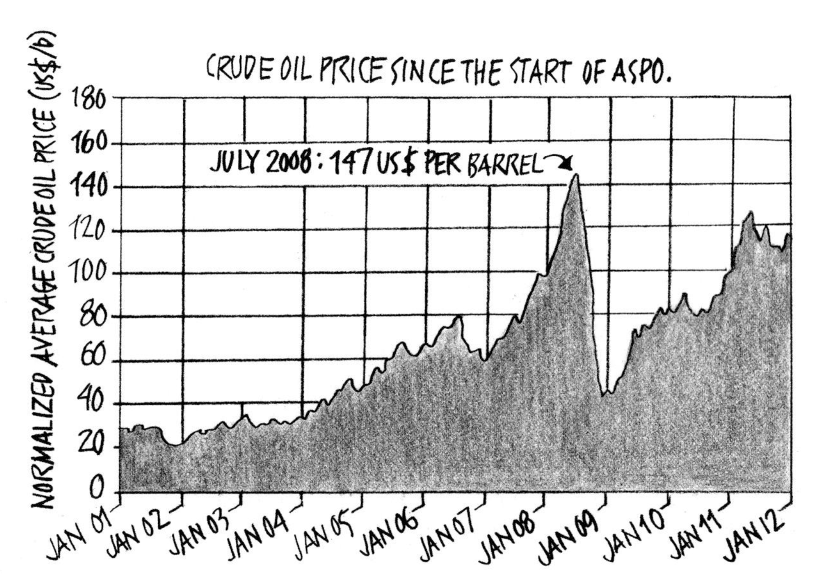

Fig. 19.9 The price of crude oil since ASPO was established (given as an average of the Dated Brent, West Texas Intermediate and Dubai Fateh benchmark oil prices). The prices shown are inflation-adjusted to the value of the US$ in July 2008 when oil reached its highest-ever price of $147. With this adjustment, the price of crude oil in 2001 (when Colin Campbell wrote his first newsletter for ASPO) was $28.50 per barrel

would pass $300 per barrel and not a newscast went by without the price of oil being discussed. When questioned about the future price of oil my standard answer then was that the price would be what people are willing to pay. In May 2008 at the International Transport Forum in Leipzig the CEO of Airbus made a comment about future oil prices that "an oil price of $200 per barrel would close down the aviation industry." This comment has influenced me greatly so that my standard answer to the question of the future price of oil is now that it cannot exceed $200. This is because an end to the commercial aviation industry would have severe impacts on globalization and, consequently, the world economy.

The fact that the price of oil increased from $100 to $147 per barrel in the first half of 2008 and oil production remained constant showed that we had reached maximal oil production. The notable exception to this observation was in July 2008 when oil production (sales of oil and oil products) increased by 1 Mb/d before it fell by 2 Mb/d in the following month. During August 2008 the Olympic Games were held in Beijing and a detailed analysis of oil sales shows that, in preparation for the Olympics, China had decided to fill its oil storages. It was this that had caused the increase in oil sales that triggered the record high oil price of July 2008. By August, the Chinese had finalized their preparatory purchases. The people who had been buying oil then spent a great deal of time in front of their TVs instead. This led to the drop in oil demand. Oil market traders around the world believed that the July spikes in oil demand and the oil price reflected oil market fundamentals. However, the fact is that it was related to the Olympic Games in Beijing. In China my friends assert that the 2008 Olympics also will prove to be the "Peak Olympic Games."

We are currently seeing a rise in oil prices similar to that before the Olympics in Beijing and it is possible that the Games of the XXX Olympiad in London will see the next "Peak Oil Price." Maybe then we will also know whether the Beijing Olympics were the Peak Games.

A nation's economy is strongly affected by the cost of oil imports. For many years it was economically beneficial for the United States to import oil because those imports generated strong economic growth that allowed payment for the oil through exports and foreign borrowings. If the price of oil is US$100 per barrel then importation of 11 Mb/d of oil into the United States costs $1,100 million per day. Today the United States is no longer able to fund this and the other imports it needs through export earnings and instead we have an increased foreign debt. Today the United States has the world's largest foreign debt and its biggest lenders are China and Japan. During the 1990s Sweden was in a similar situation and its prime minister of that time, Göran Persson, wrote a book titled, *Those Who Are in Debt Are Not Free*. Sweden's finances were then put in order and it subsequently coped better with the 2009 economic crisis than any other European nation. Economically the United States is no longer "free." China owns a large part of the US foreign debt and it can now exert a degree of control over the United States' economic future.

Norway has long been aware that its income from oil production would one day end so it has never permitted these monies to support state finances directly. Instead the income has been invested through a sovereign wealth fund and only the earnings from the fund support the state. In contrast, the United Kingdom's economy was very weak when oil and natural gas were discovered under the North Sea. The income the state received from oil production became its economic savior during the 1980s and 1990s. Now that income is declining rapidly as North Sea oil and gas production decline. For the United Kingdom, the oil that was previously a source of export income has now become an import cost. The question now is how the United Kingdom can continue to pay for oil.

Denmark is the third largest producer of oil and natural gas from the North Sea and was still a net exporter in 2010. The income from these exports contributes directly to Denmark's budget. A detailed analysis has shown that Denmark will soon become a net importer of oil and gas [21]. Oil and gas are not as great a contributor to the Danish economy as they are for the United Kingdom but Denmark can expect some economic difficulties as it adjusts to the rapid decline in production.

Russia has strengthened its position as the world's second largest oil exporter. The fear that the CIA held in the 1970s that the then-Soviet Union would become an oil importer was completely wrong (see Chap. 18). With exports of 7 Mb/d and an oil price at around $100 per barrel, Russia is currently receiving a large income from oil. In addition, Russia's income from natural gas exports is also large. Total Russian oil reserves were reported as 77 Gb in 2010 with an annual oil production of 3.7 Gb. This means that they produced approximately 5% of their reported reserves in that year. UGES has examined various future export scenarios for Russia based on its reported reserves. One scenario is based on the assumption that Russia will ultimately discover an additional 50 Gb of reserves [22]. With its current reported reserves of 77 Gb it is doubtful that Russia could continue to be an oil exporter in 2035 but with an additional 50 Gb of reserves they could maintain 50% of their current rate of exports in that year. Russia has established a fund for investment of income from oil and as Russia's oil exports decline in the future the fund will become an important contributor to the Russian economy. There is no doubt that the large income that Russia will receive from oil and gas exports in coming years will make it a world economic power.

We have discussed some of the relationships that exist among oil, national economies, and the global economy. The fact that the world as a whole has never seen economic growth without increased use of energy—and of oil in particular—means that Peak Oil requires reorganization of the world's systems for energy supply and use now. The world should already have begun many years ago to make the enormous investments required. It may already be too late to do so, but if we do not begin immediately then we will definitely encounter severe problems due to energy decline in the near future.

The Climate and Peak Oil

At the beginning of this chapter we discussed the famous *Solvay Conference of 1911*. That conference of physicists has an interesting connection to the current debate about climate change inasmuch as the work of one of its participants explained how heat from the sun could be transferred to our planet. Max Planck attended the 1911 conference and is best known for his

Fig. 19.10 The fossil fuel fire that is heating our Earth must be quenched. The IPCC asserts that it is possible for the fire to continue to grow until 2100 and UGES has estimated that Peak Oil, Peak Gas, and Peak Coal mean the fire must die down. Nobody believes the fire will go out by 2100

discovery of the "Planck constant" that is a fundamental part of quantum theory. He found a theoretical explanation for how a warm body loses heat by radiation and it is this theory that explains how infrared radiation carries heat away from the Earth. This radiation can be absorbed by greenhouse gases and reradiated back towards the Earth (see Chap. 17).

The human wellbeing equation's fourth component is the environment. The environment and climate are affected by our burning of fossil fuels for energy. Climate change politics has the aim of putting out this fossil fuel fire that is warming our planet (see Fig. 19.10). In Chap. 17 we discussed in detail what emissions of CO_2 are possible. The coming decline of fossil fuel production that our research has foreseen means that Peak Oil, Peak Gas, and Peak Coal will reduce emissions of CO_2. This is beneficial for the climate. At the same time it is important to stress that there are other systems connected to the increasing levels of CO_2 in the atmosphere that can amplify the consequent rise in global temperatures.

Fossil fuel use began with the Industrial Revolution. Before the Industrial Revolution the atmospheric concentration of CO_2 was 280 ppm (parts per million). An increase in CO_2 concentration of 50% (140 ppm) to 420 ppm is

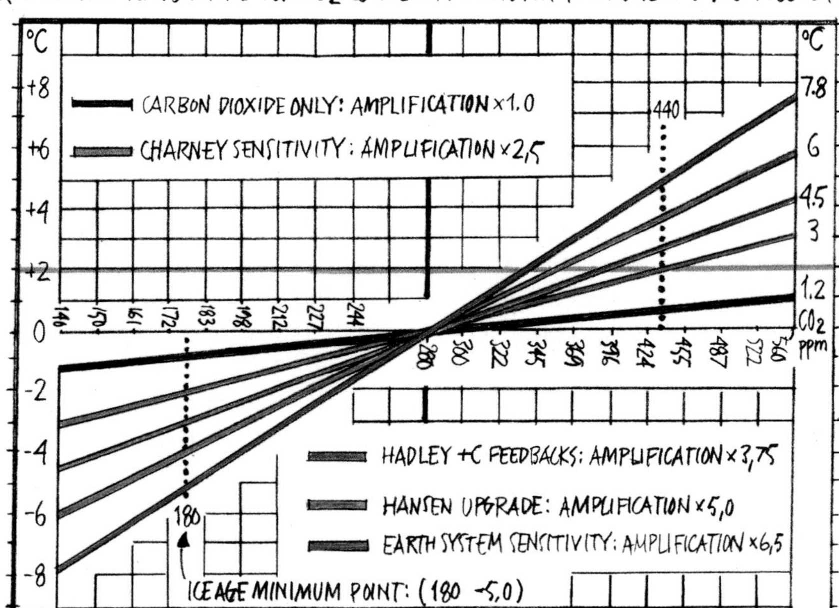

Fig. 19.11 Global average temperature changes related to different atmospheric concentrations of CO_2. The preindustrial concentration was 280 ppm. The CO_2 concentration is represented on a logarithmic scale. The amplification effect for CO_2 alone is set at 1.0. There are various systems that are affected by altered CO_2 concentrations and different assumptions about these systems lead to different amplifications of the effect that CO_2 has on temperature. Two atmospheric CO_2 concentration thresholds suggested to cause a temperature increase of 2°C are 350 and 450 ppm [23]

calculated to cause an increase in global average temperature of 0.7 degrees Celsius (°C) if one considers the effect of CO_2 alone. If we increase the CO_2 concentration by an additional 140 ppm to 560 ppm (i.e., double the preindustrial concentration of 280 ppm) then the corresponding temperature increase is not calculated to be an additional 0.7°C. Instead, it only increases by 0.5°C to reach 1.2°C above the preindustrial temperature regime. This is because, as CO_2 concentrations increase, their effect on temperature decreases logarithmically. If we plot the effect of increased CO_2 concentrations on temperature on a logarithmic scale then this relationship is seen as a straight line. This is illustrated in Fig. 19.11 where we show a diagram from an article by David Wasdell [23] that describes this relationship in detail.

If global temperature was affected only by the atmospheric concentration of CO_2 then a doubling of the preindustrial CO_2 concentration of 280–560 ppm would not lead to a 6°C increase in average temperature. It would not even lead to an increase of 2°C. However, there are other systems that are coupled to increased CO_2 concentration that amplify the temperature increase. They increase the "sensitivity" of the Earth's temperature to changes in CO_2 concentration. An example of this is found in the *IPCC Fourth Assessment Report: Climate Change 2007* where the "Charny sensitivity amplification factor" contributes to achieving an increase in global average temperature of 2°C for an atmospheric CO_2 concentration of 450 ppm. There is also a revision of the magnitude of this effect known as "Hardy plus Charney." However, the most discussed sensitivity calculation is that published by James Hansen and co workers in 2008 [24] where the atmospheric CO_2 concentration threshold required for a temperature increase of 2°C was estimated to be 350 ppm, a concentration lower than today's level. An even greater response of temperature to CO_2 concentration is given by the "Earth sensitivity factor" that is calculated using the atmospheric CO_2 concentration prevailing during the last Ice Age. If the Earth sensitivity factor represents reality then an increase in global average temperature of 6°C is feasible with a doubling of the preindustrial atmospheric CO_2 concentration to 560 ppm and we must reduce the atmospheric CO_2 concentration to 330 ppm if we are to experience only a 2°C increase. All the relationships mentioned above between CO_2 concentration and global average temperatures are illustrated in Fig. 19.11. Several people engaged in the issue of climate change have encouraged me to avoid asserting that, just because the IPCC has presented unrealistic scenarios of future CO_2 emissions, we need not be concerned about dangerous climate change. I hope that the discussion above indicates that I still regard climate change as an issue affecting humanity's wellbeing.

It is unlikely that we will see a breakthrough in the United Nation's climate change negotiations within the next few years but there are a number of subsidiary goals on which the nations of the world might be able to agree. The scenarios presented by the IPCC in 2000 envisage a huge increase in the use of methane hydrates found on the ocean floor. One goal of the negotiations could be an agreement never to use this potential resource. We currently do not know how large this resource is, what technologies could be used to produce it, or what consequences this would have for the environment. Under these circumstances reaching an agreement not to exploit methane hydrates should not be too difficult.

A worldwide discussion is currently underway about how the world's nations will decrease CO_2 emissions. In Chap. 3 we examined how we are addicted to oil. However, we could also add that we are addicted to natural gas and coal and it is use of all these "narcotics" that causes CO_2 emissions. An important part of suppressing narcotic use is punishment of producers and distributors. We can extend this metaphor to state that the largest

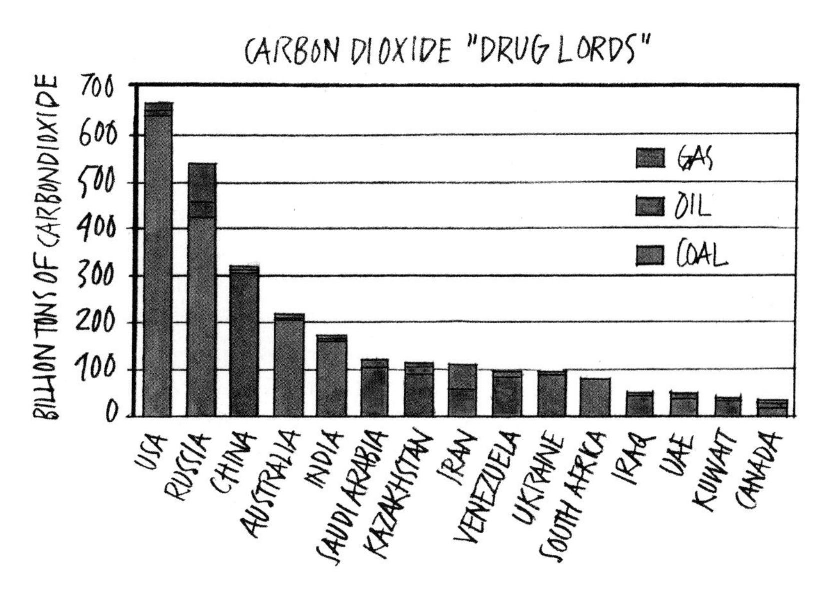

Fig. 19.12 The potential CO_2 emissions from the remaining fossil fuel reserves of the world's nations as calculated from data in the BP Statistical Review of World Energy 2011. These were calculated using the conversion factors shown in Table 15.1. The 15 nations with the largest potential to contribute to CO_2 emissions are shown. These 15 nations represent 85% of the world's total potential CO_2 emissions. We describe these nations (somewhat provocatively) as the "CO_2 drug lords"

producers of fossil fuels are "CO_2 drug lords." If we study where in the world the fossil fuels exist that can cause future CO_2 emissions when exploited then we see a very interesting situation (Fig. 19.12). When we take the fossil fuel reserves reported in the *BP Statistical Review of World Energy 2011* (*BP 2011*) and calculate the potential CO_2 emissions from these reserves then we can identify the world's largest "CO_2 drug lords."

It certainly comes as no surprise that the world's five largest producers of coal (who also have the largest reserves of coal), top the list of the CO_2 drug lords. It is interesting to note that the 15 nations shown in Fig. 19.12 together could contribute 85% of the world's future CO_2 emissions. The measures to restrict CO_2 emissions currently being negotiated by the United Nations must result in these 15 nations restricting their production of coal, oil, and natural gas. I regard it as obvious that the negotiations to reduce CO_2 emissions must include direct negotiation with these 15 nations.

In the spring of 2007 I was invited to Stockholm by the US ambassador in Sweden to discuss Peak Oil. We also discussed future emissions of CO_2 and which nations had the greatest potential to cause these. The fact that the United States and Russia are at the top of this list led me to suggest that these two nations could negotiate a bilateral treaty whereby they bind themselves never to exploit half of their remaining coal reserves. This would equate to 530 billion tonnes of CO_2 and would be equivalent in CO_2 emission potential to the world's entire remaining oil reserves. I do not know if my suggestion ever reached President Bush but I thought it worth repeating here. In practical terms it would mean that Montana's coal should remain in the ground as should the coal in central Siberia. Such a treaty would restrict future emissions of CO_2 sufficiently to limit any increase in global average temperature to 2°C according to the Charney sensitivity assumption (see Fig. 19.11). I am convinced that negotiation of such a treaty would be rewarded with the Nobel Peace Prize.

When discussing climate change I frequently tell people that "Peak Oil is a climate politician's best friend." To their list of friends they can now add Peak Gas and Peak Coal.

When the environment is discussed it is nearly always in terms of environmental destruction and threats to our future. However, I would like to present you with a positive story of environmental recovery that I hope you will find inspiring. It is a description of how ecological systems have been re-created in a large area.

At the end of 2006 I was invited to a conference in China so I took the opportunity to visit my good friend John Liu in Beijing. John arrived in Beijing more than 25 years ago as a photographer and cameraman for CBN. Since then he has documented life in China and he now produces his own documentaries. (How we met and became friends 20 years ago is an interesting story for another time.) My visit in 2006 concerned the environment and, more precisely, the world's largest ecosystem rehabilitation program, the Loess Plateau Watershed Rehabilitation Project [25].

The Loess Plateau is located in the upper and middle reaches of the Yellow River in China, and the entire plateau is approximately the size of France. It is said to be the cradle of Chinese civilization and the earlier dynasties cultivated the area thousands of years ago. Over 1,000 years ago unsustainable agricultural practices resulted in ecological degradation, and gradually the people became poorer as the desertification of the area proceeded. When the rehabilitation project began, the Loess Plateau was considered the most eroded area on Earth and its people among the world's poorest. Today, a little over a decade later, the Loess Plateau is becoming green and people in the area are experiencing an improved economy leading to a higher standard of living.

In the mid-1990s John was tasked by, among others, the World Bank, to document the Watershed Rehabilitation Project on the Loess Plateau in China. On the evening that we met in Beijing in 2006 he was organizing his documentation of the project and what he showed me was astonishing. An area that had become desert had been restored to a landscape of leafy green vegetation and streams and all this had been done in only one decade.

We discussed whether an energy system study of the project had been performed and, as far as he knew, there had not. My thoughts immediately began to revolve around this and an idea for a possible project began to take form. In the spring of 2007 I was contacted by Kersti Johansson, a student in the energy system program at Uppsala University. She inquired whether I had a suitable task that could become her Honors project. The idea of an energy system analysis of the Loess Plateau restoration reoccurred to me and Kersti thought it sounded interesting. During that summer she traveled to China and, with the help of my colleagues at the China University of Petroleum in Beijing, she succeeded in obtaining interesting data on energy system changes during the project [26].

Before the project began the population there constantly scoured the area for biomass that could be used as fuel for cooking and heating. Once the project was underway, the opportunities to do so became limited. In the statistics describing the area's coal consumption there is a clear increase during the early phase of the project. However, during the later phase of the project one can also see a clear decline in coal use when the area was generating a surplus of biomass that could be used for domestic purposes. Use of biogas by households in the area is also an important component of their energy consumption. The limited investigation that we did indicated that there is a transient phase during the rehabilitation when an increased level of fossil fuel use is needed before it finally declines.

To restore an ecosystem in a certain area is virtually the same as storing energy as biomass in the area. This also means that CO_2 is stored in the area because the vegetation and humus in the soil function as carbon stores.

In our study, the Loess Plateau was delineated as an energy system, and if the ecological restoration was successful, then energy should be retained within the boundaries of the system. The system's population (humans and other species) can be seen as part of its internal energy. A high level of biodiversity and symbiotic relationships between the species will enhance the retention of energy within the system. This is a new way to approach assessment of the outcomes of rehabilitation of an ecosystem.

John Liu has a dream that one could recreate the ecological system south of the Sahara from the Atlantic to the Indian Ocean. The government in Rwanda has accepted the idea and work there on the project has begun. I hope that there will be a small corner of the project where I can participate.

Peace, Conflict, and Peak Oil

The final component of the human wellbeing equation addresses peace on Earth. In Chap. 18 I discussed why intelligence and military organizations are interested in Peak Oil. There are several aspects of the Peace(E) component of the HWB equation that are affected by Peak Oil and conflict over resources is one of these. Michael T. Klare is Professor of Peace and World Security Studies at the Five Colleges in western Massachusetts in the United States. He has been invited to several ASPO conferences to lecture on resource wars. I do not discuss resource wars in detail here but anyone interested in that topic should refer to Klare's two books on the issue, *Resource Wars* [27] and *Blood and Oil* [28].

Many regard the invasion of Iraq as a resource war where the goal was to secure the future production of oil. If that is the case then it is interesting to analyze the monetary cost of the war in relation to the oil that exists there. In November 2008 I had the opportunity to organize a symposium in Uppsala that included Dr. Issam A. R. al-Chalabi who was the oil minister in Iraq from 1987 until 1990. In connection with his visit we discussed Iraq's oil reserves. He asserted strongly that, when he was oil minister in 1990, Iraq's oil reserves were 115 Gb [29]. Iraq has produced 12 Gb during the last 20 years and this means that at least 100 Gb remains to produce.

The US costs for the war in Iraq are enormous but so is the value of the oil in Iraq. When the symposium with al-Chalabi was organized in 2008 the United States had already spent $600 billion and now (June 30, 2011) the cost has risen to $950 billion [30]. That is $9.50 per barrel but the price of a barrel of oil on world markets is currently around $100.

The Uppsala Protocol

During ASPO's second international conference in Paris in 2003 I was elected to the position of ASPO's president. My contact details were subsequently displayed on ASPO's website. Soon afterwards I received several inquiries about what could be done to prepare for the advent of Peak Oil.

In September 2003 Colin Campbell came to Uppsala to participate in Anders Sivertsson's presentation of his Diploma work [31]. (Colin had co-supervised Anders' studies.) I then began to discuss with Colin the question of how people and nations should respond to Peak Oil. We concluded that every nation should be prepared to adapt to declines in future oil production. In his Diploma work Anders predicted that oil production would decline by approximately 2% per year after Peak Oil. Thus, a logical conclusion was that every nation should decide to adhere to a protocol to reduce its oil consumption by an equivalent amount. Colin grabbed his computer while commenting that, during his days in the oil industry, he had written many contracts and protocols and so he knew how to draft such documents. First, we needed to collect the relevant facts into a number of short statements. Then we needed to state what must be done. We named the protocol we wrote "The Uppsala Protocol" and before the day had ended it was on display on ASPO's website and later at other websites [32].

At that time Colin was associated with Uppsala University and was a member of our research group. One month later, Colin gave the first lecture that included the idea of "The Uppsala Protocol" at a conference in Rimini, Italy. Since then the idea has most commonly been referred to as "The Rimini Protocol." The protocol inspired Richard Heinberg to write the book, *The Oil Depletion Protocol*. However, he unfortunately forgot to mention that the idea was originally named, "The Uppsala Protocol".

The Uppsala Protocol

WHEREAS the passage of history has recorded an increasing pace of change, such that the demand for energy has grown rapidly over the past 200 years since the Industrial Revolution;
WHEREAS the required energy supply has come mainly from coal and petroleum formed but rarely in the geological past, such resources being inevitably subject to depletion;
WHEREAS oil provides 90% of transport fuel, essential to trade, and plays a critical role in agriculture, needed to feed an expanding population;
WHEREAS oil is unevenly distributed on the Planet for well-understood geological reasons, with much being concentrated in five countries bordering the Persian Gulf;
WHEREAS all the major productive provinces had been identified with the help of advanced technology and growing geological knowledge, it being now evident that discovery reached a peak in the 1960s;
WHEREAS the past peak of discovery inevitably leads to a corresponding peak in production during the first decade of the twenty-first century, assuming the extrapolation of past production trends and no radical decline in demand;
WHEREAS the onset of the decline of this critical resource affects all aspects of modern life, such having political and geopolitical implications;

(Continued)

(Continued)

WHEREAS it is expedient to plan an orderly transition to the new environment, making early provisions to reduce the waste of energy, stimulate the entry of substitute energies, and extend the life of the remaining oil;
WHEREAS it is desirable to meet the challenges so arising in a co-operative manner, such to address related climate change concerns, economic and financial stability and the threats of conflicts for access to critical resources.

NOW IT IS PROPOSED THAT

1. A convention of nations shall be called to consider the issue with a view to agreeing an Accord with the following objectives:
   (a) To avoid profiteering from shortage, such that oil prices may remain in reasonable relationship with production cost;
   (b) To allow poor countries to afford their imports;
   (c) To avoid destabilising financial flows arising from excessive oil prices;
   (d) To encourage consumers to avoid waste;
   (e) To stimulate the development of alternative energies.
2. Such an Accord shall having the following outline provisions:
   (a) No country shall produce oil at above its current Depletion Rate, such being defined as annual production as a percentage of the estimated amount left to produce;
   (b) Each importing country shall reduce its imports to match the current World Depletion Rate.
3. Detailed provisions shall be agreed with respect to the definition of categories of oil, exemptions and qualifications, and scientific procedures for the estimation of future discovery and production.
4. The signatory countries shall cooperate in providing information on their reserves, allowing full technical audit, such that the Depletion Rate shall be accurately determined.

Countries shall have the right to appeal their assessed Depletion Rate in the event of changed circumstances.

Proposed by
Uppsala Global Energy Systems Group
Uppsala University, Sweden

As I noted in Chap. 17, Anders' Diploma thesis was discussed in *New Scientist*. On the day that the *New Scientist* article was published online, CNN called me. We discussed the Diploma work, but I also mentioned the Uppsala Protocol [33], CNN subsequently published that, "Aleket said his team had now established what they called the "Uppsala Protocol" to initiate discussion on how the problems of declining reserves could be tackled, protecting the world economy but also addressing the problem of climate change."

Until last spring our protocol had lived a fairly quiet life but on April 6, 2011 I received the following message from our ASPO member in Portugal, Professor Rui Rosa.

> Dear Colin and Kjell:
>
> I am to inform you that the Parliament in Lisbon, today, approved a Resolution by which it recommends to the government to promote and subscribe in the national and international plans the Depletion Protocol, referring to ASPO and the Workshops held in Uppsala and Lisbon. The Protocol text is fully transcribed. It was approved by the whole House, except the Christian Democratic Party.
>
> ......
>
> Best regards,
> Rui

This is a small step forward but an important one. More nations should follow Portugal's example and decide to follow the Uppsala protocol.

Peak Oil: Final Reflections

Earlier in this chapter we stated that global oil exports reached a maximum in 2005 and then declined by 4 Mb/d by 2010. For the OECD nations the volume of imported oil declined by an even greater amount. Our calculations show that the volume of global oil exports will fall by an additional 6 Mb/d during the coming decade if global oil production remains stable. However, the decline in exports will be even greater if the current plateau of oil production ends. As shown in Fig. 9.5 global oil production and, consequently, export volumes fell around 1980. The fall generated shortages of gasoline and diesel fuels and resulted in rationing. The predicted continuing downward trend in oil export volumes may, once again, result in rationing of fuel. To avoid this the United States and Europe would need to begin immediately to reduce fuel consumption in an orderly fashion, although this process will be assisted by the economic contraction caused by falling world oil production and high oil prices. Around the world, government policies to ameliorate climate change will probably favor the use of fuel-efficient

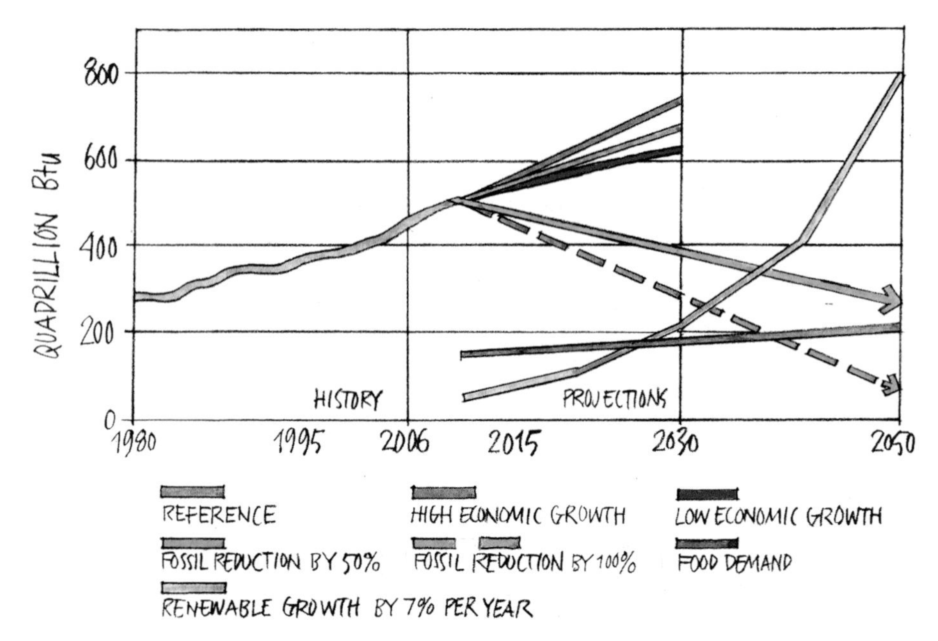

Fig. 19.13 Components of the human wellbeing equation and how they relate to future energy use. Using 2010 as a reference point we can imagine several different scenarios for our future. The world's human population will increase to over nine billion by 2050 which will require more energy for food production. A growing world economy requires more energy and when we discuss a "business-as-usual" scenario this involves increased use of fossil fuels. In contrast, climate change policies require reduced use of fossil fuels. A 100% reduction by 2050 would see us nearly totally reliant on renewable energy. If we imagine an expansion of our current renewable energy production by 7% per year then, hypothetically, renewable energy could double every decade. Nevertheless, it would be decades before renewable energy could provide the same volume of energy that fossil fuels provide today. Peak Oil constrains the future scenarios that are possible and shows us that adaptation to a future with less fossil fuel use will be necessary

diesel-powered vehicles. This could soon lead to shortages of diesel fuel and limited rationing of diesel as essential transport is prioritized.

We have discussed the conditions and thresholds for the various factors in the human wellbeing equation and Fig. 19.13 is an attempt to summarize these relative to the amount of energy required. The world's growing human population will require more food and the increase in energy shown as "food demand" in Fig. 19.13 is related to this increased food production (assuming that food production continues to be as energy-intensive as it is currently). An increased efficiency of food production means that more of the people currently experiencing hunger can receive more food.

Economic growth requires increased energy use. The curves in Fig. 19.13 for energy use under high, business-as-usual ("reference"), and low rates of economic growth are as predicted by the US Energy Information Administration [34]. These curves are based primarily on projected fossil fuel demand (and so do not accord with the predictions of UGES). However, the curves defined by political policies to address climate change point in a completely different direction. The political goal is to reduce fossil fuel consumption by at least 50% by 2050 and by 100% if at all feasible. These fossil fuel use reduction scenarios are shown as green curves in Fig. 19.13.

The energy from the sun could supply all our energy needs but the question is how to harvest it. The total global production of renewable energy (biomass, solar, wind, and hydro) is significant but is currently at quite a low level relative to other sources of energy (see the lowest point of the yellow "Renewable growth by 7% per year" curve in Fig. 19.13). During the early expansion of our oil-based economic system in the 1950s and 1960s, the annual increase in exploitation of oil was around 7% which gave a doubling of oil use every 10 years. If we regard all forms of renewable energy as part of a single energy system (which is a very simplistic assumption), and if we assume that this renewable energy system is still in an early phase of its growth, then it is not inconceivable that renewable energy production could grow by 7% annually in the same way that our oil-based energy system grew. This means that, hypothetically, total production of renewable energy could double every 10 years although this would depend on the resources (including energy resources) being available to build the renewable energy infrastructure. The yellow curve in Fig. 19.13 shows this hypothetical future expansion of renewable energy production.

We have not discussed nuclear power inasmuch as it is used almost entirely for generating electricity and so is more closely related to those issues concerning production and consumption of coal rather than oil. New nuclear technologies mean that nuclear power has the potential to be an important contributor to future energy production but the question is whether we will let it be part of our future energy system.

In this chapter I have not suggested a final solution to the human wellbeing equation but I have discussed some of the constraints acting on the global integrated energy system. There is no doubt that the factors constraining the HWB equation mean that we must work on every one of its five components in an integrated manner if a solution is to be found. I have made some proposals that can be important for this work, but what the world needs most is a global leader who understands systems thinking.

In July 2008 the BBC broadcast a TV mini-series titled, *Burn Up*. For someone uninformed about Peak Oil and the oil industry it would be difficult to follow some of the subtleties of the storyline but for a "Peak Oil insider" it was a fascinating drama [35]. The drama series was recorded in 2007 when the price of oil was still rising. In the series the (then) astonishing price of $98 per barrel was mentioned. The story opens dramatically with data being copied over to a USB memory hard disk in an oil exploration field station in a Middle Eastern desert. The field station is attacked and everyone except the person with the USB hard disk is killed. It is only in the second episode that we learn why the data that were copied onto the USB hard disk are so important that they must be kept secret at any price including murder. The secret is that only a trickle of oil will come from under the desert in western Saudi Arabia rather than the flood required during the next 60 years. (In Chap. 13 we discussed Saudi Arabia's future oil production and we concluded that their reserves are insufficient to allow constant production over the coming 60 years.) This "shocking" truth was discussed by two of the drama's characters,

- "This makes Peak Oil now!"
- "Yesterday. More oil has been used than remains but nobody knows that yet."

The truth is that we probably have a little more oil remaining than we have used so far (see the cover of this book). It is possible that if the truth about Peak Oil were widely understood then the price of oil would double, the world's economy would go into freefall and there would be conflict over the remaining oil. There are many people who try to associate such scenarios with Peak Oil but I do not accept this portrayal of the future. We humans have managed to reach our current level of development because we can adapt and innovate to find new solutions to challenges that stand in our way. I believe our clever youth (among them my students at Uppsala University) to whom we will pass on our oil-based economy can transform it into an economy where energy is used more efficiently and where renewable energy plays an important role.

At the Bibliothèque Solvay the discussion about the future of oil ended as the last grains of sand fell through the hourglass. However, important issues around Peak Oil remain and will continue to be discussed. In reading *Peeking at Peak Oil* you have followed my involvement in this issue beginning with a telephone call in December 2000 and, as I write this in June 2011, continuing most recently with a Solvay conference on the future of oil. It is unrealistic to think that this particular Solvay conference will become as famous at the 1927 conference on physics but it has definitely been an important step in shaping our understanding of the future we face.

In the film *Burn Up* public knowledge of the Peak Oil issue created chaos and panic. A foretaste of such panic could be seen during the autumn of 2008 when global financial markets, and the price of oil, crashed. My view is that it was the high oil price leading into July 2008 that was one of the factors causing the crash. My hope is that public understanding about Peak Oil and knowledge of the fact that we are actually at the peak of oil production now will draw the world's leaders together to make the decisions that the future now requires. These decisions are required by leaders in both politics and religion and must be based on the principle of co-operation by all for the good of all. The fact is that "too many have too little energy to share." [37] Grappling with the consequences of Peak Oil must become the world's greatest peace project.

The French author, playwright, and poet Victor Hugo used to carry with him a notebook in which he wrote down his thoughts and ideas as they occurred to him. On the morning when it was discovered that he had died in his sleep, a final note was found in his book that he had written during the night,

> Il y a une chose plus forte que toutes les armées du monde, c'est une idée dont le temps est venu. (More powerful than all the armies in the world is an idea whose time has come.) [33]

In *Peeking at Peak Oil* we have defined Peak Oil, we have explained how oil was formed millions of years ago, what is needed to find oil, when and where it has been discovered, how much of the world's oil has been consumed, what alternatives to oil there might be, and much more. However, the most important thing that you, the reader, must understand is that we are currently at Peak Oil.

Peak Oil Is an Idea Whose Time Has Come!

References

1. Wikipedia: The Solvay Conference. http://en.wikipedia.org/wiki/Solvay_Conference (2012)
2. Friends of Europe: The future of oil: how realistic is a post-oil society?15 June 2011. http://www.friendsofeurope.org/Contentnavigation/Library/Libraryoverview/tabid/1186/articleType/ArticleView/articleId/2839/The-future-of-oil-how-realistic-is-a-postoil-society.aspx (2011)
3. Aleklett, K., Höök, M., Jakobsson, K., Lardelli, M., Snowden, S., Söderbergh, B.: The peak of the oil age—analyzing the world oil production reference scenario in world energy outlook 2008. Energy Policy **38**(3), 1398–1414 (2010). accepted 9 November 2009, Available online 1 December 2009, http://www.tsl.uu.se/uhdsg/Publications/PeakOilAge.pdf
4. Aleklett, K.: Oil production limits mean opportunities, conservation. Oil & Gas Journal **104**(31) (2006). http://www.ogj.com/articles/print/volume-104/issue-31/general-interest/comment-oil-production-limits-mean-opportunities-conservation.html
5. BP: BP Statistical Review of World Energy. http://bp.com/statisticalreview (historical data: http://www.bp.com/assets/bp_internet/globalbp/globalbp_uk_english/reports_and_publications/statistical) (2011)
6. WEO: World Energy Outlook 2010. International Energy Agency.. http://www.worldenergyoutlook.org/2010.asp (2010)

7. CIA: The World Fact Book.https://www.cia.gov/library/publications/the-world-factbook/rankorder/2174rank.html (2011)
8. The following nations are included as importers in the China, India—SEA group; Bangladesh, Burma, Cambodia, China, China Hong Kong, Laos, India, Indonesia, Philippines, Singapore, Sri Lanka, and Thailand (2011)
9. The Economist: Barack Obama's Energy Policy; Reheated Proposals. 30 Mar 2011. http://www.economist.com/blogs/democracyinamerica/2011/03/barack_obamas_energy_policy (2011)
10. Johansson, K., Liljequist, K., Ohlander, L., Aleklett, K.: Agriculture as provider of both food and fuel. AMBIO: Roy. Swed. Acad. Sci. **39**, 91 (2010). http://ambio.allenpress.com/perlserv/?request=index-html
11. The Royal Institute of Art: Beyond oil: Shanghai. http://www.peakoil.net/files/Resources72dpi.pdf; https://www.kkh.se/index.php/en/study-programmes/mejan-arc/62-arkitektur/774-resources07-beyond-oil-shanghai (2008)
12. Redfors, A.: A Field Study of PRT in Shanghai, Uppsala University. http://www.tsl.uu.se/UHDSG/publications/Shanghai_Anton.pdf (2009)
13. The Green Momentum Group and Bristol City Council: Building a positive future for Bristol after peak oil. http://www.bristolgreencapital.org (2011)
14. Dodson, J., Sipe, N.: Oil vulnerability in the Australian city. Urban Research Program, Research Paper 6, Griffith University, December 2005. http://www.vlga.org.au/site/defaultsite/filesystem/documents/climate%20change/oil%20vulnerability%20in%20the%20australian%20city%20-%20dec%202005.pdf (2005)
15. Transition Town Network. http://www.transitionnetwork.org/ (2011)
16. Jakobsson, K., Aleklett, K.: Oil in the veins of sub-Saharan Africa. New Routes **2010**(2), 23 (2010). http://www.tsl.uu.se/uhdsg/Publications/Oil_in_SSA.pdf
17. WEO: World Energy Outlook 2008. International Energy Agency. http://www.iea.org/textbase/nppdf/free/2008/weo2008.pdf (2008)
18. WEO: World Energy Outlook 2004. International Energy Agency. http://www.iea.org/textbase/nppdf/free/2004/weo2004.pdf (2004)
19. WEO: World Energy Outlook 2006. International Energy Agency. http://www.iea.org/textbase/nppdf/free/2006/weo2006.pdf (2006)
20. WEO: World Energy Outlook 2010. International Energy Agency. http://www.iea.org/textbase/nppdf/free/2010/weo2010.pdf (2010)
21. Höök, M., Söderbergh, B., Aleklett, K.: Future Danish oil and gas export. Energy **34**(11), 1826–1834 (2009). http://www.sciencedirect.com/science/article/pii/S036054420900317X, http://www.tsl.uu.se/uhdsg/Publications/Denmark_Article.pdf
22. Mäkivierikko, A.: Russian Oil—A Depletion Rate Model Estimate of the Future Russian Oil Production and Export, Uppsala University, 2007-10-01. http://www.tsl.uu.se/uhdsg/Publications/Aram_Thesis.pdf (2007)
23. Wasdell, D.: Critical issues in the domain of climate dynamics. The Apollo-Gaia Project. http://www.apollo-gaia.org/Climate%20Sensitivity.pdf (2011)
24. Hansen, J., Sato, M., Kharecha, P., Beerling, D., Berner, R., Masson-Delmotte, V., Pagani, M., Raymo, M., Royer, D.L., Zachos, J.C.: Target atmospheric CO2: where should humanity aim? Open Atmos. Sci. J. **2**, 217–231 (2008). ISSN: 1874–2823, http://benthamscience.com/open/openaccess.php?toascj/articles/V002/217TOASCJ.htm

25. The World Bank: Loess Plateau Watershed Rehabilitation Project. http://www.worldbank.org/projects/P003540/loess-plateau-watershed-rehabilitation-project?lang=en (2011)
26. Johansson, K., Aleklett, K., Pang, X., Mei, Z., Liu, J.D.: Energy system in the Loess plateau—a case study of changes during the rehabilitation period. Global Challenges in Research Cooperation Conference Proceedings, 27–28 May 2008, Uppsala, Sweden. http://www.tsl.uu.se/uhdsg/publications/LoessConf.pdf (2008)
27. Klare, M.T.: Resource Wars: The New Landscape of Global Conflict. Metropolitan/Owl Book, New York (2002)
28. Klare, M.T.: Blood and Oil. Metropolitan/Owl Book, New York (2004)
29. Aleklett, K.: Iraq's Oil and the Future. Aleklett's Energy Mix. http://aleklett.wordpress.com/2008/11/15/iraq%e2%80%99s-oil-and-the-future/ (2008)
30. Cost of Iraq War and Nation Building. http://zfacts.com/p/447.html (2011)
31. Sivertsson, A.: Study of world oil resources with a comparison to IPCC emissions scenarios. Diploma Thesis, Uppsala University, 2004-01-01. ISSN: 1401–5765. http://www.tsl.uu.se/uhdsg/Publications/Sivertsson_Thesis.pdf (2004)
32. Campbell, C., Aleklett, K.: The Uppsala Protocol. Uppsala Global Energy Systems, Uppsala University, 3 Oct 2003. http://www.mnforsustain.org/oil_uppsala_protocol.htm (2003)
33. Jones, G.: World oil and gas 'running out'. CNN, Thursday, 2 Oct 2003. http://edition.cnn.com/2003/WORLD/europe/10/02/global.warming/index.html (2003)
34. WEO: World Energy Outlook 2009. International Energy Agency. http://www.iea.org/textbase/nppdf/free/2009/weo2009.pd (2009)
35. BBC: Burn Up. http://www.bbc.co.uk/drama/burnup/ (2008)
36. Gapminder. http://www.gapminder.org/ (2011)
37. Aleklett, K.: Peak oil and the evolving strategies of oil importing and exporting countries. Discussion Paper No. 2007–17, Dec 2007, Joint Transport Research Centre, Paris, France. http://www.internationaltransportforum.org/jtrc/DiscussionPapers/DiscussionPaper17.pdf (2007)

Chapter 20

An Inconvenient Swede

"Kjell Aleklett, a perky and persuasive physicist at Uppsala University, talks with characteristic Swedish candour," is the opening sentence of an article titled, "An Inconvenient Swede" published in the business journal *Canadian Business* in 2006 [1]. The article was written by Andrew Nikiforuk whom I met in Vancouver at the beginning of that year, shortly after I had testified before a committee of the US House of Representatives [2] (see Fig. 20.1). Andrew and I had a long conversation about oil, Canada's oil sands, why far too many people try to hide the truth about oil's future, and why I am so determined to say what I consider to be true. *Peeking at Peak Oil* is not an autobiography but I have described how my personal experiences have influenced the research of the Global Energy Systems group and vice versa. My interest in Peak Oil was ignited while preparing teaching material in December 2000. Now, a decade later in August 2011, it has led to an invitation to the EU Parliament to describe the research of my Peak Oil-focused research group, Uppsala Global Energy Systems.

Many of the most significant moments along my Peak Oil path in the past decade have concerned ASPO, The Association for the Study of Peak Oil and Gas. There was that first conversation with Colin Campbell in December 2000 that set me on this path, and the first ever Peak Oil conference in Uppsala in 2002. In December 2000 Campbell was planning an organization that was to focus on the oil peak that he and Jean H. Laherrere had described in their article, "The End of Cheap Oil," in *Scientific American*, March 1998 [3]. The proposed name was ASOP, the Association for the Study of the Oil Peak. I told Colin that the acronym ASOP was not a good one. He then proposed a change from "oil peak" to "peak oil" and the acronym ASPO. This lead to the creation of a new English term, "Peak Oil" (Chap. 2) that today is in common use internationally. Indeed, the term

K. Aleklett *Peeking at Peak Oil*, DOI 10.1007/978-1-4614-3424-5_20,

Fig. 20.1 On Wednesday December 7, 2005 Peak Oil was discussed officially in the US House of Representatives when the House Energy and Air Quality Subcommittee held an inquiry into the issue. Those invited to testify included Professor Kjell Aleklett, Uppsala University, Dr. Robert L. Hirsch, Senior Energy Program Advisor, SAIC, and Robert Esser, Senior Consultant and Director, Cambridge Energy Research [2]

Peak Oil has founded an entire family of "Peaks." We now hear discussion of "Peak Gas," "Peak Coal," "Peak Phosphorous," and so on, and, most recently, "Peak Child" (for the world's maximal rate of childbirth [4]).

In his 2006 article in *Canadian Business*, Nikiforuk mentioned "characteristic Swedish candour" and there is more to that description than is immediately apparent. During the eighteenth century King Gustav III established the Royal Academies of Sweden with the intention that professors would perform research advantageous to his kingdom. He was aware that such research might not benefit the ruling elite so he introduced a form of protected employment for academic professors. The King would have the ultimate power to appoint or dismiss professors. This meant, in effect, that professors were appointed for life and could not be dismissed. The only ground for dismissal was failure to tell the truth. During the twentieth century the power to appoint professors was transferred to the Swedish government and lay in the hands of the minister for education. Towards the end of the twentieth century the conditions of appointment for professors were altered and today they can, in practice, be dismissed. However, the old tradition lives on and truth is valued above all else.

For my appointment as a professor in the spring of 2000 the new regulations were in place but Uppsala University emphasized that their intention was to follow the traditional principles. There are certainly those who would like to see me prevented from researching Peak Oil. However, my adherence to the pursuit and promotion of the truth meant that, for example, I did not agree to remove from the Internet my analysis of the

International Energy Agency's *WEO 2004* report when the IEA's chief economist suggested I do so (see Chap. 18). Adherence to the truth also means that one must change one's viewpoint if new facts come to light. Indeed, in Chap. 13 I have described how the oil reserve numbers declared by Saudi Aramco can be realistic despite earlier having opposed this view. In fact, I have even proposed that the estimated reserves of Saudi Arabia's Abqaiq oilfield should be raised by 700 million barrels.

During the past decade of Peak Oil research there have been numerous meetings and occurrences that I have not mentioned in this book and that I will not make public. In some cases this is because the people involved have asked me not to describe them on my blog, *Aleklett's Energy Mix*. Some of the meetings I attended have been held under "Chatham House Rules" where one promises not to reveal who said what. When I have described meetings and conversations it is because I judged that doing so would benefit those interested in Peak Oil such as you, the reader.

In June 2008 I was invited as one of the plenary speakers to the *Asia Oil and Gas Conference* in Kuala Lumpur. The then acting director of BP, Tony Hayward, was also invited. Malaysia's Prime Minister Abdullah bin Haji Ahmad Badawi was present at the commencement of the conference when Tony Hayward gave the opening address. After Tony's presentation there was a reception with the prime minister at which the PM was seated center-stage, the organizers were seated to the PM's right and Tony and I were seated to his left. Most of the attention at the reception was directed to others, so Tony Hayward and I began to discuss Peak Oil. Out of the blue he suddenly said to me, "I bet you that production of oil will be higher in 10 years than today." I asked what the stakes were and he answered, "The price of a barrel of oil in 10 years time." Of course I accepted the bet and I then told him that I would describe our wager during my presentation that afternoon. Tony shot back, "But you can't do that!" I asked him whether he thought it better that I say that the CEO of BP does not back his own assertions and after a momentary silence (when I could see his mental wheels turning) he accepted.

During my presentation I described our wager and some minutes later a journalist for Bloomberg News cabled the story around the world under the headline, *BP Boss Hayward Bets Price of a Barrel over Peak Oil* [5]. Tony and I discussed what should be regarded as "oil" for the purposes of our bet and we agreed that it would be the same as BP reports as oil production in its annual *Statistical Review of World Energy* [6]. At that time in June 2008 oil production was 85.5 million barrels per day. There was commentary in the media that the price of oil could rise as high as $1,000 per barrel by 2018. Today our understanding is different. The price of oil will never rise as high as that (adjusted for inflation). Nevertheless, it will be interesting to see how large a check I can bank in June 2018.

The Nobel Prize-winner Glenn T. Seaborg worked at the University of California at Berkeley and the Lawrence Berkeley National Laboratory [7]. He had a very great influence on me regarding my engagement in issues related to the wellbeing of society. Seaborg asserted that scientists must become involved in societal issues outside their research specialty, a belief that resonates strongly with Gustav III's intention in creating the Royal Swedish Academies. This was why Seaborg accepted President Kennedy's invitation in 1961 to head the US Atomic Energy Commission. During his 7 years as head of the commission he was very actively engaged in the atomic test ban negotiations with the Soviet Union. He later described these negotiations in his book *Kennedy, Khrushchev, and the Test Ban* [8].

Professor Seaborg's father emigrated from Sweden with the name "Sjöberg" but this was partially anglicized to "Seaborg" upon his arrival in the United States. Glenn T. Seaborg's mother also immigrated to the United States from Sweden so that Glenn grew up speaking Swedish until he entered school. In May 1977 Uppsala University celebrated the 500th anniversary of its foundation in 1477 and Glenn T. Seaborg visited the university to receive an honorary doctorate and to be the main speaker at the festivities. I had just received my Ph.D. degree and my dream was to work in Berkeley with Seaborg. I traveled to Uppsala and waited outside the university hall to try to speak to him when he came out. I managed to introduce myself, give him a copy of my Ph.D. thesis and express my ambition. I was somewhat surprised when he told me he hoped it would work out. It did so, and the following year I traveled with my family to California and to Berkeley to begin a research collaboration that continued until Seaborg's death in 1999.

In March 2008 I was invited to Washington to participate in a discussion of "America's Energy Future." Some months earlier I had written the report for the OECD on Peak Oil and the world's future oil supplies (see Chap. 11). At the meeting in Washington I had the opportunity for a short conversation with the then Secretary of Energy Samuel Bodman and I gave him a copy of my OECD report. I received the impression that Bodman had heard of the term Peak Oil but had no interest in it. Bodman's successor and the current incumbent as US Secretary for Energy is the Nobel Prize-winner Professor Stephen Chu [9]. At the time Chu was appointed secretary by President Obama, he was head of the Lawrence Berkeley National Laboratory that had been the center for Seaborg's research activities. My hope was that Chu would prove to be another Seaborg and, among other things, recognize the validity of the research supporting Peak Oil so that he could push for the transformation needed in US energy supplies. My belief is that, as a researcher and Nobel Prize-winner, Stephen Chu must prioritize the truth. Anything else would be a great disappointment.

The United States' first secretary of energy was Dr. James R. Schlesinger. He was invited to the *Sixth ASPO Conference* in Cork in September 2007. At the end of his speech (that can be heard on YouTube [10]) he declared:

> And therefore to the peakists I say, you can declare victory. You are no longer the beleaguered small minority of voices crying in the wilderness. You are now mainstream. You must learn to take yes for an answer and be gracious in victory.

I think the first and latest US secretaries of energy need to sit down together and talk.

I have dedicated this book to my family: my wife, my daughters, and my grandchildren. If I live as long as an average Swedish man then I will see another 15 years of the future. If that is so then I will experience Peak Oil. Indeed, oil production is currently on a plateau that will be followed by decline and I believe the inability to increase oil production is one of the factors behind the world's current economic difficulties. It is urgent that we, as a society, realize the critical role of energy and Peak Oil in the future of our civilization (see Chap. 19). My children and grandchildren and the

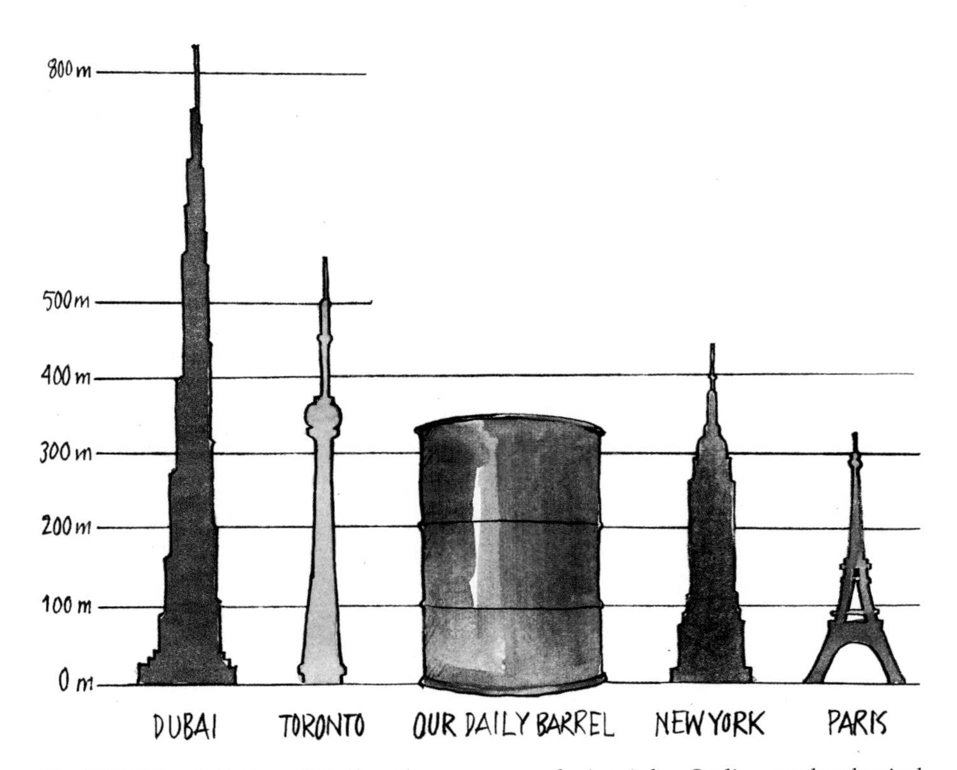

Fig. 20.2 "Our daily barrel" is the oil we consume during 1 day. Scaling up the classical 55-gal drum holding 218 L, we get a barrel that is 340 m high and 280 m wide

generations that follow them must live with the consequences of Peak Oil and my hope is that *Peeking at Peak Oil* will ease the unavoidable transition from the Oil Age into a future with limited resources.

The oil we all use each day would fill an oil barrel scaled up to 340 m high and 280 m wide. That is higher than the Eiffel Tower and more than the volume of the world's tallest building. However, our daily barrel of oil will soon begin to shrink. The world's economy quivers in fear of an uncertain future and the natural world quivers in gratitude at the coming decline in carbon dioxide emissions. I hope that knowledge of Peak Oil will hinder the conflicts that might follow in its wake and will help us take the necessary steps towards a new future where none are starving and we live in peace. If not, then at least my grandchildren will be able to say that Grandpa tried (Fig. 20.2).

References

1. Nikiforuk, A.: An inconvenient Swede. Canadian Business, 9 Oct 2006. http://www.canadianbusiness.com/article/14375--an-inconvenient-swede (2006)
2. Subcommittee on Energy and Air Quality of the Committee on Energy and Commerce: House of Representatives, Understanding the Peak Oil Theory, Hearing before the Subcommittee on Energy and Air Quality of the Committee on Energy and Commerce, 7 Dec 2005 (Aleklett page 32). http://www.peakoil.net/Aleklett/Hause_Peak_Oil_hearing_2005.pdf (also http://www.energybulletin.net/node/11621) (2005)
3. Campbell, C., Laherrere, J. H.: The end of cheap oil. Sci. Am. http://dieoff.org/page140.htm (1998)
4. Rosling, H.: Gapminder. http://www.gapminder.org/ (2011)
5. Schmollinger, C.: BP Boss Hayward Bets Price of a Barrel Over Peak Oil, 10 June 2008. http://www.bloomberg.com/apps/news?pid=newsarchive&sid=akmFgc4ikBDk (2008)
6. BP: BP Statistical Review of World Energy (2011)
7. Seaborg, G.T.: Biography, Nobel Prize winners. http://nobelprize.org/nobel_prizes/chemistry/laureates/1951/seaborg-bio.html (2011)
8. Seaborg, G.T.: Kennedy, Khrushchev and the Test Ban, Seaborg, G.T., Loeb, B., 16 Mar 1983. http://www.amazon.com/Kennedy-Khrushchev-Test-Glenn-Seaborg/dp/0520049616/ref=cm_cr_pr_orig_subj (1983)
9. Chu, S.: Biography, Nobel Prize winners. http://nobelprize.org/nobel_prizes/physics/laureates/1997/chu.html (2011)
10. Schlesinger, J.: Oral presentation at ASPO 6, Cork, Ireland (1 minute and 26 seconds into the YouTube film). http://www.youtube.com/watch?v=1Ia-sk1OqHk&eurl=http%3A%2F%2Fwww%2Easpo%2Direland%2Eorg%2F (2007)

Epilogue

The Champagne Festival

There is an atmosphere of excitement and high expectation at the International Champagne Agency's annual festivities held at the Hotel Crowne Plaza in the heart of London. The ICA is headquartered in Paris and was established in the 1970s to watch over the interests of the large champagne consumers. Every year they invite journalists, producers, market players, and champagne experts to a banquet where they present their predicted demand for champagne in the coming year and over the next 20–25 years. The members of the ICA produce some champagne themselves but far too little to satisfy their needs. The ICA's thirsty members need other producers to supply them with champagne.

As midnight approaches everyone is enjoying themselves. As one scans the function room one notices that all the large champagne consumers are crowding around the delegations from the vineyards in the Middle East. There one can see the old established customers from Europe, Japan, and the United States who for many years have purchased large quantities of those vineyards' finest champagne. However, there is also a large delegation from China that is doing everything it can to get the vineyards' attention. The world's most populous nation has developed a taste for champagne.

One can also see that many of the delegates from the European Union have gathered around the Russian producers who are accompanied by the new growers from central Asia. One of the most enthusiastic buyers is the Swedish delegation that recently realized that its neighbors and most reliable suppliers, Denmark and Norway, will have difficulty delivering champagne to them in the future. A new incurable disease afflicting the oldest

K. Aleklett *Peeking at Peak Oil*, DOI 10.1007/978-1-4614-3424-5, vineyards has recently been discovered by a group of researchers from Uppsala University in Sweden. Their studies show that the Danish and Norwegian vineyards will be the worst affected.

The delegation from Brazil is also getting a lot of attention but not from champagne consumers. Rather, the interest comes from the international champagne producers who need new land for cultivation.

Some of the officers of the ICA have noticed that champagne production has, in fact, begun to decline in many of the world's vineyards but they have been uncertain why this is happening. Now they are convinced that the world has reached its maximum possible champagne production. To keep their employers (the large consumers) happy they are now trying to find substitutes for champagne but the traditional producers forbid these being called "champagne." The substitutes will now be marketed as "sparkling wine" and no future shortages of these are foreseen.

The largest producers of sparkling wine are in Canada and Venezuela. Their representatives are also attracting many interested buyers, not the least from the new consumers in China who do not have the same requirements vis-à-vis that the champagne must come only from the finest vineyards. They are simply happy to get in on some of the sparkling action!

Every year the ICA draws attention to something new and special in the champagne branch. This year's high point is that champagne production in Iraq—and especially in the area that once held the Garden of Eden—will be ramped up with renewed enthusiasm. Some of the new tenants are also intending to plant new vine stocks to increase future production.

Also invited to this year's festivities are members of the research group from Uppsala who recently discovered the disease DRRR (depletion of remaining recoverable resources) that will limit future production. After the clock has struck midnight and the fantastic fireworks display has ended, the grey-haired professor from Uppsala approaches the microphone and begins to speak.

"Cheers and many thanks for a fantastic party! Today we celebrate most of all that champagne production in Iraq will be ramped up and many of the buyers here tonight expect it to be top-quality product. Many believe that these new harvests will be so large that we do not need to concern ourselves about future champagne shortages. But the fact is that even Iraq will be affected by DRRR and the anticipated increase in production will reach a maximum and then decline."

"Let's imagine that we have the world's largest magnum bottle and we fill it with all the future champagne that can ever be produced in Iraq. In total it amounts to about 100 billion oak barrels. Now let's represent that magnum bottle with this smaller example that I hold here in my hand. The world has been celebrating with champagne for 150 years and, in total, has emptied 11 such bottles. According to the champagne experts there are eight or nine other full bottles left in the refrigerator. Rumors are circulating that enormous quantities of sparkling wine are available but, in reality, these amount to only three bottles. In North, South, and Central America there is only one bottle of champagne left. There is one bottle in Africa, one bottle in all of Europe and Russia, half a bottle in central Asia, and an additional half in the "Asia Pacific" region. Information on the number of bottles in the Middle East varies but there are certainly four and there might even be five."

With a practiced hand the professor opens the bottle of champagne he holds and that now represents all the champagne in Iraq. On a nearby table stand three large glasses and a smaller glass. The professor fills the glasses with the golden fluid from the bottle and then lifts one of the large glasses to his lips to savor the champagne. He declares,

"This glass contains 30 billion barrels of champagne, as much as the world drinks every year. All the future champagne production from Iraq could only satisfy the world's thirst for 3 years and 4 months. The United States, the world's thirstiest nation, already has a problem because its future domestic production amounts to about the contents of this champagne glass."

"Of course the world can plant new vineyards but during the next 50 years the most we can produce from those vineyards will only be enough to fill three bottles. We also need to remember DRRR, which means that, in any one year, one can only produce a limited fraction of what remains to be produced."

"Dear Friends, of course we can continue to celebrate with champagne today and even tomorrow but in the future we will need to drink less. It is time for us all to alter our habits because, in the future, there will be less champagne to share."

Index

K. Aleklett *Peeking at Peak Oil*, DOI 10.1007/978-1-4614-3424-5,

S

T

X

Y

Z